Industrial safety handbook

Second edition

Industrial safety handbook

Second edition

Editor: William Handley, M.B.E.
North-West Regional Association
of Industrial Safety Groups

McGraw-Hill Book Company (UK) Limited

London · New York · St Louis · San Francisco · Auckland · Beirut · Bogotá
Düsseldorf · Johannesburg · Lisbon · Lucerne · Madrid · Mexico · Montreal
New Delhi · Panama · Paris · San Juan · São Paulo · Singapore · Sydney · Tokyo
Toronto

Published by

McGRAW-HILL BOOK COMPANY (UK) LIMITED

MAIDENHEAD . BERKSHIRE . ENGLAND

Library of Congress Cataloging in Publication Data

Main entry under title:

Industrial safety handbook.

    Bibliography: p.
    Includes index.
    1.   Industrial safety—Handbooks, manuals, etc.
I.   Handley, William.
T55.I52   1977   614.8′52   76–54350
ISBN 0–07–084481–X

2345   WC&S 80798

MADE AND PRINTED IN GREAT BRITAIN

Contents

Contributors

ATHERLEY, G. R. C.
Professor of Safety and Hygiene, The University of Aston in Birmingham

BARKER, G. C. E.
Independent Consultant

BENNELLICK, E. J.
Training Officer, National Radiological Protection Board, Harwell

BERRY, R.
Group Manager, British Standards Institution

BLAND, J. G. V.
Product Manager, Wolf Electric Tools Ltd

BRAMLEY-HARKER, R.
Formerly Group Safety Adviser, Wiggins-Teape Ltd

CAMERON, DR J. D.
Group Medical Adviser, Pilkington Brothers Ltd

CHEFFERS, S. J.
Formerly Home Sales Manager, Siebe Gorman and Co. Ltd

FAIRCLOUGH, J. V.
Formerly of British Engine Boiler and Electrical Insurance Company Ltd

GIBSON, DR N.
Senior Research Physicist, I.C.I. Organics Division

GOLD, W. D.
Barrister,
Senior Law Lecturer,
Blackburn College of Technology and Design

GRANDIDGE, F. E.
Ventilation Engineer, H.H. Robertson (U.K.) Ltd

HAIGH, R. D.
Chief Engineer, U.M.L. Ltd

HALL, DR M. H.
Consultant-in-Charge,
Emergency and Accident Department,
Preston Royal Infirmary

HALLIDAY, J.
Lifting and Handling Instructor,
Central Electricity Generating Board,
N.W. Division

HANDS, DR A. H.
Chief Medical Officer, British Leyland Truck & Bus UK Ltd, Leyland

HAYWARD, J. A., M.B.E.
Formerly Superintending Safety Officer,
John Laing Construction Ltd

HEARN, R. W.
Director of Training, Ro.S.P.A.

KERSHAW, DR G. R.
Formerly Medical Officer,
Rolls Royce Ltd

McCREANNEY, W. C.
Century Oils Ltd

NASH, P.
Head of Extinguishing Materials and Equipment Section,
Fire Research Station, Building Research Establishment

NEWALL, J.
Fire and Security Group, Mather and Platt Ltd

PARTRIDGE, H.
Managing Director, Fork Truck Training Ltd

PATERSON, A. G.
Formerly of Air Mover Research Syndicate, Glasgow

RAGSDALE, K. H.
County Surveyor's Department, Lancashire County Council

RIDDELL, F.
Group Safety Engineer,
Reed International Ltd (Northern)

RIGBY, H. F.
Technical Advisory Service, Flat Glass Division,
Pilkington Brothers Ltd

ROBERTS, E. L. M.
Formerly Manager, Oldbury Safety and Environmental Hygiene Unit,
Albright and Wilson Ltd

SANDERSON, P. G.
Manager, Commercial and Marketing Services,
Philips Electric Arc Welding Ltd

SMITH, S. F., O.B.E.
Formerly Chief Officer, Safety, Health and Welfare, Dunlop Ltd

SOUTHWELL, K. B.
Technical Marketing Manager, Universal Grinding Wheel Co Ltd

STANCLIFFE, I.
Head of Colour and Design Service, Crown Decorative Products Ltd

SYMONS, W. G.
Formerly H.M. Superintending Inspector of Factories,
North Western (Liverpool) Division and Midlands (Birmingham) Division

TYE, J.
Director General, British Safety Council

WARD, DR A. F. H.
Consultant, and Editor, *Safety News*

WAREING, T. H.
Textile Product Manager, Courtaulds Ltd

WARWICK, R. G.
Chief Engineer, National Vulcan Engineering Insurance Group Ltd

WHITE, M. J.
Principal Assistant, Pensions and Social Services Department,
General and Municipal Workers Union

1. The scope of occupational safety

Professor G. R. C. Atherley

Department of Safety and Hygiene,
The University of Aston
in Birmingham

Edward de Bono (1974) points to the paralysing effect which the dominance of old and apparently adequate ideas can have on scientific progress. The old idea is seen as the only stepping stone for research. Much time and effort is devoted to research always with the same point of origin. Despite all the effort the field as a whole appears to make little advance. De Bono uses hole-digging as his analogy. 'It is not possible', he wrote, 'to dig a hole in a different place by digging the same hole deeper. Digging the same hole deeper is vertical thinking; lateral thinking is trying again elsewhere.'

Vertical thinking has affected health and safety at work in at least one important respect. For a century and a half there has been emphasis on human error and the human element as a cause of accidents. Much research has been devoted to their elucidation. Hale and Hale (1972) reviewed the accident research literature. They identified 355 research studies all concerned with human error and human element as a cause of accidents. It would be unfair to say that all these studies were fruitless. Nevertheless, as Hale and Hale make clear, clear-cut findings are few and far between.

The safety movement and especially the voluntary safety movement has also been greatly concerned with the human element. Much of the safety propaganda put out by the various safety organizations is aimed at human error. A great deal of the safety training which goes on in industry is aimed at attacking human error and the human element by changing attitudes and behaviour. People in industry have often said that this educational process should begin in schools, colleges, and universities.

This idea plainly influenced the Committee of Vice-Chancellors and Principals of the Universities of the United Kingdom in their evidence to the Committee on Safety and Health at Work (1972):

> 'Universities have a responsibility not only for ensuring the protection of their staff and students but also for instilling into each succeeding generation of undergraduates a measure of safety consciousness which they can carry forward into subsequent careers in industry and elsewhere. The ultimate objective of safety arrangements in university laboratories is thus the creation in both staff and students of an attitude of mind about safe working. This is something that can be achieved, not by legislation, but by precept and example in the normal course of teaching and individual work. Indeed, the imposition of mandatory regulations from outside might have the effect of encouraging reliance on the legal minimum precautions as a substitute for intelligent appreciation and application of the principles at issue.'

Vertical thinking is involved when the various ideas outlined above are linked together in a particular way:

> *Accidents are caused by human element and human error; changes in attitudes and changes in behaviour can reduce human error and the human element in accidents. Therefore, efforts to advance health and safety at work should be directed towards altering workers' attitudes and including safe behaviour. Research into health and safety at work should be directed towards mistakes and failings on the part of workers in order to advance understanding about the causes of human error and human factor in accidents in order that they can be better prevented by effective alterations in attitudes and better safety behaviour.*

The problem with this vertical thinking is that it leads to accident prevention being seen largely as a matter of changing workers' attitudes and behaviour sometimes to the exclusion of efforts to improve plant and machinery. In most practical situations the likelihood is that accident prevention must be directed towards people and plant, and people include employers, managers, and work people. The problem is to decide where the emphasis should lie. The vertical thinking about human error and human element has so dominated the field that no one knows how to decide the priorities. We need a strong dose of lateral thinking about ways to advance health and safety at work in which we can leave human error and the human element as a problem for continuing mining operations and at the same time start excavating somewhere else. The aim of this chapter is to put human error and the human-element factors in accidents into perspective and to show that the scope of occupational health and safety at work is very much wider than these two though it nevertheless includes them.

Over the past 100 years a considerable amount has been written about the causes of accidents. Human error has always been a popular theme. An early example (quoted by Atherley, 1975) is Dr Cooke-Taylor's reference to the textile industry published in 1844:

'No-one can examine a cotton mill without seeing, not only that its operations can be conducted with perfect safety, but that there must be an utter disregard of ordinary precautions—a total want of prudence, and not a little perverted ingenuity to get into danger.'

Cooke-Taylor's book was written at a time of intense political discussion on the issue of government regulation of cotton mills. He was reflecting a particular political viewpoint: that extension of government regulation was neither necessary nor desirable.

We can see the same essential point being made with the same argument when we compare Cooke-Taylor's statement with the evidence given in 1972 to the Committee on Safety and Health at Work by the British Chemical Industry Safety Council of the Chemical Industries Association. This included the following statement:

'Reliable estimates place the proportion of accidents due to human error as 85 per cent of the total. It is difficult to visualise legislation which can have a salutory effect on human behaviour in this context. We must therefore place considerable reliance on voluntary effort by employees and employers to contain and try to reduce the number of accidents.'

The Committee of Vice-Chancellors and Principals and the Chemical Industries Association are both employers' organizations. It is not surprising, therefore, that they have in common a particular political viewpoint: opposition to legislation. They adduce quasi-scientific ideas about human behaviour and human error in support of their viewpoint. From Cooke-Taylor's writings we can see that this pattern of reasoning can be traced back a long way. The political viewpoint has not changed with the passage of the years. The quasi-scientific ideas subordinated to the political viewpoint have not changed either. Nor will they change while they retain their links with politics—they will remain impervious to research, as well as argument. This is why lateral thinking is so necessary.

New approaches to thinking about accidents

An important knight's move in thinking about the causes of accident was marked by the publication in 1974 of the Factory Inspectorate's study of accidents in factories (Department of Employment, 1974). The study was an analysis of 0·5 per cent of all accidents reported to H.M. Factory Inspectorate in 1969. Information relating to each accident was analysed so as to identify which of a range of predetermined categories it fell into. No category was titled human error.

The heading used was: *Failure to take Reasonably Practicable Precautions.* This was sub-divided into: *Failure by Management, Failure by Workers,* and *Failure by Management and Workers Jointly.* In addition, accidents were identified where there had been a breach of law. A distinction was drawn between all accidents and fatal accidents. There were other analyses which do not concern us here.

Booth (1975) has summarized certain of the key findings from the study (see Table 1.1). We can see that a proportion of accidents arising from failure to take reasonably practicable precautions range from 35 to 64 per cent for the four types of accident. Of these, failure by workers ranged from 40 to 75 per cent according to accident type.

Table 1.1 Comparison of accident statistics

Type of accident*	0·5% Sample of Reported Accidents 1969					All reported accidents 1967–71	
			Preventable accidents				
	Percentage involving breach of law	Percentage which were reasonably preventable	Failure by management	Failure by worker	Failure by both management and worker	Percentage of all accidents	Percentage of fatalities
Machinery	40(54)†	64(72)†	50	40	10	17·5	24
Transport	14	51	45	45	10	8	21·5
Handling goods	4	37·5	48·5	51·5	0	27·5	0·6
Hand tools	0	35	12	75	13	7	0·2

*Factory Inspectorate Classifications. Accidents in this list involve both continuing and contingent dangers.

†Figures in brackets: machinery in motion (6 per cent of the breaches of law relating to machinery involved breaches by workers).

Table 1.1 shows that for the four types of accident (machinery, transport, handling goods, hand tools) about a quarter can be attributed to failure to take reasonably practicable precautions on the part of workers. An incidental fact emerges from the table—the consequences of failure vary considerably according to the type of accident. The consequences of machinery and transport accidents are far more serious than those for handling goods and hand tools.

The study has been criticized on the grounds that the criterion of the reasonably practicable was too dependent on the subjective judgement of the factory inspector who carried out the investigations. There is some substance in the criticism. A different set of investigators—industrial safety officers, for example—might have used different criteria and thereby produced different results numerically. The criticism would be eliminated if the focus was shifted from the reasonably practicable on to precautions. Accidents would then be analysed according to the nature of the failure in precautions—irrespective of their practicability. Practicability could then be judged as a separate question. This, as we shall see later, is the Factory Inspectorate's crucial lateral movement in thinking which opens up the way to the idea of strategic analysis.

Before we look at strategic analysis there is a need to consider briefly the scope of what we mean by accident. Under the Health and Safety at Work Act 1974, health and safety are combined as an objective. We can, therefore, use accident to describe impairment of health as well as physical injury. Hence the aim of people practising health and safety at work is to prevent accidents in the broadest sense.

No study comparable to the Inspectorate's study is available about accidents involving impairment of health. However, experience with the carcinogen beta-naphthylamine may be indicative of what might be found. Detailed information is to be found in the judgement by J. O'Connor in the case *Thomas Cassidy v. Dunlop Rubber Co Ltd, Thomas Cassidy v. Imperial Chemical Industries Ltd, Christopher Wright v. Dunlop Rubber Co Ltd, Christopher Wright v. Imperial Chemical Industries Ltd* (*Knight's Industrial Reports*, Vol. II, p. 311).

The men started work in one of the rubber company's factories in 1947 and 1946 respectively. Until the middle of 1949 they were exposed to a substance called Nonox S which was used to prevent rubber from perishing. Nonox S contained beta-naphthylamine, amongst other chemicals. In 1949, following an investigation by Professor R. A. Case, Dunlop stopped using Nonox S and I.C.I. withdrew it from the market.

In 1966 both men were diagnosed as suffering from cancer of the bladder. The judge accepted that the cancers were due to occupational exposure to the carcinogen.

In his judgement the judge said: 'As far as the Plaintiffs are concerned it is sufficient to hold that I.C.I. ought to have withdrawn Nonox S before the end of 1946 . . . but I would, if necessary, have held that it ought to have been withdrawn by January 1st 1940.'

Dunlop told the judge that they did not know that there was any risk attached to the use of Nonox S until they were told about it by Professor Case in 1949. They contended that their ignorance was not due to any want of reasonable care on their part. The judge accepted what Dunlop said and found, in this respect, no breach of duty at common law.

The judge accepted evidence that by the autumn of 1970 some 450 cases of bladder cancer had come to light in the rubber industry. Although there is no detailed published information about these cases it is reasonable to assume that the circumstances would resemble those brought out by Cassidy and Wright.

One of the key failures in precautions in the case was highlighted in the Court of Appeal: 'There was no evidence that anyone in I.C.I. before 1948 gave thought to what effect dust and fumes from Nonox S would have on those in the rubber industry who had to use it; but there was ample evidence that they gave much thought to the risks to which their own workers were exposed from that chemical.' (Sachs, L. J., and Lawton, L. J., *Knight's Industrial Reports*, Vol. XIII, p. 255.)

As we now know the precaution of the manufacturer's warning the user is now a statutory requirement under the Health and Safety at Work Act, 1974.

There is an urgent need for research into the circumstances of industrial disease. A great deal is known about the causes of industrial disease, but relatively little is known about the circumstances in which it arises. Investigations of causes are largely a matter for pathologists and epidemiologists, but investigations of circumstances using the strategic approach described below are well within the scope of occupational health and safety advisers.

Approaches to investigations of accidents of all types

I believe that the principal conclusion to be drawn is that the search for human error as the cause of accidents is unfruitful. This is because error on the part of one of more human beings must be involved in the causes and/or circumstances of all or virtually all accidents. Errors of one kind or another are an inescapable part of living: who writes or speaks with never so much as a slip? Workpeople make errors in the ordinary performance of their duties. If a particular error results in a catastrophe, who is in error—is it the worker for making the error or his management for failing to foresee the possibility of the error? Questions such as these inevitably dominate employer's liability cases. But their pursuit beyond a certain point in research does not make much contribution to the advance of knowledge. And their pursuit by specialists in occupational health and safety may be counterproductive because people may perceive questions relating to error as attempts to attribute blame—and consequently rationalize their answers.

An alternative—the strategic approach to accidents—promises better progress. The approach starts from the proposition that accidents represent an absence of or a failure in the appropriate accident-prevention precautions. Because precaution is too narrow in concept I use instead the word strategy.

Hence, accident investigations should explore the extent of strategic failure, highlighting those strategies where overhaul is needed.

Table 1.2 shows the strategic framework. There are three classes of strategy.[1] The three categories are pre-accident, post-accident, and collateral. As before, accident is taken to include diseases of occupation as well as injuries and death caused by work.

Table 1.2 Strategies in health and safety at work

Class and sub-class of strategy	Example
Pre-accident	
Safe place	Safe premises, plant, processes, equipment, materials, systems of work, access, transport, storage, disposal of materials, adequate supervision, well-informed and competent people*
Safe person	Use, provision, and no misuse of personal protection, care of the vulnerable †, personal hygiene ‡, cautious and obedient behaviour in the face of danger
Post-accident	Contingency planning for disasters, first-aid training, feedback of information
Collateral	Compensation, many routine medical examinations §, industrial discipline, manning levels

*These ideas are taken from the Health and Safety at Work Act, which in turn stem from Lord Wright in *Wilsons and Clyde Coal Co v English (1938) AC57.*

†Early factory legislation aimed solely at young persons. Later it was extended to women and young persons. Male workers were not then considered in need of protection. See Section 75 of the Factories Act 1961 for a recent example.

‡Personal hygiene—see Section 37 of the Factories Act 1961.

§Many but not all of the routine medicals are done for collateral reasons like protection of the pension fund, safeguarding the company against claims, or as part of a company-sponsored private health scheme for employees.

Pre-accident strategies describe all the classes of action directed towards preventing the accident from happening. This class divides into two sub-classes called, for short, *safe-place* and *safe-person* strategies. The essential difference between these two is that the former aims at tackling danger at source, whereas the latter aims to protect people against danger. Obviously safe-place strategies are generally to be preferred. When these cannot be achieved, safe-person strategies are then necessary.

Post-accident strategies are concerned with actions to be taken after the accident has happened. An important sub-class consists of feedback strategies, which aim at accumulating information which can be used in the future for preventing or coping with similar patterns of accidents—the information is fed back into pre-accident strategies.

Collateral strategies are those with relevance to health and safety, but where the principal objectives are only remotely concerned with preventing or coping with accidents, or not at all. For example, the Committee on Safety and Health at Work (1972) was especially sceptical about compensation's effects on accident prevention. On this view compensation is a collateral strategy.

The framework set out in Table 1.2 can be used as a basis for accident investigations. It can readily be adapted by health and safety specialists in order to fit in with specific industries or factories. It also has the merit of being useful for the categorization of accidents where the investigation has already taken place against a different framework or even none at all.

The work of people specializing in occupational health and safety

The investigation of any accident should lead to the identification of weaknesses in strategic approach. But inquiries into strategic weakness need not—and indeed should not—await an accident to act as the trigger. The health and safety specialist specializes in detecting strategic weakness before accidents take place. He also specializes in ensuring that strategic weaknesses are avoided in the first place.

The scope of occupational health and safety can now be demonstrated. The practitioner should be expert in reviewing health and safety strategies. He should be able to identify danger in its various forms at the earliest possible stage in its development, know which strategies are necessary, and monitor their effectiveness. In addition, he should be able to lead accident investigations so that the maximum of useful, strategic information is brought out.

All health and safety strategies have human involvement. Therefore attitudes and behaviour of employers, managers, workers and specialists are crucial in the implementation of strategies. From the health and safety specialist's point of view, an understanding of human attitudes and behaviour may be the most crucial factor in getting appropriate strategies adopted and maintained. We can now see where the lateral thinking has taken us. Instead of obsessional hole-digging into human error as a cause of accident we now find ourselves wondering how we can influence attitudes and behaviour as obstacles to safety strategies. For example: why do employers always appear to believe that safety is too costly to be achieved; why do managers always seem to give priority over safety; why do some workers think that health and safety is management's responsibility; why has the safety movement been so preoccupied with human error as the cause of accidents?

References

de Bono, E. *The use of lateral thinking,* Jonathan Cape, London, 1974.
Hale, A. R. and Hale, M. A. Review of the industrial accident research literature. *Committee on Safety and Health at Work, Research paper,* H.M.S.O., London, 1972.
Committee on Safety and Health at Work 1970–72, Vol. 2, *Selected written evidence,* H.M.S.O., London, 1972.
Atherley, G. R. C. Strategies in health and safety at work. *The Production Engineer,* January 1975.
Department of Employment. *Accidents in factories.* H.M.S.O., London, 1974.

Booth, R. T. (1975). Dangers at work. Proc. INLOGOV Conference on Health and Safety at Work, University of Birmingham.
Health and Safety at Work, etc., Act 1974 c 37. H.M.S.O., London, 1974.
Knights Industrial Reports, Vol. II, p. 311, Charles Knight, 1967.
Knights Industrial Reports, Vol. XIII, p. 215, Charles Knight, 1972–73.

2. Training in workshops and laboratories in schools, colleges, and universities

Dr A. F. H. Ward

Consultant, and
Editor, Safety News

Introduction

Attention to safety in educational establishments has two main aims: to ensure that the actual working conditions for students are safe, and to inculcate in the students a knowledge of, and a respect for, the principles of accident prevention as a preparation for their later life in industry.

Teaching safety

Teaching 'safety' is quite different from teaching an academic subject. It is not just providing information, but changing a mental attitude. This is difficult, because students resent being preached to. They need to be shown how to recognize dangers, and encouraged to make personal suggestions for the improvement of the safety organization. The staff must adopt the same attitude. When every person acts as his own safety officer it is easy to make safety rules effective. This psychological side of a safety campaign is as important as the provision of safety equipment and the organization of procedures to prevent accidents and create a safe environment.

It is not generally advisable to teach 'safety' as a separate subject. Methods of safe working should be incorporated in practical instruction in the various subjects. The safe way of doing a job is the most efficient way, and if students are properly taught a good experimental technique they will have learnt a way which is not only safe but also has other advantages.

A chemistry student who is learning how to bore a hole in a cork and fit a glass tube through it would be taught in his normal instruction a safe technique designed to give a neat well-fitting product; as a result, he should avoid finishing the exercise with a broken end of tubing fitted through his hand. Similarly, when he is learning the techniques of distillation, he will realize that a volatile, flammable solvent should not be heated over a naked flame. In workshop practice a student must learn how to use hand and machine tools. He must absorb such details of techniques as not to use a cutting tool with a blunt edge, a hammer with a loose handle, a wood chisel as a screwdriver. This sort of information will help him to produce a satisfactory product and to remain undamaged himself. If students are regularly brought into contact with pro-cedures and equipment for safety, as part of a normal every-day working environment, they will become used to, and accept, safety through familiarity. A student will see around him all the equipment of fire-fighting—extinguishers, buckets, fire blankets, hose reels, etc.—brightly painted and in obvious posi-tions. The example of senior students who have already been converted to an interest in safety will have an influence. The co-operation of all the teaching staff should be enthusiastic, for much thought must be given to publicizing safety procedures throughout normal teaching.

Organization for safety

A safety officer should be appointed to take overall responsibility, but, unless the place is very small, he will be unable to cope personally with all details, especially if he has normal academic duties as well. This is the case, too, in large organizations which can employ a full-time safety officer. There should be departmental safety officers, members of the academic staff, who have a detailed knowledge of the work in their departments. When conditions justify it, non-academic sections of a college, such as a canteen department or a works depart-ment (which are likely to contribute largely to the accident record), should also have departmental safety officers.

It is important to avoid any confusion of authority between a safety officer and a department head, who is the top of the chain of command. The depart-ment head should appoint the safety officer and it should be made clear just what authority is delegated to him. In major problems which demand executive action the duty of the safety officer should be advisory. The safety officer should be of a persuasive and enthusiastic temperament since he has to interest others in safety.

The safety officers, collectively, will constitute a safety committee. It can usefully co-opt members from outside who are especially qualified to advise and help. Experts such as representatives from the local fire service, local industry, H.M. Inspectorate of Factories, and from a municipal or university medical officer's department can give valuable assistance. It is also important that the students should be represented. The most useful function of a safety committee is to allow the exchange of information and opinions. Safety problems which are causing difficulties in one department may be solved by specialized knowledge from another. A safety committee can provide safety education for its own members. A newly appointed departmental safety officer may have enthusiasm, but initially he has very little specialized knowledge of safety matters if his own academic training took place some years previously when there was little stress on safety in education. He is very likely to pick up a lot of information in a safety committee where general discussions range over many subjects.

Members of a safety committee should look outwards rather than inwards. It is easy for a committee to yield to the temptation of holding discussions for the benefit only of the committee itself. But members should always remember that knowledge and plans about safety should be spread as widely as possible throughout the establishment. If there are students who do not know the identity of the safety officers, or are unaware that any safety organization exists, then the safety committee is failing in its task.

Reporting of accidents

Whenever an accident occurs, it is an opportunity to study the deficiencies of the safety arrangements and to plan improvements. The circumstances of all accidents should be examined and discussed, and there should be a careful system of reporting all accidents. It is useful to have standardized report forms which can be torn out from a pad, leaving a permanent carbon-copy record. These pads should be available in many parts of the buildings so that report forms can be easily obtained, wherever an accident occurs. Normally the member of staff in charge of a class, who would be responsible for dealing with the accident in the first place, would fill in the form and pass it on to the safety officer. The information to be given on the form should include time, place, name, details of what happened and any useful comments, what action was taken, and who dealt with it. It should be stressed that all accidents and injuries should be reported, however slight. It is also useful to report 'near misses', where an incident might have been potentially dangerous, but where no one was actually injured.

Statistics prepared from accident reports over a long period of time will enable the trend of accidents to be followed and show whether preventive measures have had any effect. Such statistics must be used with care. It is frequently found that when a safety campaign has been started and well publicized there appears to be an immediate *increase* in accidents. In fact this is because all accidents, even those with no resulting personal injury, are now being reported.

Safety equipment

Students are likely to co-operate in a campaign for safety if they see that it is being taken seriously by the authorities. Safety equipment has to be maintained in a state of efficiency. The extent of this equipment will depend on the complexity of the work done, and the requirements of by-laws.

Fire

Any educational building will be provided with fire extinguishers; the types and number of these must depend on the risk. Work carried out in laboratories and workshops may involve a high fire risk, and the staff should see that appropriate fire-fighting equipment is available. Officers of the local fire brigade will always be pleased to give technical advice. (See Chapters 7 and 8 for advice on manual and automatic fire control.)

Fire alarms vary greatly in complexity from hand-operated bells to electric sirens or hooters operated by a break-glass press or automatically by the heat generated. Whatever the type of the alarm, the sound should not be similar to any other signal used in the establishment. With an electrical system there should be numerous actuation points, so that there is one within easy reach in any part of the building. They should have a power supply from batteries so that they are still usable if the mains supply is cut off. In small establishments the alarm may call for the immediate complete evacuation of the building. In modern buildings with a high fire resistance there may be two types of alarm, one indicating only a local fire outbreak (which may be controlled without disturbing people in other parts of the building), and a different sound to call for a complete evacuation.

There should be a good provision of fire notices telling people what to do if a fire starts or if they hear an alarm. There should also be direction signs in corridors and stairs towards exits, all of which should open outwards and be clearly marked, and all passages should be kept unobstructed. Everyone using the building should receive periodical training in the action they should take in case of fire.

First aid

The essence of first aid is that it can be carried out quickly. There should be a wide distribution of first-aid boxes in laboratories and workshops, or wherever hazard exists. For minor injuries (e.g., small cuts, burns, and scalds) first-aid treatment may suffice. For more serious injuries the casualty should be taken as soon as possible to the surgery or hospital. The procedure for this should be known thoroughly in advance and instructions should be shown inside the first-aid box (with any telephone numbers required). It should be clear to which hospital the patient should be taken by car, or what procedure should be used to summon an ambulance. Such details should be discussed beforehand with the authorities.

Eye protection

At least two eye-wash bottles filled with clean water should be available at each first-aid box site, and also near the source of specific danger, and the water should be changed at monthly intervals. For splashes of corrosive substances in the eye there must be no delay in washing the eye with a continuous stream of water. It is strongly recommended that after any eye injury the patient should be seen by a doctor. Even though the eye may seem undamaged after the first-aid treatment trouble may develop afterwards, and a teacher who takes the responsibility of saying that there is nothing wrong with the eye may find himself later in a very difficult position if he has made a mistake. (See also Chapter 29.)

Since injury to the eye can so easily be serious, it is normal practice for safety spectacles or some other form of eye protection to be worn where there is a hazard. This applies in workshops where flying pieces of solid may strike the eye. In chemical laboratories the eyes may be threatened not only in advanced work where dangerous substances may cause an apparatus to explode, but also in elementary work where liquids may be shot out of test tubes in irregular boiling. As well as the experimenter, people standing near by or walking past may receive eye damage.

Machinery guards

This is a large problem, and is dealt with fully in other sections of the book. Essentially it must be made impossible to put any part of the body in a dangerous place when a machine is working.

Some special hazards

Not all educational establishments will be subject to the same hazards. Where advanced work is carried out there may be very serious dangers, necessitating elaborate precautions. Every school, college, or university should assess its own possible dangers carefully. Some problems which frequently occur are discussed in the following paragraphs.

Mercury vapour

Mercury is used for many purposes: collection of gases and experiments with them, in pressure gauges, seals, thermometers, etc. Although most people know that soluble mercury salts are dangerously poisonous, fewer are aware that the vapour can have serious effects and acts as a cumulative poison. The signs can include tremor, stammering, tension, agitation, outbursts of anger, headaches, loss of memory, insomnia, jerky movements, and odd behaviour (the expression 'mad as a hatter' originated from the use of mercury in the hat-manufacturing trade). Later there may be abscesses in the mouth, a blue edge to the gums, loosening of teeth and fillings, vomiting, diarrhoea, and nephritis. The safe limiting concentration of mercury vapour in the air is taken to be about 50 $\mu g/m^3$. If air is allowed to become saturated with mercury vapour at 25°C the concentration will rise to about 300 times the safe limit. This indicates the importance of good ventilation wherever mercury is used, and the avoidance of spillages and large exposed surfaces of mercury. Regular monitoring with a direct-reading mercury vapour meter should be carried out. Spilled mercury should be collected by suction into a receiver, or can be frozen with solid carbon dioxide and brushed up. Inaccessible droplets in floors or woodwork can be brushed with a paste containing lime and sulphur. (See also Chapter 4.)

Waste solvent disposal

Arrangements are often made with a local authority for removal of waste solvents. There should be warning notices on waste solvent bottles in a laboratory against including poisonous or otherwise dangerous liquids. Also, any acid should be neutralized as this may cause corrosion and leakage in metal drums used subsequently for removal. Consult Control of Pollution Act 1974.

Carcinogenic substances

'The Carcinogenic Substances Regulations 1967' came into force in December 1967. These regulations drew attention to the action of certain chemicals in causing tumours of the urinary tract, when inhaled, absorbed through the skin from contact with contaminated clothes, benches or apparatus. The use of some compounds was **prohibited** in industry (beta-naphthylamine, benzidine, 4-aminodiphenyl, 4-nitrodiphenyl, and their salts) and others were only to be used under carefully controlled conditions (alpha-naphthylamine, orthotolidine, dianisidine, dichlorbenzidine, and their salts, and auramine and magenta). In addition to those compounds mentioned in the regulations, diazomethane is similarly dangerous, and many nitroso compounds used in its generation, particularly nitrosomethylurethane. Most of these compounds have been used in educational work. Publicity should now be given to their danger.

Benzene

This widely used compound is treated with great familiarity, but it is now realized that its vapour has an insidious effect in the bone-marrow and can lead to serious blood disturbances. Its maximum safe concentration in air is 10 p.p.m. People who are frequently exposed to it should have regular medical examinations.

Radiation

Lasers. These give a very highly concentrated beam of radiation which can be dangerous to the person and especially to the retina of the eye. What is not often realized is that a reflected beam can be equally dangerous. In using them great care must be taken to contain the beam and warning notices should be displayed.

X-Ray equipment and radioactive isotopes. Workers have to be protected from dangerous radiations. The equipment has to be properly shielded and stray radiation monitored. The worker should wear a film badge to ensure that the maximum permissible cumulative dose is not exceeded. There are detailed legal requirements for this work and specialist advice should be obtained.

Flame-retardant fabrics for amateur dramatics

Fire brigades receive more enquiries from educational establishments about this than any other single topic. Local by-laws usually require that fabrics must be treated to be fire-resistant. Combustion should not be self-supporting, but it should die away and char only. A popular solution to apply is 10 oz of borax and 8 oz of boric acid in a gallon of water, applied to deposit at least seven per cent added solids to the fabric weight, but there are many alternative recipes. The treated fabric should be tested by applying a match for 30 s to a hanging strip. The flame should be self-extinguishing and leave no after-glow.

Legal liability for accidents

The provisions of the Factories Acts do not apply to schools, colleges, and universities. However, these establishments would certainly wish to maintain at least as high a standard of safe working as is required by these Acts. If a student is injured as a result of negligence by a teacher or by his employer, he can sue for damages which could be very heavy. If the injury is caused by a state of affairs for which the teacher is not responsible, such as faulty construction of a building, a claim may be made against the authority. But if the teacher is negligent, both he and the authority may be liable if the negligent act is within the scope of the teacher's employment. However, the teacher alone may be personally liable if his negligence arises from behaviour not consistent with the terms of his employment. Negligence may be failing to do a thing or not doing it properly. The duty owed by schoolmasters to pupils has been defined as 'to take such care of his pupils as a careful father would take of his children'. This duty may not be interpreted so strictly with older students in colleges and universities, but a duty still exists.

Authorities normally take out insurance to cover their liability for negligence. A very small percentage of accidents occur and cause serious injury which are not the fault of the teacher, the authority, or the student. A college can take out a group insurance to cover accidents, however they are caused, for all students. This is not expensive and is well worth doing.

The provisions of the Health and Safety at Work etc Act, 1974, and also Regulations issued under this, apply to educational establishments. Attention should also be given to requirements arising from other recent pieces of legislation, such as The Highly Flammable Liquids and Liquefied Petroleum Gases Regulations 1972, The Protection of Eyes Regulations 1973, The Abrasive Wheels Regulations 1970, Employers Liability (Defective Equipment) Act 1969. In recent years there has been a rapid increase in interest in safety among members of educational establishments. The Department of Education and Science have recently issued an excellent series of booklets (D.E.S. Safety Series) on various aspects of safety in schools and colleges; and the Committee of Vice-Chancellors and Principals have published a *Code of Practice for Safety in Universities*. Many Universities have also produced their own Safety Manuals, some in considerable detail. The Royal Institute of Chemistry have also published a leaflet on *Some Sources of Information on Safety* for members of educational institutions.

3. Communicating the safety message

J. Tye

Director-General, British Safety Council

Introduction

In almost every city or town, in workshops, colleges and schools, steelworks and mines, on television and radio, in magazines, newspapers, on poster sites and, in fact, almost everywhere the safety message is hopefully beamed at the public. In many cases, this message just does not get through. Badly conceived safety messages which raise anxiety levels in nervous workers and drivers can even contribute to accidents.

Conversely, when safety communications are properly planned and directed at the right target group in the right language, and supported by illustrations which attract attention, they have a big part to play in reducing accidents.

Preparation

Over 80 per cent of accidents can be directly attributed to human error and very few of them have anything to do with machinery. Therefore it follows that if we are going to make any impression on the tremendous accident toll, we have to communicate safety as part of our educational process.

Let us lay down some guide lines for better communication and consider improvements in transmitting our safety message with greater impact to motivate the kind of behaviour required. In communicating, we should know:

(a) at whom we are aiming,

(b) what we are trying to say,

(c) why we are saying it,

(d) when is the best time to communicate,

(e) how and at what cost,

(f) where it will take place, and, above all,

(g) what action we are aiming to bring about.

Definition

The object of safety communication is to transfer ideas and knowledge from one person to another so that the message will remain in the memory and motivate certain action.

To whom the message should be communicated

Safety communication should be not only between the safety officer and the workers on the shop floor. It should start by indicating to the chairman, the managing director, and the board room the need for safety, not only on human grounds but also on grounds of finance. No insurance covers the cost of reduced productivity that inevitably follows accidents. After a fatality on a building site in the centre of London, over a thousand workers downed tools for the rest of the day; the cost of this was £5000, which was not covered by insurance. These are the kinds of costs that should be mentioned to top management to make them support fully the safety officer's efforts. At middle and supervisory management level it is vital to prove that the right way to do a job must inevitably be the safe way. When middle management agree to see safety 'built in' to the job, then this line of communication is being a success. Supervisory management has to be convinced of the need for safety; the supervisor, foreman, or charge hand is the nearest management representative to the man on the shop floor. Finally, the facts of safe working have to be communicated over and over again to transmit the safety message to the individual worker.

How people learn

We learn through our five senses, and it therefore follows that the greater the number of senses we can use the greater the impact. We learn through seeing 80 per cent of the time, and yet even today people will talk endlessly without any visual aid and are surprised when they fail to communicate.

We learn by:

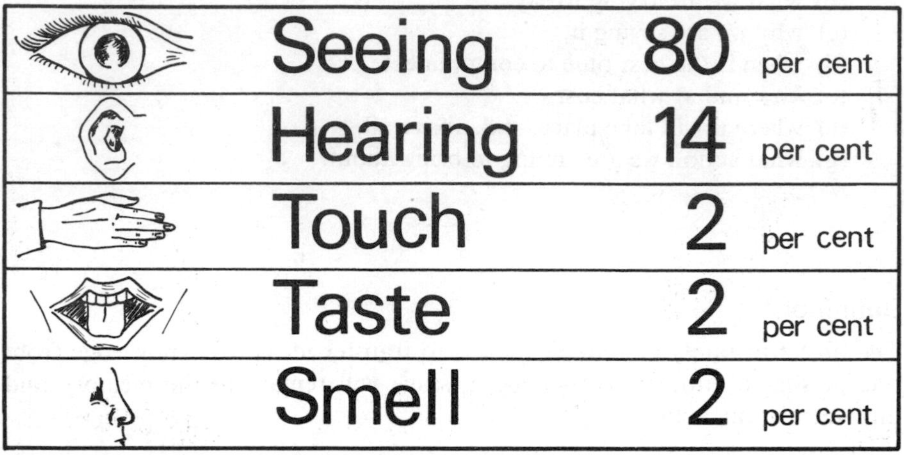

	Seeing	80	per cent
	Hearing	14	per cent
	Touch	2	per cent
	Taste	2	per cent
	Smell	2	per cent

Fig. 3.1 How we learn.

The spoken word is not a good means of communication because it is very frequently misunderstood. It is difficult with an average vocabulary to express oneself adequately and even professional communicators can be misunderstood. If anyone has doubts as to the effectiveness of the spoken word, I suggest that he stands in front of an audience and, with hands firmly clasped behind the back, describes some complex piece of equipment, say the movement of a sleeve in a rotary engine, or even something simple like a spiral staircase! Most experts agree that there is 30 per cent more comprehension and probably 50 per cent more retention when multisensory channels of communication are used.

Choosing your words

If you want to communicate to nuclear physicists, by all means use technical language; but if you want to be understood on the factory floor, then use simple short words which convey the meaning and which are impossible to misinterpret. 'Beanz meanz Heinz' may be terrible English, but the message certainly gets over.

Be specific—do not be vague

Of all the thousands of posters now being displayed, not only in factories but on roadsides and public buildings, 50 per cent fail because the reader simply does not understand the message. The 'Stop Accidents' and other general campaigns, unless motivated by a clear and specific call for action, are simply a waste of money. Sit down and study your accident statistics. Sort out a specific problem and then use a specific slogan. Do not refer to 'lifting safely' but give the specific instruction 'bend your knees and not your back'.

The safety message communicated to an assembled audience by means of the spoken word with mechanical aids

Thirteen guide lines for safety communicators

1. Visuals, diagrams, and charts used in front of an audience should preferably start in the bottom left-hand corner and work up to the top right-hand side. This indicates progress (think of sales graphs) and falls into line with what we might expect.
2. Disjointed layouts on any form of presentation indicate disjointed thinking and illogical thoughts and ideas.
3. Simple straightforward symbols should be aimed at when using visual aids: the watchwords should be simplicity, clarity, and understanding.
4. Use signs and symbols which are readily understood by all your audience.
5. Words used in visuals must be clear, short, and simple.
6. Photographs must be blown up large enough to be seen: any competent artist will 'white out' anything in the picture which is irrelevant to the point. Nothing must be allowed to get between the communicator and his audience.
7. Sketches and drawings must be ruthlessly concentrated on the point to be demonstrated.
8. Technical sketches must focus on the point: the machine as a whole need not be in great detail.
9. Cartoons and caricatures have their place in visuals, but beware that the point of the safety communication is not lost in the resultant laughter. Note that national advertisers are very chary of using humour to promote their products.
10. 'Sick humour' with funny tombstones, blinded workers, or bleeding bodies, are never well received by those who have had close connection with a bad accident. Keep off it.
11. 'Horror' or shock photographs or films, unless very carefully worked out, are of very limited value. If you use them, make absolutely sure that they have a clear message and a course of action.
12. Complex diagrams, organization charts, etc., are often defeated by their own complexity. If you must use complicated drawings, try to break them down into understandable units or groups.
13. Colour is sometimes used on visuals; make sure that symbols in colour conform to the appropriate British Standards.

What methods are best?

There can be no one method of safety communication. The method you use will depend upon the size of the group you are aiming at, the amount of material you wish to put over, the age and experience of the group, whether you wish to implant a completely new thought or to present 'recall material', and the amount of money available.

Sometimes a black-board will be better than a film show, other times a built-up 'flannel board' will get home, but I suggest that unless your message is both seen *and* heard you are no more than 50 per cent effective.

Methods in order of priority

When selecting the main means of communication to an assembled audience, the order of priority should be:

(a) moving pictures with realistic sound-track and/or commentary;
(b) still pictures with spoken commentary;
(c) still pictures with written commentary;
(d) oral communication;
(e) written communication.

Planning your safety presentation

Diagnose your problem

Why is it necessary to communicate and at whom are you aiming the communication? The answers to some of the following questions can be a guide. 'What is the real need?', 'what financial difficulties are there?', 'what other barriers are there?', 'can you obtain active support as distinct from "lip service" from the board room?', 'can you get trade unions' support?', 'is the safety communication aimed at passing on information quickly, or at the changing of ideas and of behaviour?'.

Prepare the brief

Who is likely to be able to contribute worthwhile suggestions? Certainly management might advise you on any budget limitations, but, if your diagnosis has been thoroughly prepared, even budgets can be extended. Preparing the brief takes into account the time necessary for preparation, the time to write a script, to take photographs, to edit, and to revise.

Write your script

Collect your information and then do the actual writing. Start by putting all your thoughts down on paper. Just put down as much as you possibly can which is relevant to the problem. Having done this, sort out the ideas into a logical sequence. You should now be able to group together the various points in some progressive order. It might help at this stage to ask yourself the journalist's six basic questions: who?, why?, what?, when?, where?, how?, and add at what cost? Separate each idea and plan the most effective way to convey the point, bearing in mind that it has to be seen as well as heard.

Plan your script with the audience in mind. To illustrate a machine guard the 'flannel-board treatment' might be best for apprentices, a sketch for engineers, a photograph for actual operatives, and for management a diagram with a comparative cost chart.

No matter what visual aids you propose to use, prepare your script as though it were for a film or a television show. It helps to divide the page into three columns and mark the first column 'visual', and the two right-hand columns 'spoken'.

Assemble your audience

When you have completed the script, visuals, diagrams, photographs, blow-ups, and working models, you are ready to present them, but where? Is it to be in a lecture room, canteen, theatre, cinema, or first-aid post? It is advisable to have a look at the place to see if it is light enough, too hot, or too cold.

Run through the following check list before the audience gathers.

Check list.

Have instructions for attendance been issued?
How many will attend, and are there enough chairs?
Are tables and chairs set out as planned (horseshoe shape or straight)?
Are the name plates and working materials in front of every place?
Are registration forms or other services required?
Is there a registration table by the entrance?
Are suitable badges supplied and printed?
Are seats for the speaker and following speakers located as required?
Is the speaker plainly visible to all the audience?
Are blotters, ashtrays and water on hand for speakers?
Have the signals between speaker and projector operator been worked out?
Is the projection screen high enough to be seen at the back of the hall?
Displays and exhibits—are they clearly visible to all the audience?
Are all models in good order?
Audio-visual equipment—is it correctly located and in good order?
Are aids dependent upon light circuits? Is there a spare fuse?
Have lights been tested?
Do the lights distract, and are the switches readily available?
Who is responsible for light and power control, and is voltage correct?
Has the power point been checked and is it readily available?
Have the lighting arrangements been rehearsed?
Are the projector lenses clean and are there spare bulbs?
Are all scenery and props on stage and readily available?
Will the platform curtain need to be opened and closed?

Microphones should be readily available; are they correctly placed?
Is there a lectern or raised table for speaker's notes?
Is the room too hot, or too cold, or is it adequately ventilated?
Will a black-board, chalk, duster, rubber erasers, or crayons be required?
Has a handout or précis to be given out afterwards?
Who, in case of breakdown or other crisis, makes immediate decisions?
Has a timetable or order of the day been set out?
Have the press (and photographer) been briefed?
Are exit doors free of obstruction?

Rehearsal

The more you rehearse and polish up your presentation the stronger will be the impact of your safety communication. The better you know (and can handle) your props, speech, and cues, the more effective you will be. By freeing your mind of mundane things you will be able to establish a stronger contact with the audience.

The props to the script

Using the black-board

Even though black-boards are to an extent overshadowed by more advanced visual aids, they are still invaluable to clarify a quick point. They are now made in different colours, able to take different coloured chalks and crayons, move freely up, down, and horizontally along classroom walls. They are cheap to buy, independent of electricity and require only a piece of chalk to perform a useful visual function. The greatest drawback is that when you are actually writing on the board you lose contact with the audience because your face is turned away from them.

Hints for using the black-board. Decide well in advance for what you are proposing to use the black-board and then plan each usage. (The reverse side of the board can be prepared in advance.)

Use only capital letters which are sufficiently large to be seen by everybody in the room. Do not bother about details. Draw a bus in outline, not with the driver and conductor in position. The same thing applies with a machine guard, a cross hatch will do; it does not, for the purpose of black-board illustration, have to be a work of art.

Use coloured chalks if these can be useful.

Having written or drawn on the board, stand to one side so that people can see it, or even better, train yourself to stand to one side as you actually write on the board.

Using flip-over charts

A flip-over chart can be used like a black-board but has the great advantage that your various points can be 'flipped over' and just as easily 'flipped back' to restate, reinforce, and summarize your point, and can be used again at a later date. Briefly, a flip-over chart is made from a plywood back, to which 30 or 40 sheets of cheap newsprint paper are clipped, and it either stands on an easel or is held on any convenient table. Minimum size should be 768 mm × 504 mm which is metric double crown size. In the Safety Council we write on the charts with broad felt pens in a size easily readable in a room of a hundred people, but others prefer black grease crayons.

There is a word of warning about charts in the hands of lazy communicators. There are a number of information sheets and charts produced by governmental and business concerns which contain a mass of information which needs to be dealt with point by point, section by section. It is no use putting a complex information chart on display without explaining each particular section.

Flip-over chart—previous tracing

It sometimes happens that an item of machinery, an article of safety equipment, or even a part of the anatomy needs to be carefully illustrated and 'built up' on the spot. For this you may pre-plan your drawing by using a pale yellow wax pencil with which you trace on your flip-over chart any design required. This is quite indistinguishable to members of the audience and all you have to do at the time is to follow your previous tracings with a thick black crayon.

Using flannel board

Anyone can make a flannel board or felt board at a cost of about £2, together with the various models to go with it, to illustrate almost any safety problem. The board may be covered with a piece of flannel or felt, and any model/symbol/design which is backed by flannel or flock wallpaper will stick to it. All the material can be previously prepared and can easily be pressed onto the flannel board with the fingers.

Hints for using flannel board. Plan in advance and have your cutouts ready for placing on the flannel board and keyed to your script.

Draw your models in simple but understandable forms; cut them out and stick them onto flock paper (available at any wallpaper shop). The cost will be negligible.

Build up your presentation logically so that one step reasonably follows another.

Keep away from detail—keep it simple.

Using other display boards

There are other types of boards serving much the same purpose as flannel boards, but with the use of other adhesives. You will have seen magnetic boards in use on television; sometimes these have iron pellets embedded in the board itself, and others have the faceboard treated with iron filings which will retain models with a series of small magnets at the back.

There are plastic boards upon which cut-outs of clear plastic will stick so that a visual presentation can be built up. The sheets of plastic may be as large as 1 m × 1 m and as many as 15 sheets can be superimposed one on top of another.

You will have seen peg boards and simple working models (possibly powered by a small battery) in shop window displays. The cost of these in total may be only a few pounds and they can often be adapted for a safety communication purpose.

Remember that whatever your safety problem, there is always a method of conveying it to your audience providing that you are willing to break it down into ideas which can be assimilated, and to use accessories to the spoken word.

Working models and demonstration

There can be very few methods of communicating safety quite as effective as the actual demonstration of some type of protective equipment. Some manufacturers have models for demonstration purposes, showing the effectiveness of their goods.

Certainly one of the most effective models is that showing what happens to the spine when a load is lifted wrongly. The spine breaks dramatically, and, if the model is used in conjunction with a skeleton, and quoted statistics (including financial cost), it is a potent means of conversion to proper lifting.

The only limiting factors against effective models are those of your own skill, money, and ingenuity. Furthermore, a good model will stand up to being handled by the persons to whom you are demonstrating. As with other visuals, they must be big enough to be seen at the back of the room.

Demonstration units and mock-ups

It is often quite impossible for reasons of space, time, preparation, and mechanical or electrical complexity to make a working model of something to be put over visually, but a demonstration unit or mock-up can be built to illustrate one particular point.

I have seen the principle of photo-electric guards quite clearly demonstrated with mock-ups. A wiring system can be easily shown on a board complete with switches and relays when it simply cannot be seen completely in a car or even an aircraft. Barrier and cleansing creams can best be demonstrated actually on the hands, particularly when the main function is the removal of oil or grease.

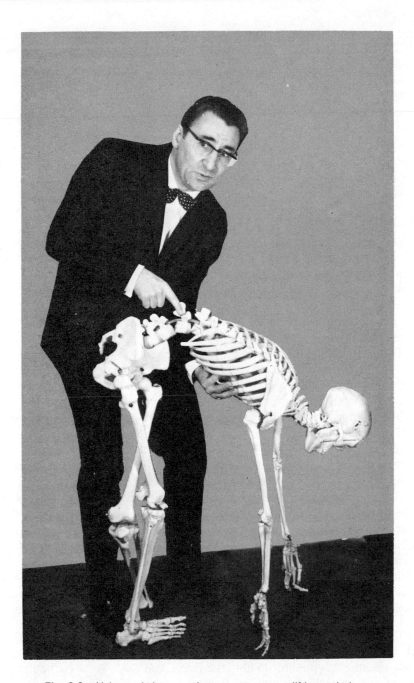

Fig. 3.2 Using a skeleton to demonstrate correct lifting technique.

Using films

There are several hundred safety films available for hire in the U.K. The techniques used vary from the documentary single colour approach to the colour cartoon of a few minutes' duration.

Space does not allow film making to be dealt with here but there are several excellent publications available, and many safety officers are able to communicate safety by using films they have made themselves, with spoken commentaries dubbed on.

I suggest you follow these ten rules when using films:

1. Regard the film as just one of the media of communication and fit it into your programme—do not make space for it unless it integrates reasonably well and performs an effective communication function.
2. Put your screen up only when the film is to be used, otherwise its presence will distract attention from the rest of the session.
3. Very few films have *exactly* the kind of message you require; therefore you will need to direct attention towards the point you want made both in an introduction and also in a summary afterwards.
4. If you are showing a film make sure that people can see it, stagger the seating, and fix the screen high.
5. Film-showing should not be 'dozing sessions'—therefore check ventilation and ensure that the room is not too hot.
6. If you can link up the film with a working model to be shown afterwards to illustrate visually the point you wish to make, your success is guaranteed. For instance, a 'lifting film', backed up by a working model to be used in the discussion and summary afterwards.
7. Do not mix up the idea of safety communication with entertainment. Humorous films shown to break the monotony have no place in a communications programme. They serve only to distract.
8. Note-taking during the actual showing of a film is difficult and lessens the impact so it should not be encouraged.
9. If special names, terms, or even slogans are to be used, then viewers must be briefed beforehand. Nothing is worse than not understanding a key word at the beginning of a film and puzzling over it for half an hour.
10. Never rely upon a volunteer who 'happens to be around' and says he will run the film.

Finally, it is vital to understand that safety films are purely instructional aids to complement or supplement the 'safety communication'. They are only as effective as the foundation which has been previously prepared by the instructor.

Using filmstrips and slides

Both filmstrips and slides are particularly useful in safety communication, mainly because it is possible to hold the frame and study some particular point in detail in a way which cannot be done with the moving film.

The use of filmstrips demands just as much preparation as any other visual aid because every successive slide or film shot should be slotted in to form a pattern. Although silent film strips and slides are easily available, visual material makes deeper impact if supplemented by the voice. Some strips run for 45 minutes or even an hour, but if one cannot convey a point in 20 minutes, then the session should be broken into two.

The criticism is made that filmstrips produced in the U.S.A. or Russia or anywhere else do not apply to the U.K. This should be expected and dealt with properly and may even provide a lead in a dull discussion period.

Hints for using filmstrips and slides. Use the filmstrip as a tool in the overall safety communication rather than make your presentation around the filmstrip.

As a general rule, keep the time of showing to about half the length of your total time, and thus allow time to introduce your subject, to brief the audience on the various points you want to raise, and to summarize— allowing discussion of the salient points afterwards.

Always do a run-through, mark your slides and get them located the right way up; audience concentration is too valuable to allow it to disappear amidst laughter if a slide is shown out of sequence or upside down.

Do not use 'The End' on a slide; this breaks up the class too easily and people tend to start moving off.

Actual case histories shown on filmstrips are tremendously valuable, but they must always be accompanied by a positive message. Just running through a series of 'after' pictures may shock people but does not tell them what to do to avoid being victims.

If one slide can effectively make your point then it is not necessary to use more. Ask yourself the question 'Do they add anything to what has already been shown or said in support?'

Using overhead projectors

One of the most recent and valuable tools for visually communicating safety is the overhead projector, of which there are at least a dozen models on the market. It has a tremendous advantage in that the speaker can face his audience at the same time as he places transparencies right in front of him and they are immediately projected on a screen or wall behind him.

A cellophane film is fed across the two rollers which will take a wax or grease crayon. The projection is much about the same as with a 16 mm film and can be used in a normal class of up to 100 people.

As with films and filmstrips, colour can be used most effectively, but with the added advantage that the slides can be written upon, areas traced, safety points and features marked. These can be retained for a later date or discarded.

Much the same rules apply as with films and filmstrips and it is not necessary to repeat them here.

Using tape recorders

Although we learn more effectively by looking than by listening, there are times when tape recorders can be used alone, and times when they can be used to supplement the visual message to make greater impact.

With the most professional sound recordists in TV and radio there is always a 'voice level' test; when this simple precaution is ignored amateurs are surprised that the subsequent recording booms, blurs, whispers, or is completely inaudible. As with other equipment, become familiar with it beforehand, and have at least sufficient technical knowledge for adequate operation.

Hints for using microphones. Make sure beforehand that the microphone will adjust in height.

Become accustomed to speaking in a normal speaking voice when using a microphone.

A good distance to be from the microphone is 0·5 m, this allows room for more than one person to speak. Avoid the pop-singer's microphone-hugging technique.

If you are speaking to an audience through a microphone, concentrate on the audience and do not be hypnotized by the microphone.

A few people find that using a microphone limits their movement in front of an audience when using visual aids. This is a very real difficulty which must be overcome and can only be defeated by preparation and practice.

Closed circuit television

Undoubtedly, the communication medium of the future will be closed circuit television. Travelling units are complete television studios in themselves, and have up-to-date equipment with the control room housed in a single vehicle. There are fixed and moving cameras, telecine equipment, a video tape recorder, and monitor screens which can be installed up to 200 m away from the studio. Using both inside and outside cameras, extensive safety programmes can be broadcast through the closed network.

Because it is live, closed circuit television has the same attraction as a non-scripted national television programme and when accompanying graphic illustrations are shown this adds to the dramatic impact.

Setting up a closed circuit television studio (which might well cost £25 000) is outside the scope of most safety officers' budgets but it is now possible to hire these units weekly for approximately £120, and certainly there is a case for trade associations or trade unions to think about co-operating to purchase one.

The safety message communicated within the works during work operation

The guide lines on page 21 should be borne in mind.

The purpose of safety posters

As with any other visual aid tool, the user of a safety poster should have in mind quite clearly the specific job he wants it to do, which will probably be one of the following:

1. To remind employees of some specific safety practice that they have previously been taught, or of a hazard they have laid themselves open to.
2. To remind employees of human traits that cause accidents.
3. To underline the basic need of using safe working practices and to stress that accidents are a sign of an unskilled worker.
4. To encourage the right sort of behaviour to reduce accidents, and perhaps stress the need to wear safety clothing and footwear.
5. To encourage employees to work and think as a team in preventing accidents.

There is plenty of evidence to show that safety posters aimed at specific hazards, with a clear message and sited in the right place, do a good and effective job. However, because they require very little effort to select and put up, they have frequently been used mistakenly as a complete safety programme rather than just one of the many visual aids available.

Categories of posters

Every poster you decide to use, or design, or in any way become associated with, should be evaluated on the basis of the British Safety Council's seven-point formula, which goes under the mnemonic of STAPIMA.

Strategy. Your safety strategy must be to make the best use possible of the money, time, materials, and talent you have at your disposal, towards a specific objective.

Target. If the strategy is the overall plan, then we must have a specific target for a specific safety communication: the target group. Ask at this stage:

    (a) who is it aimed at?
    (b) is it understandable?
    (c) does the picture clarify and not confuse?
    (d) is there a clear call to action?

It is better to aim your communication to hit the target group 50 per cent of the time than to pepper the message in every direction, missing the target all of the time.

Attract attention. This is absolutely imperative; unless the communication will be read the whole enterprise is a waste of time. Translated into British Safety Council terms, this means dramatic photographs, strong colours, eye-catching artwork, and sometimes the use of celebrities; but no horrific shock posters which only raise anxiety levels.

Positive and attainable slogan or message. At this point you have decided at whom you are aiming. You have worked out the means whereby you direct attention and it is vital now to make sure that the message you are going to put over is positive, and that the action desired is attainable.

Interest. The poster must be interesting to the reader. There are so many ways of presenting material which might attract attention, which might have a positive slogan, but which are boringly uninteresting if the subject matter is not applicable to the target group.

Memorability. The average person is subjected to more than a thousand persuading messages every day and although your poster safety message is only momentarily seen, it will be seen repeatedly. If the picture is arresting and to the point, the message will gain a fraction more impact each time.

If it can be immediately memorable it will have even greater impact. Perhaps the memorability is brought about by a dramatic illustration, a photograph which appeals, a rhyming couplet which 'clicks' or a warranted trick of presentation.

Action. Unless the safety communication is designed to promote some kind of action or response in the mind of the reader, then it is useless. Therefore we must aim that people *will wear their seat belts, lift with the knees bent, wear protective helmets, practise tidiness at work,* and *observe all the rules of safe working.* We *must motivate action.*

Using the press and other mass media

Safety is worthy of promotion as *NEWS*. There are many ways in which this can be done, if you exploit the circumstances by dressing the subject up in a suitably publishable announcement. Perhaps it was the case that your 'safe month' was the result of your using a new and original method of work; again, perhaps it came as a result of an employee's brainwave. Whatever the reason it should be passed on and publicized as much as possible, for the benefit of your own firm and others.

The actual writing of news stories and feature articles, together with the taking of photographs suitable for publication, are, strictly speaking, jobs for professional journalists and photographers. However, there is no reason why the safety officer, or the firm's director responsible for safety, should not supply the press regularly with facts and figures on his firm's safety record and achievements. Regular publication of such material is of inestimable value to the overall safety campaign; it keeps the theme of industrial safety constantly before the public (and your factory workers are of course members of the public).

In the writing of such material, be brief, factual and accurate. Supply as many names as possible (properly spelt and with first names). Where some aspect of the story calls for a picture, invite the local press to send a photographer. If your firm has a house magazine or wall newspaper, these are admirable vehicles for safety propaganda as well.

Conclusion

Wherever we see a company with really low accident rates, we see behind it the safety officer doing his job efficiently, and, in the process of so doing, communicating his safety message effectively.

He uses all the tools of communication to which I have referred, and keeps them well 'honed' by using the most up-to-date materials available in the form of books,[1] such as *Safety Management* by Dan Petersen and *Management Introduction to Total Loss Control*, film guides,[2] and slides, cassettes, and other visual aids.[3]

In addition, he remembers the need to exchange ideas and information through meetings with his fellow professionals at courses and conferences.

By communicating the safety message he makes a tremendous contribution for good within, and without, the industrial environment.

References

1. Both available from The British Safety Council.
2. Available from such sources as Central Film Library, Guild Sound & Vision, Sorel Films Ltd, and British Safety Council.
3. Obtainable from Camera Talks Ltd, Rank Audio Visual Ltd, and BBC Enterprises.

4. Chemical handling

E. L. M. Roberts

Manager, Oldbury Safety and Environmental Hygiene Unit,
Albright and Wilson Ltd

Introduction

In this chapter, an attempt is made to give guide lines on which plans for the safe handling, i.e., the safe receipt, storage and movement of chemicals, can be based. It does not purport to deal extensively or comprehensively with any one aspect of chemical safety; its terms are general, not specific or technical, and its theme is concerned more with the user of chemicals than with the manufacturer.

There are thousands of chemical compounds and mixtures which present some form of hazard—either major or minor—to the unwise user, but, just as the hazard of the exposed gear wheel, or of the unguarded circular saw, can be controlled by correct study and application, so can the hazard of any given chemical compound be controlled. Nevertheless, with the increasing use of chemicals in non-chemical industries, the incidence of what may be loosely termed 'chemical accidents' increases year by year. Many of these accidents are due to ignorance of the properties of the chemicals involved in the process. In the chemical industry itself the safety problems involved in the use of chemicals, by reason of both size and complexity, demand the systematic evaluation of all known hazardous properties of the chemical, e.g., toxicity, flammability. By such evaluation, plant can be correctly designed, hazards eliminated or controlled, wastage of valuable chemicals avoided, and the maiming of operators prevented. Hence, in the chemical industry today, large quantities of chemicals can be manipulated without risk, reactions releasing great amounts of energy can be controlled, and the accident frequency rates can be contained at a figure below that of other industries. Furthermore, current legislation introduces an even more stringent need for hazard pre-recognition, and there is an implicit demand to inform employees of the hazards of their occupation.

During recent years it has come to be an axiom of good management in the chemical industry that the safe manufacture of chemicals is dependent upon three main essentials:

(a) knowledge of the chemical;
(b) knowledge of the processing plant;
(c) knowledge of the operator.

This maxim equally applies wherever chemicals are used in any industry, whether in small or large quantities.

It is not the intention of this chapter to deal with the last two of these essentials (both demand separate thought and study), but to concentrate on the first essential, namely 'knowledge of the chemical'. Knowledge of this first essential largely influences the requirements of the other two; it governs plant design and equipment, and determines the degree of skill and attention required by the operator.

Assessing the dangers of a chemical

Although the general knowledge of how to handle a particular chemical safely is often sufficient, if some chemical changes take place within the plant or process involving this chemical it is necessary to understand these too. For instance, if hydrochloric acid is being used in pickling steel, it is not sufficient just to know how to handle the acid. We must be alert to the evolution of hydrogen in the process and to traces of more unpleasant materials due to trace impurities in the steel. Every producer and every user of any chemical already makes some form of assessment of its properties, varying from 'I know it's corrosive' to an 84-question sheet as used by one of the larger chemical firms, but the standard safety 'assessment sheet' (Table 4.1) which the writer advocates should be used in all instances, irrespective of the amounts of chemical involved. It is one which has been used over a period of years with

Table 4.1   Safety—initial assessment

1. Name of chemical   . . . .

2. What is its physical state?   . . . .

3. Is it toxic?   . . . .
    acute?   . . . .
    chronic?   . . . .
    by ingestion?   . . . .
    by inhalation?   . . . .
    by absorption?   . . . .
    can its toxicity readily be determined?   . . . .
    threshold limit value   . . . .

4. What is its:
    vapour density? ....
    vapour pressure? ....
    freezing point? ....
    specific gravity? ....
    miscibility with water? ....

5. What is its incompatibility? ....
    ....
    ....

6. Is it flammable, highly flammable? ....
    what is its:
        flash point? ....
        explosive limits? ....
        ignition temperature? ....

7. What type of fire appliance should be used? ....
    ....

8. What type of protective clothing should be used? ....
    ....
    ....

9. Other special safeguards ....
    ....
    ....

                                         Compiled by ....
                                         Date   ....

considerable success. The assessment should be made as soon as it is known that a new chemical is to be used, and a reassessment should be made when there is any change in site, plant, or supplier.

Why are the questions on the assessment sheet so framed? What is the value of the information so compiled and from what sources can this information be obtained? The following sections will attempt to clarify these points.

Name of chemical

The need for this question is obvious, but the popular name should be given as well as the chemical name, so that in a situation where acetyl salicylic acid means aspirin to the chemist, the operator experiences no confusion or conflict. Generally steel is steel and cast iron is cast iron but the use of different systems of nomenclature in chemistry tends to lead to confusion. For instance, the chemist talks of H_2S whereas the engineer talks of hydrogen sulphide, calcium hypochlorite is chlorinated lime, phenol becomes carbolic acid and simple baking soda becomes bicarbonate of soda, or acid carbonate of soda, or sodium hydrocarbonate, or sodium hydrogen carbonate, or perhaps simply $NaHCO_3$.

What is its physical state?

The object here is simply to determine whether the chemical is received as a solid, liquid, or gas—not, at this stage, to ascertain its general physical properties. We should also consider in this section under what conditions the chemical is a solid, liquid, or gas. Sodium hydroxide (caustic soda) can be purchased as a solid flake in drums or as a strong solution in road tankers or drums; carbon dioxide can be purchased as a solid, liquid or a gas. Generally, heat in or heat out is required for any change of state, so that this assessment determines how and where the chemical should be stored. Will sunlight and heat affect it? Will it freeze if left in the open? If it is a solid, is it a dusty powder? Is dust suppression or extraction necessary? Consideration must be given if chemical is delivered and stored in an unstable state, for example, solid carbon dioxide, liquid air, liquid ammonia. Dangers to be considered with these items are, say, rapid temperature rise due to proximity of fire, rapid emission due to leakage, etc. If it is a liquid, where will the spillage run to? Can the drums rot if left on damp ground, or corrode internally if stored over a long period of time?

Is it toxic?

The clarification between toxicity and hazard should here be clearly understood. The toxicity of a chemical is one of its inherent properties which cannot be removed if the chemical is to remain true to its formula, but hazard is determined by the frequency and duration of exposure and the concentration of the chemical. Injury cannot occur without exposure to a given concentration and the design and operation of any chemical process determines the amount of exposure, concentration, etc. Therefore, by correct design and safe handling, the hazard can be removed or its potential markedly reduced.

Because of its common usage, it is almost universally known that concentrated sulphuric acid is a corrosive liquid which rapidly destroys the tissues of the body and causes severe burns on contact. Nevertheless, hundreds of tons of sulphuric acid are manipulated, transported, and stored daily without great hazard. This is because its toxic properties are known and understood and the full accident prevention measures are enforced. Hence the determination of the toxic properties of any chemical which is to be used is of paramount importance, particularly where the toxicity is of an insidious form leading to chronic poisoning. The action of sulphuric acid on the skin is immediate and acute, as is the effect on inspiration of ammonia fumes, but although benzene liquid on the skin in small quantities is not particularly objectionable, the accumulative effects of its insidious properties can lead to a serious form of anaemia and death. In a similar manner nitrous fumes are extremely dangerous because of their insidious character. Often there are no immediate or acute effects and the first indication that there has been an exposure may be the latent development of the characteristic symptoms of nitrous poisoning some hours later. The determination of the manner in which the toxic substance or poison enters the body, i.e., by inhalation, ingestion, or absorption, indicate the tests to be applied and the preventive measures to be adopted.

A further aspect of the question of toxicity is whether the toxicity can readily be determined and what is its threshold limit value (TLV) expressed in parts per million, which represents conditions to which it is believed that nearly all workers may be repeatedly exposed day after day without adverse effect. However, warning should be given that the TLV, in its proper context, should be interpreted only by a person trained in industrial hygiene, and should not be used:

(a) as a relative index of hazard or toxicity;
(b) in the evaluation of air pollution nuisance;
(c) in estimating the toxic potential of continuous uninterrupted exposure.

The written advice of the supplier, or of H.M. Factory Inspectorate, in this matter is desirable if expert opinion is lacking within the particular organization.

Although the dangers of reliance for detection on the sense of smell cannot be over-emphasized, there is no doubt that the characteristic odour of certain chemicals is a clear indication of their presence, though not of their concentration. Herein lies a danger in human monitoring. For instance, the odour of chlorine can be recognized by smell in very small concentrations and, as there may be no immediate irritation or adverse effects, no remedial action may be taken. However, the maximum permissible concentration of chlorine by volume of air should not exceed one part chlorine per million parts of air for an eight-hour exposure, and as small concentrations detectable by smell generally indicate three or four parts chlorine per million parts air, anything more than a brief exposure to such a concentration could be hazardous. If chlorine can be detected by smell something is wrong with the installation.

Notwithstanding such contra-indications, the recording that chlorine does have a characteristic odour is of value in the case of accident or leakage. Similarly, the ready identification of toxic concentrations by means of indicator papers and either monitoring equipment for spot checks, or, where applicable, continuous detection equipment, with a warning device indicating a concentration rise to danger level should be recorded. The recording of the TLV is of real value to the laboratory technician or other person who is assigned to carry out atmosphere sampling where this is necessary, and it is also a fair indication of the toxic property of the chemical, e.g. chlorine 1 p.p.m., carbon monoxide 50 p.p.m,, carbon dioxide 5 000 p.p.m., arsine 0·05 p.p.m., hydrogen chloride 5 p.p.m.

Physical characteristics

Knowledge of the five characteristics, vapour pressure, vapour density, specific gravity, freezing point and miscibility with water, provides separate and valuable facts and information. All exposed liquids vaporize, but the rate at which they vaporize depends on temperature and pressure; generally, hot liquids vaporize more rapidly than cold liquids. The vapour pressure of liquids and solutions should be quoted, preferably at ambient temperatures. This is extremely important when storing drums containing a hazardous liquid. A leak from such a chemical, say, from a drum in a warehouse or workshop, would present a danger. A comparison of the density of a vapour/gas with that of air (standardized as 1·0) shows whether the vapour at normal temperature (0°C or 32°F) and normal pressure (76 cm or 29·92 in. Hg) is more dense or less dense than air, viz. it will continuously rise up into the atmosphere or it will sink to the lowest possible level. Petroleum, for instance, has a density in the region of 2·5. Hence a leak of petrol, readily vaporizing at normal temperatures, creates a vapour which tends to creep along the ground. Such conditions are, of course, affected by wind velocities, ambient temperatures, etc., but there have been a number of instances where petrol has been spilled at some distance from a garage inspection pit but the spillage vapours have moved along the ground into the pit, creating an explosive atmosphere requiring only a spark to cause a disaster.

Such information is of particular value where work is done on a plant having several levels of platforms and where spillages or leakages of a toxic nature could occur at lower levels. When considering means of escape from such plants it might be that escape routes should enable an operator to escape upwards and outwards rather than downwards and outwards. For instance, chlorine has a vapour density of 2·5, so that it becomes obvious that a person working above the point of leakage should be able to make his way out of the plant without moving down to a lower platform level, that is, into the dense concentrations of fume.

The value of knowing the specific gravity of a chemical becomes readily obvious when determining the action to be taken in the case of a major spillage. This comparison of density with the density of water (1·0) shows whether the chemical will float on the top of the water or sink below it. All liquid spillages eventually reach the drains, and underground explosions due to contamination by highly flammable liquids can cause costly damage over a wide area. The major value of this knowledge is concerned with flammable chemicals, e.g. petrol. Petrol has a specific gravity of 0·80, so that spillages float on water. Hence water is not recommended as a fire extinguishing agent for liquid petrol fires, as water sinks below the petrol and, by the increase in the volume of the liquids, tends to spread the area of fire. Allowing the petrol to run away into the drains merely increases the hazard and possible area of involvement.

Conversely, burning carbon disulphide, a very highly flammable liquid, having a low flash point and low ignition temperature, has a specific gravity of 1·26 and can be controlled by the gentle application of water.

If the chemical is miscible in water, any spillage can be more readily dealt with as it can be saturated with water and, after appropriate precautions have been taken, swilled down into the effluent system. Incidentally, miscible is a word which tends to confuse the non-technical operator. Although it is a word which often appears on chemical labels, the operator is never quite sure whether to ignore it because he does not recognize it, or to relate it with the verb mix, thereby concluding that a miscible chemical will mix with water. The correct interpretation of this word should be clearly explained to all concerned with particular emphasis on any converse property, immiscible. An immiscible liquid poured down a drain will probably stay in the traps and other recesses and may become hazardous if, at some later date, a second immiscible liquid which reacts violently with the first is disposed of in a similar manner.

Whilst considering the miscibility of a particular chemical, it is worth considering also the action of water or water vapour upon the chemical. There are many case histories of serious injuries arising from the simple introduction of water into partially emptied containers of a variety of chemicals causing violent reactions. For instance, phosphorus chlorides are not of themselves corrosive but, in contact with water or water vapour, react violently, giving off heat and corrosive fumes of hydrogen chloride. Similarly, some years ago, a quantity of sodium cyanide was spilled from a container and the spillage was swilled with water into the drains. The reaction between the sodium cyanide and water in the drains gave rise to volumes of deadly hydrogen cyanide gas. Again, some chemicals reacting with water liberate heat: for example, concentrated sulphuric acid mixed with water generates considerable heat, and under certain conditions produces steam sufficient to cause ejection of material. Consideration therefore must be given to water–chemical mixtures. A danger may arise, for example, if water and a liquid chemical are inadequately mixed, and stratification, or layering, occurs; the layers may then start to mix progressively, with possible violent reaction. Because water is nearly always readily available and has almost universal usage, the danger of its indiscriminate use should be clearly explained to all concerned.

To know the freezing point of a chemical is of value in determining its storage. Obviously a liquid chemical solidifying at, say, 30°F must not be stored in the open during the winter period as, apart from the fact that it may crystallize and become difficult to use, there is a danger that it will burst its container.

What is its incompatibility?

Certain chemicals react violently with other chemicals and the chemicals involved are said to be incompatible. For instance, acetylene reacts violently with chlorine, so that the accidental bringing together of these chemicals must be prevented at all times. Similarly, concentrated nitric acid should not be brought into contact with flammable liquids.

The real danger of 'incompatibility' lies in the failure to 'make an assessment', so that when certain chemicals are inadvertently brought together, a violent reaction occurs, and plant and personnel are damaged. The enquiry into the reasons then readily shows the reactive nature of the chemicals involved. Providing the incompatibility of the chemical is known and understood, i.e. 'knowledge of the chemical', it can often be controlled and is in fact occasionally used in chemical processes. These instances are, of course, 'planned' reactions, generally taking place in the presence of a third chemical acting as a diluent. The possibility of unplanned mixing of incompatible chemicals should always be guarded against.

Under this heading 'incompatibility' the related properties 'oxidizing agent' or 'reducing agent' should be determined and considered. Many chemicals which are not flammable strongly support combustion by readily combining with other chemicals, giving off large quantities of oxygen. Not only does the immediate atmosphere become enriched with oxygen, but the 'heat of reaction' may be sufficient to cause ignition and a fire can occur. Oxidation is the chemical combination of oxygen with another substance; it can be rapid or slow, and any substance which readily supplies oxygen to another substance can be considered as an 'oxidizing agent', e.g. nitric acid (HNO_3), manganese oxide (MnO_2), hydrogen peroxide (H_2O_2), chromic acid (CrO_3).

Conversely, a substance which takes oxygen from a compound and combines with it, is known as a reducing agent, e.g. hydrogen, carbon, hydrocarbons, organic substances, etc.

Oxidation and reduction are opposite processes which always occur together, and the incompatibility of potassium permanganate ($KMnO_4$), a strong oxidizing agent, compared with the incompatibility of aluminium powder, a strong reducing agent, readily indicates the converse nature of these properties and immediately shows why the two substances should not be stored alongside each other.

Having determined the incompatibility of the chemical with which you are immediately concerned, it follows that a check must be made to determine whether any 'incompatibles' are used, not only on a given plant, but also anywhere else on the premises. Bitter experience shows how easy it is for the 'wrong chemical' to be delivered by internal transport, and the process operator, long accustomed to handling a given size of package, does not always stop to read the labels before emptying the contents into a vat or vessel. One drum looks very much like another; only the label gives a guide to its contents.

Flammability

This question deals with the flammable properties of the chemical, its flash point, explosive limits and ignition temperatures. All these properties are more ably discussed elsewhere in this handbook, so that it is only necessary here to state briefly their relationship in this 'initial assessment'.

The determination of whether a chemical is flammable or highly flammable is usually governed by the arbitrary flash point figures of 66°C and 32°C, but rigid adherence to these figures is not advocated here. Knowledge of the flashpoint, flammable limits, and ignition temperature markedly influences the storage and usage of the chemical wherein weak and rich mixture control, flameproof equipment, static control, ignition control and the like are all being fully considered.

Remaining points

The last three questions (7–9) are not actually assessment questions but are a natural follow-up to the previous questions. Compilation of answers to these questions at the time of making the assessment does much to ensure that both fire and protective equipment will be available as soon as the chemical is received and not at some later date—perhaps after a fire or injury has occurred. The final question allows space for special safety precautions which the assessment may have shown to be necessary.

Sources of information

The previous sections give the essential factors which should be known about any chemical which is to be used in the plants for which the 'assessor' has some responsibility. Whilst this assessment may at first appear complex, all the relevant information can be obtained with little difficulty from some or all of the following sources:

(a) supplier or manufacturer;
(b) labels on the containers;
(c) reference libraries and handbooks;
(d) H.M. Factory Inspectorate;
(e) The Royal Society for the Prevention of Accidents.

Although many chemicals are sold as proprietary compounds, the manufacturer, who also affixes the labels, usually gives sufficient information to ensure safe transport, storage, and usage. However, it should be remembered that although the manufacturer knows all about his particular chemical he does not necessarily know all the purposes for which the customer wishes to use it, so that such information must be relative. For instance, the writer once informed a customer that nickel and lead wire were suitable materials to use in conjunction with a certain acid, and this advice was taken too far, inasmuch as the customer made his tank coupling pipe of lead. As this pipe had to be 'made up' every time a delivery was made, the inevitable happened—the frequent movement of the lead pipe caused it to become fatigued so that it ultimately broke under pressure. It was not thought necessary to advise the customer that lead pipe was not suitable for this type of application irrespective of what liquid was passed through it.

Labels on containers, as issued by most reputable manufacturers, give valuable precise information, which is unfortunately often rendered void by the failure of the customer to read and understand the label and to ensure that all those in his works who move, manipulate, or empty the container also read and understand the label attached. It will be noted that many chemical manufacturers of repute can give valuable after-sales service in the form of safety-handling data sheets and other useful safety-handling information. Forthcoming legislation will make a more rigorous demand on manufacturers to be more forthcoming about the potential dangers, corrective procedure and safe handling of chemicals supplied. To obtain a full understanding of such instructions it is good practice to read the basis by which such information is governed in a publication entitled *Marking Containers of Hazardous Chemicals* published by the Chemical Industries Association Ltd, containing an explanation of different classifications, e.g., harmful, extremely harmful.

Reference to libraries is not often necessary, but the use of several handbooks can often give all the data required. In particular, the following publications are useful additions to the library of non-chemical users of chemicals, and supply many of the data for a safety assessment sheet:

(a) *Marking Containers of Hazardous Chemicals,* published by Chemical Industries Association Ltd.
(b) *Road Transport of Hazardous Chemicals*, published by the Chemical Industries Association Ltd.
(c) *Dangerous Properties of Industrial Materials*, Van Nostrand Reinhold, London, 1968.
(d) Technical Data Note 2—*Threshold Limit Values,* published by H.M. Factory Inspectorate.
(e) *Encyclopaedia of Occupational Health and Safety,* I.L.O., Geneva, 1972.

The help given by H.M. Factory Inspectorate, the Chemical Industries Association and the Royal Society for the Prevention of Accidents on all problems relative to safety, health and welfare, is already well known and, wherever any doubt exists or the data are incomplete, reference to these organizations would bring ready assistance.

Receiving, storage and transport of chemicals

On completion of the safety assessment sheet the receipt, storage, and transport of all chemicals within a works should be carefully planned. The layout of stores should follow the well-established principles of orderly stacking, clear gangways, good drainage, good lighting, etc. and the following additional points should receive consideration.

Receiving

The first essential is to ensure that you receive what you ordered before the delivery man leaves your premises. This appears to be so obvious that it is taken for granted and no emphasis is made on instructions to the receiving departments to make an immediate check on all chemicals received. With small packages in particular, where the delivery is by a contractor's vehicle, making perhaps 10 or 20 deliveries from the same load, it is easy for the wrong package to be delivered. Persons receiving chemicals should be instructed to read the label on the package and cross-check this with the order; where there can be confusion, e.g. potassium chlorate instead of potassium chloride, very clear instructions should be given. The man in the receiving stores may not even know that there is in fact a chlorate and a chloride—why should he know unless instructed? At the same time the package should be examined for damage.

Apart from the inconvenience and waste of time and money in making an exchange for the correct package, there is always the possibility that the wrong package could be the one that is most 'incompatible' with a chemical you normally receive and store. Hence an unconsidered potential can arise and, although a chain of causes is generally necessary for an accident to take place, such chains do occur and numerous incidents could be quoted wherein the so-called unknown or unanticipated factor has closed the links in the chain.

Even bulk deliveries can be wrongly delivered. There is a classical example of a road tanker driver, who, arriving late at a works in bad weather, mistook certain instructions which, together with a chain of errors, resulted in a quantity of sulphuric acid being wrongly discharged into a storage tank containing a sodium bisulphite solution, so that there was an evolution and escape of sulphur dioxide gas. To avoid such incidents, a written procedure should be adopted which has as its starting point the arrival of the tanker at the factory gates and, as its closure, the departure of the tanker from the same place. This procedure should include labelling of the storage tank with its intended contents, and this labelling should also apply to the filling inlet pipe. Guidance in the formation of such procedures can be obtained by studying the requirements of the Petroleum Spirit (Conveyance by Road) Regulations 1957 and from information given in the publication by the Chemical Industries Association entitled *Road Transport of Hazardous Chemicals*. The procedure should also govern the uncoupling of the tanker; there is a record of a tanker driver who drove his lorry off the loading pad before the delivery hose had been uncoupled. In particular, consideration should be given to the route taken by the tanker, both in and out of the works, where an accident involving a tanker could have disaster proportions, e.g. a petrol tanker passing close to a large furnace house, a bulk oxygen tanker stopping alongside oil storage tanks. Where alternative routes are available these should obviously be used. During the off-loading there are generally workmen employed by two different employers, the supplier and the customer, and it may well prevent accidents if 'who does what' is clearly determined in the written procedure.

Before leaving the matter of bulk storage of chemicals, and bearing in mind that, apart from considerations of health and safety, such installations represent high capital outlay, a warning should be given about using the correct materials of construction. It must be emphasized that any small change in the impurity content of a chemical can make a vast change in storage and tank corrosion life. For example, small quantities of chloride in acid solution can give rise to rapid attack on stainless steel, with the liberation of hydrogen and the possibility of an explosion in the gas space.

Sampling

At some time between receipt and actual usage it may be necessary for samples to be taken to ensure that the chemical is to an agreed specification. Often the procedure for sampling is haphazard and no procedure is laid down concerning when, where and how. As a check of this statement do you know if samples are taken at your works; by whom is it done and how? Does the sampler have a safe means of access; is he qualified to do what he has to do; does the 'person in charge' know when the sampler is taking a sample; is proper sampling equipment available? Suitable carriers for glassware and for other breakable containers of samples should be provided and their use enforced. Often such carriers are scorned and disregarded on the assumption that 'it cannot happen to me', but it should be pointed out that the most agile of persons has been known to slip on an icy roadway or stairway during wintry weather. Similarly, the patch of oil in a darkened passageway is no respecter of persons. Falling or slipping is still one of the largest single causes of accidents, and a small amount of concentrated acid flowing into a severely lacerated hand from a broken sample jar can cause irreparable damage. When a sample of a toxic chemical has to be taken, consideration must be given to the ready availability of protective equipment, e.g. respirators, in case of accident. Ideally, the taking of a sample should be planned to be without risk so far as is practical, but the possibility of breakage and spillage cannot be overlooked.

Storage

The actual storage of the chemicals is governed by many factors—not the least being cost and space—and the usual recommendations that they should be stored in a cool, dry, well-ventilated, well-drained building is an ideal always to be aimed at, even though adverse factors may prevent its attainment. These factors include the quantity to be stored, the properties of the chemical, the package in which it is received, the method of internal transport, the lifting appliances available, the method of discharge at the point of usage. Governed by these factors, a storage site or area, be it large or small, must be clearly established. This may be within an existing general store or a special site, but unless this specification is made the odd drum will be tucked away with little regard for its contents and flammable or oxidizing substances will be mixed up with other goods.

The manner in which a package or container should be stored must also be determined. This determination should not be confined to such problems as 'palletization methods', 'drums on end stacking', etc., but must include the safe limits of any stack. For instance, certain hazardous chemicals, e.g. bromine, may be received in bottles in crates and, both by reason of the extreme hazard if spilled and the strength of the case, it may be necessary to limit stacking to, say, three cases high. If this rule is not made and enforced, a temporary shortage of floor space may arise and the storeman may endeavour to make do by stacking higher than a true safety margin would permit.

A system should be devised whereby there is a continuous clearance of stock—first in, first out—thereby limiting any corrosion of package or, in some cases, build-up of drum pressure or decomposition of the chemical due to extended storage.

Drums and barrels may be stored end vertical, end horizontal, on pallets, in racks, stacks, or piles, but whatever the method used they should be off the floor—particularly on open ground. This enables a speedy identification of a leaking container, the early recognition of corrosion or pressure build-up, limits the amount of corrosion, improves ventilation and dispersion of flammable or toxic vapours which may arise from the leakage and, if on open ground, keeps the container clean and free from foreign matter which could be carried into the process. In large stocks there should be adequate gangways or avenues to enable this inspection to be carried out. Where a defective container could give rise to hazardous conditions, the period from one inspection to another should be determined and a procedure enforced to see that the inspection is in fact carried out. Experience shows that because hazardous leakages from stock containers are not a very frequent happening, the need for periodical inspection is readily and conveniently overlooked.

Several new types of containers of up to 45 l liquid measure are now in use and are generally of some plastic material. It is advisable to bear in mind that some chemicals extract the stabilizer from plastic containers, thus reducing their rigidity and strength of the seams. Certain of these containers, being of very robust construction, do not require a protective metal basket and are ideally suitable for stacking. Because of this factor there is a tendency to stack the containers on pallets and then to stack pallet on pallet with little regard to the actual strength of the lower containers which carry the load. The fact that a stack looks quite safe when first built can be very misleading; the plastic container can gradually distort under a load—as a snowman melts in the morning sun—so that some hours after an apparently safe stack has been made there is some distortion at the lower levels and a failure of the stack. Generally, such distortion gradually disappears after the load is removed and the container regains its normal shape. In the absence of suitable racks or load-bearing pallets, or unless advice to the contrary has been received from an authoritative source, no container of liquid chemicals should be stacked more than two high. Even here a bonding method should be used, viz., two on three, three on four, four on five and so on.

Consideration should also be given to the unauthorized use of chemicals and, if necessary, a locked compound should be provided. The writer well remembers an elderly labourer who was told to clean the floor of a machine shop superintendent's office which was very oily. Determined to make a good job of the cleaning the labourer obtained a quantity of a certain proprietary cleaning powder from a drum left at the end of a workshop which he had seen used on the oil shop floors during weekend cleaning. This cleaning powder had a very high caustic soda content and the men normally cleaning the shop floor were well equipped with protective clothing. This point escaped the labourer who put the powder into a bucket of warm water and literally hand-scrubbed the office floor—he made a first class job of the cleaning. When questioned some days later he said that he felt a 'tingling sensation' on his hands and arms when he was scrubbing the floor; the same night he could not sleep because of pain, and the following day his hands had 'broken out'. Some 24 hours after using this compound his hands and forearms were pockmarked with small ulcer like holes, similar in appearance to chrome ulcers, and he was off work for many weeks. This was an incident which should never have happened—the powder was known to be extremely corrosive and should not have been left lying about.

Handling

Perhaps the greatest single hazard in the use of a drum or other container is that created by a loose stopper—particularly where only part of the contents have been used and the stopper appears to have been correctly replaced. A combination of a loose stopper and a fairly hefty bump as the drum is put down on the floor causes a forceful ejection of some of its contents which is likely to splash over the operators with most serious results. Most stoppers are of the screw-in or screw-on pattern and, unless fully secured, can gradually work loose by vibration, as when on a vehicle or truck. Unfortunately, in many instances, little or no vibration is necessary, the user merely giving the stopper no more than an initial twist so that it becomes a man-trap for the unwary handlers.

Whilst every effort has to be made to educate users of such containers to secure the stoppers fully, each and every person who has to move or handle drums or similarly constructed containers must be taught 'before the drum is moved, always ensure that the stopper is secure'. This check should become completely automatic. If one watches the experienced delivery man or the experienced handler of drums, it will be seen that this placing of the hand on the stopper to ensure it is tight before the drum is moved is as automatic as switching on a motor before pressing the starter. They do this on every drum— full, part-full, or empty. As a further safeguard they ensure that the drum is never tilted during the whole of the vertical lift, and that it is placed, not dropped, onto a floor or platform.

Drums and other containers must always be handled with care and not subjected to physical shocks and, except for a vertical lift, they should always be moved by some form of special drum truck or by a suitable sack truck. There are many efficient and relatively cheap patterns available and, by careful selection, the same two-wheeled truck can be used for moving small drums, plastic containers, sacks, bags, etc.; this is particularly useful when a variety of packages is used in the smaller works. If the drum has to be moved manually, this should again be a two-man operation, and it should not be lifted any higher than is absolutely necessary—if the bracket breaks, the closer the drum is to the ground the less shock it will suffer.

An insidious hazard created by the clumsy and ignorant operator should be guarded against. Here he drives his truck at the drum to load it instead of gently tilting it away from him whilst placing the foot iron of the truck underneath it. Even if he is successful in loading at the first attempt the drum receives a heavy physical shock and may thus become damaged whilst still retaining its contents. During some later movement, perhaps during the actual emptying of the drum, the damage is worsened and the contents are spilled onto the handler.

When one or more drums are moved across factory yards or similar places by means of a steel platform truck, e.g. Lister, Brush, etc. the drum must be prevented from moving about on the platform. Blocks, wedges and similar devices should be used. Metal-to-metal contact on the platform provides little frictional resistance and the drum creeps along the platform and may fall off before the driver is aware of any danger. Standing the drum on a sack bag on the platform is helpful in limiting this movement. Similarly, although running over a small pothole might not seriously disturb a wood or cardboard container on a platform truck, the same event could completely dislodge a drum. These circumstances should be clearly explained to the drivers of such trucks as they may not otherwise be aware of these fractional changes and of the additional forces created by heavy liquids in a container.

In use at this time are several containers of new design, some apparently still in the experimental stages, which are gradually replacing the old glass carboys and metal drums. These containers vary from a simple plastic 45 l container protected by a wire basket to a sophisticated free-standing polythene container fitted with moveable spout and handles. Polythene-lined drums up to 205 l capacity are popular for several of the lower strength acids and are usually classed as a returnable package. Most of these containers provide considerable improvements in handling; there are also several new types of drum trucks, drum carriers, and drum tilters, all having a most useful application, and these should be used wherever possible. Manual handling should be restricted to the absolute minimum; where it has to be used it should be in conformity with the principles of correct manual lifting and handling which are excellently dealt with in the booklet *Lifting in Industry* published by the Chartered Society of Physiotherapy.

Emptying

There are several variables in the use of drums and other containers; hence, there are several methods of emptying them. The use of compressed air or any compressed gas should be strictly forbidden; the carboy is not designed as a pressure container. A method used for the infrequent 'topping up' from a single drum, whereby the requisite quantity is tipped from a free-standing drum into a bucket or similar receptacle, is not advocated here even though several of the new plastic containers which are replacing drums have handles and pourers to facilitate such usage. A preferable and safer method is to use one of the special drum holder and tilter appliances. Although the tilting mechanisms on a few appliances require the operator to tilt the drum towards himself when emptying—and this in itself can create a hazard—continuous and complete control can be secured by the exercise of care and common sense. Ideally, the bucket or receptacle used in this method should be fitted with a lid which can be easily placed and secured in position whilst the bucket is being carried. This is particularly important when the operator has to ascend one or more steps to empty the bucket. To ensure good housekeeping the bucket or container should be placed in a shallow tray when being filled so that any splashes or spills due to misalignment or careless handling are prevented from running across the floor. The tray itself should be maintained in a clean condition.

Syphonic, gravitational or pumping methods of emptying are well known and should be used when dealing with the more hazardous chemicals (refer to the label on the drum to assess the hazard). However, there are still many applications where the drum can be manually emptied directly into a vat or open-topped vessel, but this method should be reserved for the less hazardous chemicals. The following method of manual emptying, which two safety-minded experienced operators had themselves devised and, most important, had invariably used, has much to recommend it.

The drum is brought by means of a drum truck and is put down at the nearest point alongside the vessel for a vertical lift. The operators, wearing appropriate protective clothing, i.e. knee-length rubber boots, gauntlet PVC gloves, PVC apron and chemical workers' goggles, stand facing each other, sideways to the vessel which has its rim 1 m from floor level. One operator—the same one every time—removes the stopper completely and then replaces it a *full half turn*. In this instance the partially placed stopper creates no hazard because the circumstances are known. On the instruction, lift—again from the same operator every time—each operator lifts with one hand on the top rim of the bracket and one hand underneath until the drum is resting on its side in an angled position on the rim of the vessel, the neck being tilted away from the operators. The drum is then tipped until its contents are barely touching the stopper and on the word, hold, the operator who first moved the stopper now removes it completely leaving three hands holding the drum. He places the stopper on a convenient ledge and again taking hold of the drum assists the other operator to empty its contents gently. After emptying, the stopper is then securely fixed.

This procedure can be brought to a fine degree of precision, and the inevitable splashing when the first gallon or so is poured reduced to the absolute minimum. Users of drums will know that until about a gallon of liquid has been poured it is almost impossible to prevent some splashing as the incoming air competes with the outgoing liquid at the junction of the neck and body of the drum. Splashing is also limited if the liquid can be poured down a flat surface running into the receiving vessel, e.g. an electrode, or down the inside wall of the vessel.

There can also be a number of variations of the method of emptying quoted, for instance, where the drum and its contents are heavy, it might be well to make the lift in two parts, the first being onto a small stand at, say, 0·5 m high. Legislation demands that the edge of any vessel containing dangerous material must be at least 1 m above the ground level or working platform; this should determine the maximum height to which a drum has to be lifted for emptying, and where a vat is of greater height, consideration should be given to the building of a working platform to reduce the effective height to 1 m. A word of caution, however, because such work generally requires the wearing of rubber boots, and because the platform will get wet, it must be wider than the normally accepted working platform; 1 m is a good width. Furthermore, if the difference in height is small, remember that a person can become accustomed to duck-board height or to a normal step height—anything between these two leads to misjudgement in stepping up and stepping down. Hence the merits and demerits of a platform less than say 18 cm high should be carefully considered.

The most important factor in emptying drums, as in fact with any other container, is that the method must be considered, determined, put into practice and enforced. Unless this is done, the operators will devise their own methods, and although in a few instances a safe and efficient method may be devised, case histories show that the converse is true. Most operators quite naturally seek the easiest way of doing a job—not necessarily the safest.

The basic principles which govern the emptying of drums should be applied when considering the methods of emptying any other container which replaces it. Generally, the new containers are of a more robust construction and easier to handle, so that the basic method can be modified accordingly. The emptying of larger drums usually demands some form of mechanical handling. As with all containers of chemicals, air pressure should not be applied to drums unless the supplier has agreed that such methods can be used. In certain special circumstances chemicals are supplied in very strong drums which are designed to allow air pressures of up to about 0·5 bar to be applied for the purpose of emptying. In all instances the suppliers' recommendations must be strictly observed. Where a pressure build-up can occur, particularly in the larger size drums, the operator will have to be trained in a safe method of venting the container. Guidance in such procedures will be given by the supplier.

A number of accidents have been caused by the misuse of empty chemical containers. For example, an apparently empty 45 gal drum was used to receive residues from a process; a slow reaction took place and a considerable pressure built up inside the sealed drum. This pressure did not burst the drum but the release of pressure when the drum was opened some months later in order to discharge the residues finally caused a gross splashing and burning to the handler. This man was a contractors' labourer working at a tip remote from the works who did not realize that there was a pressure in the drum.

The dangers arising from the application of heat to empty drums which have contained flammable materials, are still not fully appreciated. These containers are still used from time to time—often by contractors—as temporary workbenches for holding lengths of steel whilst cutting and welding. Serious injury and fatalities have resulted from this misuse.

To complete the advice on handling chemicals in drums and other types of containers, another less obvious danger must be considered, and that is the drum with 'unknown contents'. This can arise due to defacement of markings, or the drum 'left in the corner for years'. Hence, dangers arise due to unknown contents, the possibility of pressure within the container which can arise due to corrosion with evolution of flammable gases, or reaction of the contents within itself. Such dangers, which would include the possibility of bursting, fire, splashing of corrosive or toxic materials, evolution of toxic fumes, etc., do present a real danger. A procedure, therefore, should be written to the effect that such drums must only be handled for examination by a competent person. It is advisable to move the said drum into the open air prior to opening, and the competent person would be well advised to wear suitable protective equipment, and to be accompanied by a second man, as safety watchman.

Finally, spillages present potential dangers, not only to personnel, but environmental hazards. It is a good rule to ensure that any spillages should be notified to a competent person.

Accidents involving the receipt, storage, and use of liquid chemicals in bulk, although not large in number, are of great potential by reason of the quantities involved. Because of the greater capital expenditure involved 'chemicals in bulk' generally receive a more than proportionate study of all the mechanical and chemical factors involved, so that most of the accidents are caused mainly by unsound methods of operation or unsafe procedures. The attempt to deliver 9 000 l to a tank that can only receive 7 000 l is not unheard of; nor are the many failures of delivery hoses and fittings to withstand slight excess pressures due to some human error in the omission to open a delivery valve on the receiving installation.

Although all preventive measures should be adopted, it should be accepted that a spillage can occur and remedial action, even if only paper planned, should be prepared beforehand. Most major spillages ultimately find their way into the drains, so that the planning should include an immediate notification to the drainage authority, stating the extent of the spillage and of the proposals to deal with it. This notification enables the authority to make its own assessment of the spillage on the drainage and sewerage disposal plants. Failure to give this information could cause untold damage to the installations, endanger the lives of men working in manholes in the area, and result in heavy litigation claims against the perpetrator.

The handling and emptying of chemicals in the solid form present fewer serious hazards except in the case of dusty solids—particularly of a flammable nature. Spillages are obviously more easily dealt with but, as previously shown, the indiscriminate use of water should be carefully avoided. This will particularly apply where a reaction with water may give rise to considerable heat or a toxic fume. It is also as well to remember that certain oxidizing chemicals on drying out after swilling down a spillage can cause a hazard; evaporated residues of sodium chlorate for instance become highly flammable. Solid spillages should be collected into a clean container and only after the contaminated area has been brushed clean should a final flushing with water be permitted.

In conclusion, emphasis is laid upon one single fact; that nearly all accidents are a result of ignorance of simple safety laws, ignorance of human failings, ignorance of known chemical reactions, ignorance of the need to give constructive thought and attention to accident prevention. Except in the newest and most complex of chemical operations, someone, somewhere, knows the answer to any particular safety problem; with a little diligence he can be found. The standard instruction stamped on many working drawings 'if in doubt, ask', is equally applicable to the safe handling of chemicals.

5. Ionizing radiation

E. J. Bennellick

Training Officer,
National Radiological Protection Board,
Harwell

Ionizing radiations, either generated by equipment such as X-ray apparatus, or spontaneously emitted by radioactive materials, are widely and increasingly used in industry. At present in the U.K., in addition to large and specialist organizations such as the United Kingdom Atomic Energy Authority, British Nuclear Fuels Ltd, and the Central Electricity Generating Board, there are several thousand premises subject to the Factories Act in which radiation generators and radioactive materials[1] are used.

Ionizing radiations have been recognized as a potential cause of injury since the end of the nineteenth century, following the discovery of X-rays by Röntgen in 1895 and of natural radioactivity by Becquerel in 1896. Insufficient knowledge and insufficient precautions produced injuries, many of them severe or fatal, amongst the pioneers and early users. Forewarned of the risks, the modern atomic energy industry instituted the requisite precautions from the outset of its expansion. In this chapter we are concerned with the risks associated with, and the precautions necessary in, the use of:

(a) X-ray machines and other apparatus which emit ionizing radiations, and
(b) sealed and unsealed radioactive materials.

Primarily, the needs of users in general industry, particularly within factories, are considered. Only passing reference is made to the problems associated with nuclear reactors, particle accelerators, and criticality control, which are special subjects outside the scope of the chapter; nor are the questions of transport of radioactive materials and the disposal of radioactive waste considered in any great detail.

Types of exposure

External and internal radiation exposure

The two possible types of exposure are generally referred to as 'external radiation exposure' and 'internal radiation exposure', according to whether the source of radiation is outside or inside the human body. In practice, for external radiation exposure, only those radiations may be considered which are sufficiently penetrating to reach at least the basal layer[2] of the epidermis which lies beneath the inert surface layer of skin at a depth conventionally taken to be between 0·05 and 0·1 mm over most parts of the body. Internal radiation exposure arises following the intake of radioactive materials into the body by inhalation, ingestion, or absorption through the intact or injured body surface. With radiation generating equipment, such as X-ray apparatus, only the problem of external radiation exposure would arise. With radioactive materials, precautions may be required against one or the other or both types of exposure; radioactive materials may therefore be regarded as toxins which may also be capable of producing effects whilst remaining outside the body. Certain types of particle accelerators are used for producing radioactive materials or they produce such materials incidentally to their operation; with such machines it is obvious that both types of exposure have to be considered.

Radiation from natural sources

Radiation has always been part of man's environment. For 'normal areas' the sum of the doses for external and internal exposure from naturally occurring sources averages about 90 mrad a year.[3, 4] However, in some parts of the world which have highly radioactive soils, e.g., the Kerala State of India, and the town of Guarapari in Brazil, doses of up to 3000 mrads/year have been recorded.

Biological effects of ionizing radiation and concept of maximum permissible dose

Evaluation of risks of exposure[5, 6, 7, 8]

A variety of harmful effects can be caused by exposure to ionizing radiation. These effects may appear in the individual exposed (somatic effects). Where the reproductive organs have been exposed, deleterious effects may manifest themselves in the descendants of the irradiated person as a result of alterations in the genetic material (genetic effects).

Somatic effects. Information about the clinical effects of ionizing radiation comes from four main sources: from the medical uses of X-rays and radium;

from the study of a few occupational overexposures and accidents; from observations on those exposed to nuclear weapons explosions; and, by analogy, from experiments on animals.

The probability and severity of radiation injury depends on many factors including the dose, dose-rate, the extent and part of the body irradiated and the type of radiation. The early effects of acute exposure (in which a large dose is delivered in a period of hours or less) are those which develop within a few hours to a few weeks; any delayed effects from acute or chronic exposure may have a latent period of tens of years. With very few exceptions, radiation effects are non-specific and cannot be distinguished from similar effects caused by other agents. For *acute whole body irradiation,* no early clinical response would be expected below doses in the region of 50 to 100 rads. At higher exposures up to about 200 rads, a proportion might be expected to exhibit symptoms of radiation sickness. At increasingly higher doses of acute whole body exposure, severe symptoms are more probable with a small risk of fatality at a month or so after exposure, the deaths increasing to about half of the untreated cases whose exposures were of the order of 300–400 rads or more. At much higher levels, survival would be improbable and very early death could occur. For *acute local external exposure* limited to a relatively small area of the body, the local signs and symptoms may be more significant, the commonly observed ones being loss of hair and erythema (with doses of several hundred rads), and in more severe cases, radiation burns.

The delayed effects of main importance studied in man include the increased risk of various forms of cancer, and the possible damage to the developing embryo in the early stages of pregnancy. For evaluating the risks of late effects from radiation exposure, much of the human evidence again comes from large doses, delivered often at high dose rates. For example, quantitative estimates of the increased risk of leukaemia induction have been made from studies of hospital patients who received large exposures in radiotherapy and from observations on survivors of nuclear weapons explosions. Neither of the studies provides significant evidence of an effect with whole body irradiation of less than about 100 rads. Assuming a linear relationship down to very low doses leads to the risk estimate that if a population of one million persons was exposed to 1 rad of whole body radiation, a total of up to about 20 additional cases of leukaemia might occur, spread over the following 10 to 20 years. The additional incidence of all other malignancies, not all fatal, might add a further 80 cases per million persons exposed. These figures refer to the increased incidence over and above the natural incidence in the U.K. of all types of fatal cancer (including leukaemia), the figure for which is at present about 2 500 cases per million persons per year.

Genetic effects. The frequency of genetic effects of radiation is even more difficult to assess than that of the somatic effects mentioned above. There are no human data yet available that are comparable to those on the somatic effects. Surveys have been made of the offspring of irradiated parents, but none of these has revealed any unequivocal evidence of harmful hereditary effects of radiation. Any assessment at the present time is still dependent mainly on inference from animal experiments. Tentative assessments have been made of the numbers of the more serious physical defects which might be expected to occur in the first generation of offspring of irradiated parents. The exposure of parents to 1 rad is estimated as being likely to produce in the first generation an increased incidence of about 20 cases of such defects per million live births. These defects would be expected to occur during a score or more years. A comparable figure for the natural incidence of similar hereditary abnormality is about 15 000 per million live births. (The U.K. live births rate approaches about one million per year.) The total effect per generation would be greatest in the first generation, decreasing progressively in subsequent generations.

Basic maximum permissible doses

Maximum permissible doses for occupational exposure. Table 5.1 is a summary of the principal recommendations of the International Commission on Radiological Protection (I.C.R.P.)[9, 10] for occupational exposure.

Table 5.1   Maximum permissible doses* for male workers aged 18 years and above

Organ or part of body	Long-term limit	Quarterly limit
Gonads and red bone-marrow (and, in the case of uniform irradiation, the whole body)	5 rems in a year† or 5 $(N-18)$ rems to date‡	3 rems in a quarter
Skin, thyroid, and bone	30 rems in a year	15 rems in a quarter
Hands, forearms, feet, and ankles	75 rems in a year	40 rems in a quarter
All other organs	15 rems in a year	8 rems in a quarter

    * The doses comprise the contribution from external and internal radiation exposure; they are additional to the dose from natural background radiation, to the dose received as patients in medical procedures or from other exposures received by the individual as a member of the public. All unnecessary exposure is to be avoided. All doses are to be kept as low as is reasonably achievable. The values are modified in a number of cases, e.g., for women because the foetus may be somewhat more sensitive in certain respects than children or adults, for special or abnormal exposure and for persons starting work at an age of less than 18 years.
    † The annual figure should be applied except when the greater flexibility of the formula is necessary.
    ‡ This is the total dose accumulated at any age over 18 years. N is the age in years.

The maximum permissible doses represent extrapolations; as indicated above, risk evaluations are based on excessive exposures received often under acute conditions. For chronic exposure to low doses there is really no evidence to make it possible to measure the risks quantitatively with any degree of confidence. For exposure within the limits of the maximum permissible doses there would be no risk of acute radiation effects. For possible late-effects such as radiation induced malignancy, it is unknown whether a threshold exists below which there would be no risk. As the existence of a threshold is unknown it has been assumed that even the smallest doses involve a proportionately small risk, and the object of the maximum permissible doses becomes therefore to limit the risk of late effects, if any, to a negligibly small level, taking due account of the other risks of life and the benefits[11] of the uses of ionizing radiation.

To safeguard the health of industrial workers, methods of protection and control have to be based upon measurement and control of the working environment and of the radiation dose to persons, and not on the more conventional methods of occupational medicine. One reason for this is the inability of the body to feel exposure to radiation except in grossly supralethal doses. Another is the absence of early warning clinical signs of radiation injury; by the time these signs are detectable or become apparent to the individual, the degree of injury is often severe and sometimes irreparable.

Dose limits for exposure of members of the public and of the whole population. The dose limits recommended by I.C.R.P. for members of the public are one-tenth of the corresponding maximum permissible annual dose for workers, except in the case of the dose to the thyroid of young persons, for which a lower figure of one-twentieth applies. The I.C.R.P. recommended limit on genetic dose for the whole population is 5 rems per generation.

The control of occupational exposure to external radiation

Principal radiations involved and corresponding maximum permissible doses

In industrial use, the radiations[12] chiefly encountered are α-, β-, γ-, and X-rays, bremsstrahlung, and neutrons.

α-radiation. Epidermal regeneration depends on cellular division in the basal cell layer. α-radiation of at least 7·5 MeV energy would be required to penetrate the outer layers of skin where there are no dividing cells and reach the basal layer of the epidermis at a mean depth of 0·07 mm in parts of the body such as the arm (0·40 mm for palmar surfaces of the hand). Generally, α-radiation from radioactive materials has energies of 4 MeV to 5 MeV with corresponding ranges in air and soft tissue of about 3 cm and 0·04 mm respectively. Therefore, for practical protection purposes, α-radiation from radioactive materials is normally excluded from consideration in the control of exposure to external radiation. This exclusion would not necessarily apply to a direct beam of α-radiation from an accelerator, where the energy might be substantially greater.

β-radiation. For β-radiation, energies of about 0·07 MeV only would be required to penetrate the skin to the depth of the basal layer. There are some β-emitting radionuclides whose maximum β-energy is well below this value, for example tritium with a β E_{max} of 0·018 MeV, but since there may be significant bremsstrahlung emission, the general rule should be to include such sources in assessing external radiation exposure risks until it is established that they may be excluded. Table 5.2 illustrates the range of β-radiation (and, of course, electrons of similar energy) in air and in water; in this respect water may be taken to represent soft tissue.

Table 5.2   Range of β-radiation in air and water

β-energy MeV	Range in air (approx.) m	Range in water (or soft tissue) (approx.) cm
0·5	1·5	0·18
1	4	0·44
2	8	0·98
2·5	11	1·25

If the values in the first column of Table 5.2 are taken to represent β E_{max} the corresponding average ranges would naturally be less than those in the second and third columns. For exposure to external β-radiation from radioactive materials, the critical parts of the body would be the superficial tissues including the lens of the eye which is taken to be at a depth of 3 mm[13] below the surface of the eye. Therefore, for controlling exposure to this type of radiation the general practice is to apply the I.C.R.P. recommended maximum permissible doses for the skin. For the extremities, i.e., the hands, forearms, feet, and ankles, the I.C.R.P. values as shown in Table 5.1 would apply. For the lens of the eye the values recommended for other organs are taken to be the relevant ones; however, because of absorption losses in the soft tissue superficial to the eye lens, for conditions involving exposure to β-radiation alone, the skin dose would be limiting and could be used for control purposes. (*Note:* the skin dose could not necessarily be taken to be limiting for a direct beam of electrons from an accelerator.)

γ-radiation, X-rays, and bremsstrahlung. Electromagnetic radiation does not have a finite range in matter and accordingly its penetration is expressed in relative terms. Thus for γ-radiation of energy 1 MeV the half-value layer for water (taken as equivalent to soft tissue) is about 10 cm; for energies of a few keV the half-value layer for water falls to around 1 mm or less. The values for γ-radiation would also apply to X-rays and bremsstrahlung of similar energy. Because of the ability of these types of radiations to penetrate the body, the normal practice is to apply the most restrictive of the I.C.R.P. recommended maximum permissible doses, which are those for the gonads, red bone-marrow, and whole body irradiation. If this seems unduly restrictive it may be noted that U.K. factory regulations[14, 15] exempt machines or apparatus with an accelerating voltage of less than 5 kV which emit ionizing radiations adventitiously. However, for the extremities listed in Table 5.1, the most permissive of the I.C.R.P. values of 75 rems/year and 40 rems/quarter would apply as they pertain to all tissues in these parts of the body.

Neutrons. Like γ-rays, neutrons have great penetrating power and the maximum permissible doses (in rems) applied in their case are the same as those for electromagnetic radiation.

Exposure to mixed beams. All maximum permissible doses refer to the total dose in the tissue or organ under consideration, as a result of simultaneous or successive exposures to one or more types of radiation. This concept is reflected in the manner in which maximum permissible doses are specified in statutory regulations.

Maximum permissible doses due to external radiation for factory employees

To prevent difficulties of interpretation the statutory regulations specify the maximum permissible doses in terms of exposure to different types of radiation. The doses relate only to ionizing radiations originating from sources outside the body and do not include radiation for the prevention, diagnosis, or treatment of illness or injury.

Table 5.3 is based on the maximum permissible doses specified in U.K.

Table 5.3 Maximum permissible doses due to external radiation for factory employees

Organ or parts of body	Radiation	All employees		Additional restrictions for women employees
		Long-term limit rems in a year	Quarterly limit rems in a quarter	
Hands, forearms, feet, ankles	All*	75	40	
Lens of eye	All	15	8	—
Other parts of body	All (except X, γ, neutrons)	30	15	—
	X, γ, neutrons	5 (N − 18) †	3	1·3 rems in a quarter
For pregnant persons	X, γ, neutrons	—	—	1 rem for remainder of pregnancy

* X, γ, α (excluding α-radiation from radioactive materials), β, electrons, positrons, protons, neutrons, heavy particles.

† This is the total dose accumulated at any age over 18 years. N is the age in years, fractions of a year being counted as a whole year.

statutory regulations[14, 15] applying to premises under the Factories Act.

The restrictions imposed by the cumulative dose formula imply an average maximum permissible dose to the whole body of 5 rems/year. Assuming working periods of 40 h/week for 50 weeks/year, i.e., 2 000 h/year, the implied annual value of 5 rems is equivalent to an average of 2·5 mrems/h. The definition of 'adequate shielding' in the factory regulations for areas occupied by classified workers turns on this level of hourly dose rate.

The factory regulations also have relaxations and restrictions depending respectively on whether doses incurred are unlikely or likely to exceed three-tenths of the appropriate annual figures in Table 5.3. Applying this fraction to the value of 5 rems/year gives a figure of 0·75 mrems/h which is the other level of hourly dose rate specified in the regulations in defining 'adequate shielding' for areas which may also include employees other than classified workers.

Estimation of external radiation dose rates

For estimating dose rates for protection purposes, most configurations may be dealt with using rules of thumb[16] and simple formulae,[17] as illustrated below for 'point' sources. The approximations usually enable the dose rates to be estimated in advance within a factor of 2 or 3; at a later stage they should be checked by measurement.

β-radiation dose rate. With β-radiation, air absorption is important even at distances of the order of 1 m or less. Because of the rapid variation over short distances, there is not much purpose in attempting to make accurate calculations of dose rate. Also, except with thin sources, self-absorption within the source material itself may be significant. However, for a 'point' source, the dose rate may be estimated from the formula:

$$D = 3\,000C$$

where D = rems/h at 10 cm distance, neglecting air and self-absorption
          C = source strength in curies

The formula holds fairly well for distances of 10 cm to 1 m for β-energies of E_{max} between 0·5 MeV and 3 MeV. It is worth noting that dose rates can be high even with relatively low strength sources. If unshielded, a millicurie β-source produces around 3 rems/h at a distance of 10 cm, and several thousand rems per hour at distances of a few millimetres. The β-radiation dose rate at the surface of massive forms of bare (unclad) natural and depleted uranium metal is often of interest; it is in the region of 200–230 mrems/h.[18]

γ-radiation dose rate. Published data for γ-emitting nuclides are often expressed in terms of the specific γ-ray constant, i.e., the exposure rate in roentgens per unit time per unit activity at a distance from an unshielded point source. Table 5.4 gives values for the more commonly used nuclides.

Table 5.4[17]   Values of specific γ-ray constant

Nuclide	Roentgens/h/Ci at 1 m
Sodium-24	1·84
Cobalt-60	1·32
Krypton-85	0·0021
Antimony-124	0·98
Caesium-137	0·33
Thulium-170	0·0025 (including X-ray contribution)
Tantalum-182	0·66 (including X-ray contribution)
Iridium-192	0·48
Gold-198	0·23
Radium-226 (in equilibrium with daughter products)	0·84 (0·5 mm platinum filter)

For practical purposes the numbers of roentgens, rads and rems may be taken as numerically equal. The dose rate for other activities may be obtained by simple proportion; for other distances the inverse square law may be applied.

When published data are not available, the dose rate for an unshielded 'point' source of strength C (curies) and total γ-energy E (MeV) emitted per disintegration may be estimated from the formula:

$$\text{rems/h at 30 cm distance} = 6\ CE.$$

The formula applies fairly well over the energy range 0·07 MeV to 3 MeV. For other distances the inverse square law may be applied; unlike in the case of β-radiation, absorption in the intervening air is ignored as it is insignificant over distances of practical importance. To illustrate the rise in dose rate with proximity, 1 mCi of a γ-emitting material at 3 mm produces the same dose rate as 10 Ci at 30 cm.

It may be noted that a 1 mCi point source of 1 MeV γ-radiation produces about 50 mrems/h at 10 cm, i.e., about 2 per cent of the β-dose rate from a β-emitting source of the same activity. Thus, in controlling external radiation exposure with beta-gamma emitters, it is the β-radiation which usually is the more important at short distances.

X-rays and bremsstrahlung dose rates. The most important sources of X-rays are high-voltage equipment which is either specifically designed to produce them or does so as an incidental by-product of its operation. The output of X-ray apparatus at a given potential when measured in roentgens per minute per milliamp at a fixed distance from the target can be predicted with reasonable precision. It is, however, a function of the type of generator, the target material, the type of target and the total beam filtration. Typical outputs are shown in Table 5.5 but in practice it is essential that they be measured.

Table 5.5[17] Typical radiation output from X-ray sets

Operating voltage kV	Output R/min/mA at 1 m	Operating current mA	Dose rate at 1 m R/min
50	0·3	10	3
200	3	5	15
500	8	3	24

As regards adventitious sources of X-rays, any thermionic valve or similar device which operates at more than a few kV may emit a substantial intensity of X-rays. Such equipment is found in radio transmission, television, radar, power rectification and in a number of other minor spheres. In most cases where the X-ray production is adventitious, the intensities are likely to be much less than with X-ray generators, but if the hazard is not properly understood the risk of overexposure may be greater. The U.K. factory regulations do not apply to adventitious sources operated at less than 5 kV (for television receivers the exemption extends to 20 kV).

Bremsstrahlung[17] is the name given to the electromagnetic radiation emitted when a charged particle undergoes acceleration or deceleration. It is of practical significance only in the case of electrons brought rapidly to rest in an absorber. In practice, the term is frequently restricted to the electromagnetic radiation emitted when the beta radiation from a radioactive material is brought to rest. The amount of electromagnetic radiation generated increases with β-energy and the atomic number of the stopping material. At energies of 1–2 MeV and with light elements such as aluminium and water, less than 1 per cent of the energy is dissipated as bremsstrahlung. The proportion of radiation which is emitted from the absorber depends critically on the absorbing material and on the geometric arrangement of the beta source and absorber.

Neutron dose rate. For an unshielded 'point' source of output N neutrons/s, the flux density at a distance of d cm $= N/4\pi d^2$ neutrons/cm^2/s. Published values[19]

Table 5.6 Relationship between dose and neutron flux density

Neutron energy keV	Effective Quality factor	Flux density equivalent to 2·5 mrems/h (n/cm^2/s)
Thermal	2·3	650
1	2	675
100	7·4	120
500	11	35
1 000	10·6	21
5 000	7·8	17
10 000	6·8	17

of neutron output and energy are available. Having calculated the flux density the dose rate may then be estimated using the relationship shown in Table 5.6. With neutron sources, γ-radiation also is invariably present. This γ-radiation may come from a constituent of the source as in antimony-124/Be mixtures or be produced by the interaction of neutrons with matter; the γ-radiation output from the latter process is difficult to calculate and is best determined by measurement.

Protection against and control of external radiation exposure[19, 20, 21]

Planning. The first and essential stage in the control of external radiation is that of planning, and includes the choice of suitable techniques and equipment and the forecasting of radiation doses to be expected during the operations. The first step in such planning is to choose the weakest practicable source for the job in hand, followed by assessing how closely the source will have to be approached to enable estimates to be made of the dose rate at the working distances. Next, an assessment is made of the amount of time the operators will be exposed so that the expected doses may be forecast. Any shielding requirements and restrictions on exposure times can then be considered. Finally, attention must be paid to other safeguards which may be needed, such as the siting and design of the working area, monitoring requirements, operating instructions and the allocation of responsibilities. For work involving large sources, special provision may need to be made for accident prevention and detection, and procedures to deal with emergencies.

In simple cases planning may involve no more than a few minutes' thought and a quick estimation of the doses likely to be incurred. In others it may involve the design of special equipment and a full review of operating methods.

Methods of protection against external radiation. One of the aims of planning is to make the best use of the three main methods of reducing radiation exposure: distance, time, and shielding.

(a) Distance is usually achieved by the use of remote handling devices ranging from tweezers and laboratory tongs to long-handled manipulating equipment. Distance is a powerful method of protection because with physically small sources, the radiation dose rates fall off approximately as the square of the distance from the source. Conversely, dose rates at close distances can be extremely high, even for small sources. Therefore, sources should never be handled directly.

(b) Time is a useful control because both I.C.R.P. and statutory regulations (see Tables 5.1 and 5.3) express maximum permissible doses in terms of doses over substantial times, the minimum period specified being 3 months. Therefore, high dose rates can be accepted for short periods of time provided that the integrated doses remain low. However, if there is an excessive dependence on restricting exposure time, strict discipline is especially important and the risk of the operation not going according to plan should be taken into account. A trial run with inactive specimens is usually worthwhile before operations involving high activity sources with correspondingly short permitted exposure times.

(c) Distance and time together are usually satisfactory for operations involving small sealed radioactive sources. However, with large sources and generally for the control of X-rays, the radiation dose rates require to be reduced by the use of shielding. Shielding has the advantage that it can be used to provide intrinsic protection without reliance upon restricting exposure time. This is emphasized in the factory regulations[14, 15] which specify that where reasonably practicable, all sources must be adequately shielded; it should be noted that the presence or absence of adequate shielding will often determine whether persons working with ionizing radiations are to be classified. Radiation in interaction with matter may be scattered (in the forward and other directions) or absorbed. The effectiveness of a shield[17] depends upon its geometrical configuration as well as its thickness and material.

β-radiation has a finite range in matter and shielding is simple because in many situations complete absorption can be obtained by a few mm of metal or about $\frac{1}{2}$ in. of glass or plastic materials. With solutions, sufficient protection usually results from absorption in the water and the walls of the containing vessel. As an exception to the general rule against the direct handling of radioactive sources, thick gloves[18] are used to reduce the hand dose from β-radiation for work with unclad natural (or depleted) uranium metal, which is a relatively weak radioactive material. It may be noted that the factory regulations pertaining to sealed sources,[14] only prohibit direct contact with the bare hand. Eye shields[22] can be used when they are necessary to keep the dose to the lens within permissible limits. Shielding against bremsstrahlung produced by β-active radioactive materials should not normally have to be considered except with large sources of several hundred mCi or greater.

For attenuating X-rays and γ-radiation, materials of high density should be used, the common ones being lead, iron and concrete; in some situations depleted uranium has considerable advantages because of its very high density of about 19 g/cm^3. For vision, a combination of mirrors and handling equipment which operates through or over the shield is usually adequate, but high-density lead-glass (Fig. 5.1) and liquid shields[23] also can be used. The precise degree of shielding required depends on the material and the energy of the radiation, as illustrated for γ-radiation in Table 5.7.

The data in Table 5.7 are for broad-beam geometry which usually applies in practical cases. It may be noted that if narrow-beam data (derived for collimated beams) are applied to broad-beam conditions, the thickness of the shield will be underestimated and so also will the expected dose rates. Information on shielding[17] is readily available for X-rays and γ-radiation. Because of the penetrating power of X-rays and γ-radiation it is not practicable to design clothing as a shield, although an important exception is the use of lead-impreg-

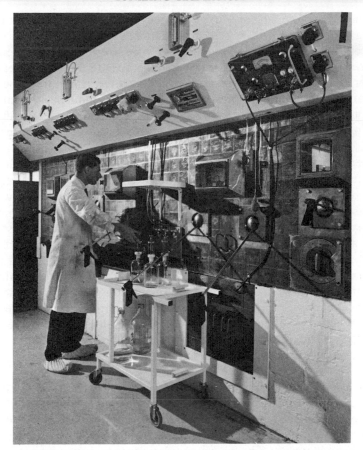

Fig. 5.1 Lead shielded facility with remote handling tongs and lead glass observation windows. Part of a radioisotope production line at the Radiochemical Centre, Amersham.

(Photograph by courtesy of U.K.A.E.A.)

Table 5.7[24] Shielding against γ-radiation

Material	Attenuation factor	Approximate thickness (cm)			
		0·5 MeV	1 MeV	2 MeV	3 MeV
Concrete	10	23	28	36	44
	100	38	49	66	79
	1 000	53	69	92	110
Iron	10	7	9	11	12
	100	11	15	19	22
	1 000	16	21	28	32
Lead	10	2	4	6	7
	100	4	8	12	12
	1 000	5	11	16	18

nated rubber aprons and gloves to reduce the dose to the body and hands when working with low energy X-rays in medical diagnostic radiology.[25]

Aspects of shield construction which may be overlooked are: the strength of flooring needed to support a lead wall; the dose to the legs or head of the operator and the dose to persons in surrounding areas arising from the absence of all-round shielding; the presence of narrow, high-intensity beams from radiation streaming through small gaps in the shield. Also, significant air scattering[17, 26] into the working area may occur when large sources are manipulated in 'open top' shielded enclosures.

In neutron shielding,[17] one of the principal factors is the reduction of their energy, since the permissible flux density increases progressively as the energy decreases. Fast neutrons are slowed down by interaction with light elements, especially hydrogen. Therefore materials such as water, paraffin wax and plastics are very effective; about 25 cm thickness will reduce the fast neutron dose by a factor of ten or more. Concrete is useful for permanent shields around neutron generators and reactors. Neutrons which have been slowed down to thermal energies are readily captured and completely absorbed by cadmium or boron. If the dimensions of the shield are small enough for cadmium to be used economically, even a thin sheet will give a very useful factor of protection. In all cases where large attenuations are necessary, shielding may also be required against the secondary γ-radiation arising from scattering and capture processes in the interaction between neutrons and matter.

External radiation monitoring. Radiation monitoring is necessary to check that the planning is adequate and that the actual doses incurred are satisfactorily low. Two well-defined methods are used, namely, the measurement of dose rates in and around the working area[27, 28] and the measurement of the doses received by persons individually. The factory regulations specify requirements for both these types of monitoring.

Radiation dose rates are normally measured by portable electronic instruments of a more or less simple type. Installations and equipment are monitored initially to ensure that ambient radiation dose rates are satisfactory and, where reasonably practicable, below the figure of 0·75 or 2·5 mrems/h, as applicable, adopted in the factory regulations as the criterion of adequate shielding. Subsequent to this, spot checks are made to indicate whether there has been any significant change. Additionally, monitoring is done for controlling working times in specific operations. In selecting instruments,[29, 30] it is important to take into account not only the advantages, but also the limitations of a particular type for the circumstances of its application. Advice on the choice of suitable instruments is available from the National Radiological Protection Board.

The individual doses can be obtained most easily by making the wearing of a personal dosemeter compulsory in radiation areas; such a requirement is specified in factory regulations which also require that the dosemeters must be obtained from and examined by an 'approved' laboratory. The most convenient way of making this dose measurement for β-, X-, γ-radiation, and slow neutrons is by the use of a film badge or thermoluminescent dose meter worn on the body; this can be supplemented as necessary by additional films or thermoluminescent dosemeters where hands and fingers are preferentially exposed. It is sometimes necessary to have a continuous indication of the dose accumulated during some particular operation and for this purpose quartz fibre dosemeters which may be read at any time and other types which also sound an audible warning at a predetermined dose, may be used in addition to the film badge or thermoluminescent dosemeter. Such direct reading dosemeters have the disadvantage that those of robust construction are suitable for γ-radiation measurements only; also, they may fail to indicate the magnitude of gross overexposures. Fast neutrons can be measured using special photographic emulsions. The National Radiological Protection Board provides a personal monitoring service for external radiation.

The control of contamination

Definition of the problem

In the handling of unsealed sources of radioactive material, apart from any problem of external radiation, there is the problem of contamination control. Whenever such work is performed, there will be some transfer of radioactive contamination from the immediate working area into the working environment. This contamination is spread not only by accidents and spills, but also by normal operating procedures, however carefully these are planned. Part of this contamination may enter the bodies of exposed persons by ingestion, inhalation, or absorption through broken or intact skin; the absorption of HTO through skin is noteworthy.[13] Contamination may also contribute to the external radiation. The other grounds for objecting to contamination is that it may interfere with the technical work of the laboratory. In tracer work, this latter aspect can impose more stringent standards of precautions than do the safety problems.

Scale of the problem

The scale of the problem is reflected in the basic recommendations of the I.C.R.P. and limiting values derived from them, as illustrated below by the limits imposed for internally deposited radioactive material and levels of environmental contamination.

Maximum permissible body burdens and maximum permissible concentrations in air and water of radioactive materials for occupational exposure. A selection of values recommended by I.C.R.P. is given in Table 5.8.

Table 5.8[13, 31]   Maximum permissible body burdens and concentration in water and air for occupational exposure (40-h week) for soluble forms of some commonly used nuclides

Nuclide (soluble form)	Critical organ	Maximum permissible body burden μCi	Maximum permissible concentrations $\mu Ci/cm^3$	
			Water	Air
^3H (HTO or T_2O)	Body tissue	10^3	0.1	5×10^{-6}
^{14}C	Fat	3×10^2	0.02	4×10^{-6}
^{32}P	Bone	6	5×10^{-4}	7×10^{-8}
^{90}Sr	Bone	2	10^{-5}	10^{-9}
^{131}I	Thyroid	0.7	6×10^{-5}	9×10^{-9}
^{226}Ra	Bone	0.1	4×10^{-7}	3×10^{-11}
Natural uranium	Kidney	5×10^{-3}	2×10^{-5}	7×10^{-11}
^{239}Pu	Bone	0.04	10^{-4}	2×10^{-12}

To illustrate the enormous differences that can exist between the values for radioactive materials and industrial toxins, the figure for strontium-90 in breathing air is equivalent to about 10^{-6} $\mu g/m^3$ which is of the order of a hundred miliion times less than the value of 200 $\mu g/m^3$ adopted for stable lead. However, in terms of the amounts normally handled in laboratories, many radioactive materials are effectively no more toxic than many conventional laboratory chemicals.[24, 29] For a few nuclides, chemical toxicity is limiting, an example being soluble natural uranium[18] whose maximum permissible concentration in terms of mass per unit volume of air is about the same as in the case of stable lead.

Derived working limits of surface contamination.[10] Radioactive contamination on a surface presents the following risks:[32] inhalation of radioactive material

Table 5.9   Derived working limits for surface contamination*

Type of surface	Principal α-emitters $\mu Ci/cm^2$	Low toxicity α-emitters $\mu Ci/cm^2$	β-emitters $\mu Ci/cm^2$
Inactive and low activity areas	10^{-5}	10^{-4}	10^{-4}
Active areas†	10^{-4}	10^{-3}	10^{-3}
Personal clothing	10^{-5}	10^{-4}	10^{-4}
Personal protective clothing not normally worn in inactive areas	10^{-4}	10^{-3}	10^{-3}
Surfaces of the body	10^{-5}	10^{-5}	10^{-4}

* Averaging is permitted over inanimate areas of up to 300 cm², or, for floors, walls and ceiling, 1 000 cm². Averaging is permitted over 100 cm² for skin or, for the hands, over the whole area of the hand.

† The limits for active areas do not apply in the case of surfaces or equipment inside fume cupboards and total enclosures such as glove boxes. In such cases, the levels should be kept to the minimum that is reasonably practicable.

through its resuspension in breathing air, ingestion of radioactive material by direct or indirect pathways, and, with β-active contamination, external radiation exposure also.

The basic derived working limits for surfaces specified in factory regulations are summarized in Table 5.9.

Fundamental principles of contamination control

The control of contamination is a complex subject because any action taken in a radioactive area may cause or spread contamination; also, decontamination is an essential part of the control measures and decontamination is often intrinsically difficult, and in some cases, impossible.

The fundamental principles of contamination control are containment and cleanliness. As with the control of external radiation, the first step is that of planning. This planning stage should, where appropriate, take into account external radiation problems in the same way as for sealed sources. Attention should then be paid to the problems of containment.

Containment is achieved by the careful selection of materials and operating techniques and the provision of adequately designed workplaces aimed at preventing contamination from spreading into the working area. The total quantity of radioactivity should be as small as practicable. Where there is a choice, solutions should be used rather than solids, especially powders. Preference should be given to materials of low specific activity and low relative toxicity. Techniques should be selected which produce the minimum of spray or dust and involve the minimum of transfer between vessels. Reliance should not be placed on single containers; the use of secondary containers should be the rule. A distinction should be made between operations which may be carried out without risk on the open bench and those which should be enclosed in fume cupboards and glove boxes.

Because containment is never absolute, some radioactive material may still reach the working area where its build-up is prevented by regular cleaning and its further spread checked by the use of clothing change, ventilation arrangements and control over all classes of movements (personnel, equipment, sources, waste, etc.) between the process or laboratory areas and others. Much of the cleaning in the working area can be done by conventional means provided that care is taken to avoid methods which give rise to dust.

In brief, all radioactive materials should be treated as poisons and conventional laboratory techniques and procedures should be modified accordingly. For example, pipetting and glass blowing should never be done by mouth; no mouth-operated equipment of any kind should even be present. 'Good housekeeping' is vital. All actions should be conducted with the thought of preventing contamination, both of the person and the environment. Apart from laboratory clothing[22] including gloves which are used in the appropriate situations for normal operations, special protective clothing, ranging from respiratory protective apparatus to pressurized suits[33] (Fig. 5.2) may be needed for special operations, abnormal conditions and emergencies. The limitations of respirators[34] and airline equipment[35, 36] should be noted. Restrictions may have to be imposed in respect of such actions as eating, smoking, etc. in the area.

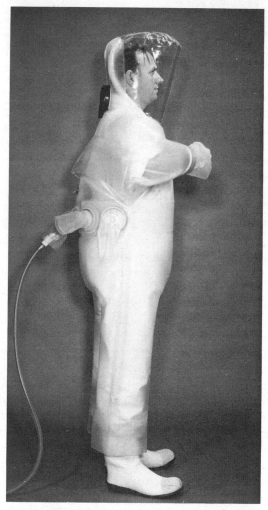

Fig. 5.2   U.K.A.E.A. standard P.V.C. pressurized suit.
(Photograph by courtesy of U.K.A.E.A.)

Contamination monitoring

As in the case of external radiation control, monitoring is applied both to the working environment and to the individual workers.

A programme of routine monitoring for surface contamination is desirable.

Monitoring instruments,[37] which may be mains or battery operated, should be available to the operators as part of their normal equipment to enable them to check their own operations as they proceed. Equipment leaving the working area should be specially checked.

Air sampling[28] does not give a precise indication of the exposure of workers to environmental radioactivity. The usual aim, therefore, is to keep the measured concentrations below one-tenth[12] of the I.C.R.P. recommended limits for the long-term average of the results from environmental sampling. Another factor is that allowance may have to be made for any external radiation exposure that may also occur. In many situations, instead of routine air sampling, reliance may be placed upon surface contamination monitoring as an index of control over radioactive material. However, some confirmatory sampling may be required, particularly for high-activity work and materials of very high toxicity. Fig. 5.3 shows a portable sampler suitable for environmental control sampling. It should rarely be necessary to institute personal exposure sampling.

As regards personal monitoring, everyone working in a contamination-risk area should wash and monitor their hands before leaving; this is particularly important at the end of the day and before meal breaks. Radioactivity deposited inside the body may be estimated[38] by direct measurement of the exposed person or by indirect methods. It should not be necessary to arrange for such monitoring unless a significant intake is suspected or unless large quantities of radioactivity are being handled. Personal monitoring for internal contamination is another of the services available from the National Radiological Protection Board.

As stated on page 56, the basic maximum permissible doses comprise the sum of the doses from external and internal radiation exposure. Whilst the forecasting, measurement, and control of personal dose from external radiation is relatively straightforward, the estimation of internal dose is often difficult and complicated, and liable to delay, important errors and omissions. Therefore, for protection in work with unsealed sources, the underlying principle is to prevent significant intake by the adoption of high standards of containment and cleanliness rather than to plan exposures within the limits recommended by I.C.R.P. One consequence of this philosophy has been mentioned, namely the application of a restrictive factor in air monitoring. It may be noted that the factory regulations[14, 15] specify maximum permissible doses for external radiation exposure only. Exposure to internal radiation is not specified in terms of numerical limits such as maximum permissible body burdens or concentrations in air of radioactive materials. Instead, the onus is on the employer to do all that is reasonably practicable to prevent the intake of radioactive materials; the worker is responsible for avoiding unnecessary exposure. The factory regulations, however, do specify requirements for invironmental and personal monitoring including, in special circumstances, estimations of body burdens also.

Laboratory standards

A classification of laboratories has been devised as guidance to the design of laboratories for the handling of unsealed sources. By classifying nuclides into a number of groups based on relative radiotoxicity as illustrated in Table 5.10, the standards of design required for broad levels of radioactivity handled in different ways can be indicated as shown in Table 5.11. The three grades of laboratory are described briefly below.

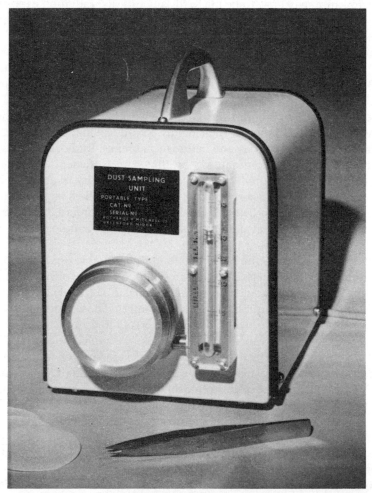

Fig. 5.3 A portable air sampler. A mains-driven positive displacement pump draws air at a known rate through a glass-fibre filter paper which is removed to conventional counting equipment for estimation of the activity of the collected dust. Charcoal impregnated filter papers can be used for the sampling of radioiodine vapour in air.

(Photograph by courtesy of Rotheroe and Mitchell Ltd.)

Table 5.10[15] Classification of nuclides according to relative radiotoxicity per unit activity

(The nuclides in each class are listed in order of increasing atomic number. For other nuclides, comprehensive lists should be consulted.)

Toxicity class	Nuclide
High toxicity (Class I)	^{210}Po, ^{226}Ra, ^{227}Ac, ^{228}Th, ^{230}Th, ^{233}U, ^{234}U, ^{239}Pu, ^{241}Am
Medium toxicity: Upper sub-group A (Class 2)	^{22}Na, ^{45}Ca, ^{60}Co, ^{90}Sr, ^{124}Sb, ^{131}I, ^{137}Cs, ^{170}Tm, ^{192}Ir
Medium toxicity: Lower sub-group B (Class 3)	^{14}C, ^{24}Na, ^{32}P, ^{59}Fe, ^{85m}Kr, ^{132}I, ^{135}Xe, ^{198}Au, ^{222}Rn
Low toxicity: (Class 4)	^{3}H, ^{232}Th, Th-Nat, ^{235}U, ^{238}U, U-Nat

Table 5.11[39, 40] Grades of laboratory required for unsealed radionuclides

Relative radiotoxicity of nuclide	Grade of laboratory required for levels of activity specified below*		
	C	B	A
High toxicity (Class 1)	<10 µCi	10 µCi–1 mCi	>1 mCi
Medium toxicity: Upper sub-group A (Class 2)	<1 mCi	1 mCi–100 mCi	>100 mCi
Medium toxicity: Lower sub-group B (Class 3)	<100 mCi	100 mCi–10 Ci	>10 Ci
Low toxicity (Class 4)	<10 Ci	10 Ci–1 000 Ci	>1 000 Ci

* Modifying factors may be applied, e.g., $\times 100$ for simple storage, $\times 1$ for normal chemical operations and $\times 0.01$ for dry and dusty operations.

Grade A. These are laboratories which are specifically designed for high-level work. They should be supported in the same building by change rooms, preferably with showers, and by separate counting rooms, etc. The extract air is exhausted through high-efficiency filters. The drainage is separate and leads through delay tanks.

Grade B. These are high-quality laboratories with generous provision of adequately ventilated fume cupboards. Glove boxes may also be required. The discharge points for exhaust air, which may require to be filtered, should be selected with care. Particular attention should be given to surface finishes. Great care in design and operating techniques is necessary to facilitate the control of contamination. Facilities should be provided at the laboratory entrance for clothing change, hand washing and the monitoring of skin, clothing, and equipment.

Grade C. Few modifications should be needed to any modern conventional chemical laboratories. The special needs are very little beyond good ventilation, at least one fume cupboard with induced draught exhausting outside the building, and easily cleaned surfaces.

Emergency procedures: immediate action drills

Typical emergencies with radioactive materials are cuts and wounds involving contamination, spills, the damaging of a sealed source, the dropping of a sealed γ-emitting source from its shield, and fire involving or threatening radioactive material. Even where only small amounts of radioactive material are handled, it is convenient to have prearranged procedures.[41] For injuries, immediate washing with copious water is the preferred method of first aid and the implement causing the wound should be retained for monitoring. In the other cases mentioned above, the primary concern is to minimize exposures and the secondary need is to confine any contamination to the area directly affected. For other than minor incidents, the area should be cleared of people as rapidly as possible and remedial work involving re-entry should be undertaken only after an assessment of the precautions needed; these can be quite simple as illustrated in Fig. 5.4. For spills,[39, 40] the persons should be monitored before they move far from the scene of the accident to avoid the unnecessary spread of contamination to other areas. Immediate attention should be given to injuries involving contamination and severe skin or clothing contamination. In the case of fire, contamination control gives precedence to fire-fighting but should not be ignored. As part of preplanning procedures, the local fire service may wish to discuss fire-fighting arrangements and may require a special marking[42] on the door of laboratories, storage areas, and source containers. The importance of minimizing the risk of accidents, particularly fire and explosion, is greatly enhanced because the effects of the release of radioactivity not only complicate local emergency procedures but may have repercussions outside the range of the initial accident. Aspects of accident prevention liable to be overlooked are that spills and explosions can also be caused by the high pressures that may develop as a result of radiation decomposition of solutions of radioactive materials stored in closed and unvented containers. Also, sufficient importance may not always be attached to preventing skin injuries; serious amounts of radioactivity can easily be introduced directly into the bloodstream from a single incident involving a puncture wound received while working with unsealed sources.

In the U.K. there are generalized arrangements on a national basis (N.A.I.R.) to enable expert advice to be made quickly available to the police when they are confronted with an incident which may involve the public being exposed to radioactivity and no other expert is otherwise available. The country is divided into zones which accord with police administrative areas. A source of radiological advice and assistance to the police is nominated for each zone. The operation of the scheme is based on the assumption that the police will normally be alerted at the outset in the event of a transport accident or other incident in a public place. Details of the scheme are obtainable from the National Radiological Protection Board which has responsibility for coordinating these arrangements.

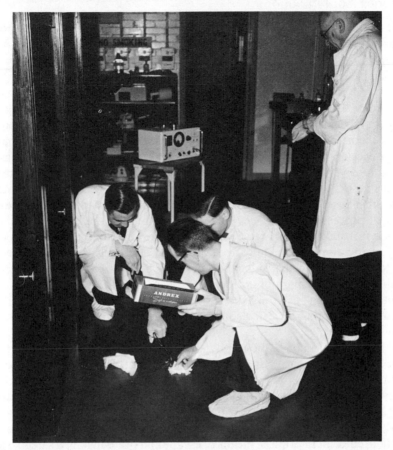

Fig. 5.4 Control of a small liquid spill. Following the breakage of an ampoule containing 1 mCi of phosphorus-32 solution, only tongs, paper tissues, rubber gloves and overshoes are needed for the initial cleaning. The first absorbing of the liquid was done with tissues held in tongs to reduce the beta dose.

(Photograph by courtesy of U.K.A.E.A.)

Legal requirements and special safety aspects of the principal industrial applications of ionizing radiations

For persons using ionizing radiations in industrial circumstances, the principal legislation is outlined below.

The information merely outlines certain of the statutory provisions and must not be considered as authoritative. To ensure compliance with the law the relevant parent legislation and the orders, etc., made thereunder, must be consulted.

Factories Act regulations

Legal requirements for the protection of persons against ionizing radiations in premises subject to the Factories Act 1961 are contained in the Ionizing Radiations (Sealed Sources) Regulations 1969, and the Ionizing Radiations (Unsealed Radioactive Substances) Regulations 1968.

The Ionizing Radiations (Sealed Sources) Regulations 1969.[14] These regulations apply to: (a) sealed radioactive sources, (b) machines or apparatus designed to produce ionizing radiation, such as X-ray sets, and (c) machines or apparatus with an accelerating voltage of not less than 5 kV which emit ionizing radiations adventitiously. Their object is to ensure that the workers incur the minimum practicable exposure to ionizing radiations and that the specified maximum permissible doses are not exceeded. Apart from X- and γ-radiography, the other principal processes and devices subject to the regulations are: high-intensity irradiation installations; X-ray fluoroscopy; X-ray crystallography and spectroscopy; static elimination; thickness, level and density gauging; and, package monitoring. The principal safety aspects have been well covered in an official handbook.[12] A summary of accidental exposures and the lessons to be drawn from them has also been published.[38] Some of the special features for the safe design and operation of sealed sources and radiation generators are described below.

X-radiography and γ-radiography should be carried out where practicable inside a walled enclosure. The operating voltages of the majority of X-ray sets lie in the range 100–400 kV, although they can range up to 2 MV (even higher energy beams are available from particle accelerators). For γ-emitters, activities up to around 20 Ci are common but can be in excess of 1 000 Ci for special applications. There must be adequate shielding against direct and scattered radiation, efficient interlocks, means of escape and other installed safeguards for persons who may be accidentally shut inside the enclosure, and audio-visual warning signs. Where construction of a working enclosure is impracticable, radiography should be isolated from other work. For dose rate surveys in areas around the enclosure or equipment, instruments should be selected for uniform response over the appropriate range of energies, ability to maintain full-scale deflection at very high dose rates and maintenance of calibration in collimated beams or pulsed fields.

High-energy, high-intensity particle accelerator installations such as betatrons[43] and neutron generators must be installed in a special enclosure. They can give rise to very specialized and difficult safety problems in respect of: the primary beam; secondary radiation; concentrations of noxious and radioactive gases in radiation rooms; radioactivities induced in the target material and also in shielding and machine components; radiation damage to insulating materials and radiation paralysis of electrical safety devices; fire hazards in irradiated materials; and high voltage electrical hazards. Monitoring is complicated by the fact that radiation fields in the neighbourhood of these installations are very complex.

High-intensity irradiation,[1] as used for example in sterilization and disinfestation processes, utilizes very large radioactive sources of up to several hundred thousand curies of cobalt-60 (or high-energy machines). As in the case of particle accelerators, these plants are highly specialized and expert advice must be sought in their design and operation.

In fluoroscopy, the X-ray apparatus should be enclosed in an adequately shielded cabinet fitted with effective interlocking devices. The useful beam should not fall outside the screen and this should be checked by monitoring. Viewing should be indirect; but if this is not practicable the window must be adequately shielded. Dark adaptation prior to viewing will reduce the dose to the operator.

Particular care is needed with X-ray crystallographic equipment. Although the relatively soft X-radiation generated at around 50 kV is readily attenuated, the very high dose rates near the target, often amounting to several thousand roentgens/s, can cause injury in a very short time. Apart from hand injuries, several accidents have also occurred in which the tube collimator has been aligned by eye, with the tube operating. It is particularly important that the fingers and other parts of the body cannot be inserted in the useful beam. To prevent the eyes from being exposed, 'lining-up' can be effected with the aid of a small fluorescent screen. The most satisfactory design is to enclose the X-ray apparatus completely in a shielded cabinet with interlocks provided for the access door and panels. Where total enclosure is not possible it is necessary to have automatic beam shutters and preferably double interlocks. For equipment of open-type construction, additional needs include local shielding, proximity switches and beam traps. Warning systems must be clear, unambiguous, close to the apparatus and properly visible. For warning light assemblies, two lamps in parallel should be used. Special precautions are essential during maintenance and adjustments which have to be made with the shielding removed or the interlocks rendered inoperative.

For the remaining processes such as thickness gauging, package monitoring and static elimination, the safety problems should be much less severe, chiefly because of the form and amount of the source material involved. Nevertheless, since many of these sources are continuously employed during the processes, are sometimes hand-operated and somewhat fragile, it is just as important to apply the prescribed care and precautions.[12]

Normally, with sealed sources, it is necessary to take precautions against external radiation only. The possibility remains, however, that the sources might suffer damage or corrosion leading to a spread of contamination. Therefore, apart from care to reduce this possibility, the regulations specify periodical leakage tests and the action to be taken in the event of breakage of a sealed source.

The Ionizing Radiations (Unsealed Radioactive Substances) Regulations 1968[15]

With unsealed radioactive substances, apart from the risk of internal exposure, there may also be the problems of external radiation. The regulations are concerned primarily with controlling internal radiation exposure by the application of the principles of containment and cleanliness. The general aim of the measures is to avoid, where practicable, direct contact between the worker and the radioactive material, to maintain surface and personal contamination below the specified levels and to ensure that no significant concentrations of radioactive substances in air occur. Two points of particular interest may be noted. First, the level of surface contamination is used as a major control parameter. Second, although maximum permissible body burdens and concentrations in air of radioactive material are not specified in numerical terms, the parameters laid down for determining what constitutes a 'significant' amount of radioactive substance would obviously require current I.C.R.P.[9, 13, 31] recommendations to be taken into consideration.

Storage of radioactive materials when not in use. The regulations specify requirements for the marking and ventilation of storage areas and care and custody arrangements. Stores should be so located[39, 40] as to simplify transport problems and minimize exposure of workers resulting from the transport of sources. Consideration should also be given to: (a) shielding, (b) contamination control, and (c) precautions against flooding, fire, and theft and to any special requirements such as the venting of solutions.

Medical supervision. Although, primarily, the radiological protection of workers has necessarily to be based on controlling the working environment and the doses to persons, medical supervision has a complementary though distinct role in the safeguarding of health. This is reflected in the requirements in the factory regulations for pre-employment examination, routine re-examination and special supervision in the event of 'over-exposure'.

Responsibility and training; sources of advice and assistance. Underlying any procedures for the control of any occupational hazard is the allocation of responsibilities. Although the principal responsibilities in practice and in law fall on the management, the worker must recognize his own responsibility for safety, and without close collaboration no satisfactory programme of protection can be operated. The factory regulations require appropriate training to be given to employees on the nature of radiological hazards and the precautions to be observed. Advantage should be taken of the courses available at colleges and other centres, particularly for the training of any full-time radiation safety staff and the nominated persons invested with special responsibilities for safety as specified in the regulations. Full use should be made of the advisory, training and other services available to users of radiation sources from the National Radiological Protection Board.

The Health and Safety at Work etc. Act 1974

The Act extends to *all* people at work in Great Britain except domestic workers in private employment. Unlike previous health and safety legislation it is aimed at people and their activities instead of at premises and processes. Its intention is to secure the health, safety and welfare of persons at work (employers, self-employed and employees) and to protect persons other than those at work from risks to health and safety arising from or in connection with working activities. General duties of care are laid down and a Health and Safety Commission created to help people fulfill their obligations under the Act and specific Regulations etc. The Commission will ensure that research and training is undertaken and information provided. The Commission will operate mainly through a Health and Safety Executive who will also be responsible for enforcement action. Several inspectorates created by earlier legislation have been brought together in the Executive. The new obligations under the Act are in addition to those already included in existing legislation which will remain in existence until it is progressively repealed and replaced by regulations under the Act and approved codes of practice. Under the Act, in addition to powers of prosecution for contraventions, inspectors can issue prohibition or improvement notices and also seize, render harmless or destroy substances or articles.

Radioactive Waste and the Radioactive Substances Act 1960

The handling and disposal[39] of contaminated wastes gives rise to two groups of problems. First, there are the problems internal to the premises giving rise to the waste; these are largely covered by handling the waste in the same way as any other radioactive material. The second and more complex group relates to public health problems. Both aspects must be considered at the planning stage, with the aim of keeping contaminated waste to a minimum and in a form suitable for the intended method of disposal. There are two basic methods of disposal. First, controlled dispersal into the surrounding environment, using air or water as the diluting medium to reduce the radioactivity to a completely safe level. The alternative method is to store the radioactivity until it decays to a safe level or, if the rate of decay is too slow, indefinitely. Many of the principles of internal and external radiation control operate favourably for the waste management services, e.g., using as little radioactive material as necessary and choosing nuclides of low relative toxicity.

The Radioactive Substances Act 1960 is concerned essentially with the public health aspects of the storage or discharge of radioactive waste. The Act requires, with certain specified exceptions, the registration of all holders of radioactive material (and mobile radioactive apparatus), and prohibits the disposal or accumulation of radioactive wastes except in accordance with an authorization granted by the appropriate Minister. The Act also allows the Minister to provide or arrange for the provision of facilities for the safe disposal of the waste in cases where no suitable local method is available. Users should discuss their waste disposal problems at an early stage with the Governmental Inspectors or the National Radiological Protection Board. Certain classes of radioactive waste may be exempted from authorizations, local disposal would be the authorized method for the great majority of users and only in exceptional cases should the National Disposal Service be required.

The Nuclear Installations Act 1965 and 1969

Nuclear reactors and criticality control are special subjects outside the scope of this chapter. Under the Nuclear Installations Act, a licence granted by the Health and Safety Executive is required before installing or operating a nuclear reactor or other installation covered by the Act, such as sub-critical assemblies and plants manufacturing or producing nuclear fuel elements, alloys and chemical compounds containing enriched uranium or plutonium. The conditions of the licence require inspection and testing during construction and the submission of detailed operating procedures in the interests of safety.

Regulations affecting the transport of radioactive materials[44]

For the transport of radioactive substances within the premises, the principles pertaining to internal and external radiation control should be applied with special care. The carriage of radioactive substances outside the premises by road, rail, post, sea, or air is governed by Regulations, Rules, Conditions of Acceptance etc., in force or under preparation, according to the means of transport. Advice should be sought from the appropriate Ministry or Department. The persons to be safeguarded include transport workers, members of the public and damage-control teams. U.K. regulations are in harmony with the international recommendations formulated by the International Atomic Energy Agency.[45]

The National Radiological Protection Board

The Board is a public authority whose members are appointed by the Health Ministers of the United Kingdom. It was established by the Radiological Protection Act, 1970, to advance the acquisition of knowledge about the protection of mankind from radiation hazards and to provide information, advice and services to persons with responsibility in radiological protection. The purpose of the Act was to create a point of authoritative reference in radiological protection. The Board engages in research and undertakes special assessments and studies. Its headquarters are at Harwell, and it has established centres at Glasgow, Leeds and Harwell for the provision of advice and services.

Work with radiation sources and radioactive materials covers a very wide range, and involves industry, hospitals, universities, colleges and schools, as well as individuals in medical, dental and veterinary practice. The Board can help in all these fields and in a variety of ways by providing advisory and practical services. The latter include: Reviews of Design, Monitoring of Premises, Personal Monitoring, Record Keeping, Instrument Tests, Testing of Materials and Equipment, Leakage Tests on Sealed Sources, and Assistance in the event of Incidents and Accidents. The Board has an extensive training programme for various levels of operational staff, supervisors and safety staff. Details of the services and charges are obtainable from the Board's headquarters and service centres.

Conclusion

The past 25 years have seen a considerable growth in the attention given to the study of biological effects of ionizing radiation and the refinement of methods of protection and control. Therefore, despite the spectacular increase in the use of radiation generating apparatus and of radioactive materials in factories in the past decade, experience so far has indicated that the health of work people can be adequately safeguarded. There is no justification for fear and, equally, none for carelessness or complacency. With relatively straightforward precautions, sources of ionizing radiation can be handled with a high standard of safety, without at the same time restricting the benefits which its use can bring to industrial work in all fields.

Acknowledgements

The author wishes to thank Mr H. J. Dunster and Mr L. D. G. Richings, Assistant Director (Operations) and Secretary respectively of the National Radiological Protection Board, for reading the draft of this chapter and for their helpful suggestions and criticisms.

Appendix
Explanation of terms

Ionizing radiations

Electromagnetic radiation (X-rays, γ-rays), or corpuscular radiation (α-particles, β-particles, electrons, positrons, protons, neutrons, and heavy particles) capable of producing ions.

Roentgen (R)

This is the unit of exposure of X- or γ-radiation, based upon the capacity of the radiation to produce ionization in air. Over the energy range 10 keV to 3 MeV,

1 R is associated with an energy dissipation of about: 87 ergs/g of air, 87–97 ergs/g of water (or soft tissue).

Rad

The energy deposited by radiation in material is called the absorbed dose. The unit of absorbed dose is the rad, which is equal to an energy deposition of 100 ergs/g of any material, irrespective of the type of radiation to which the material is exposed. From the foregoing it will be seen that 1 rad is approximately equal to the absorbed dose delivered when soft tissue is exposed to 1 R of X- or γ-radiation, i.e., that the numbers of roentgens and rads are numerically equal.

Rem

Biological effects depend not only upon the energy deposited (rads) but also upon the type of radiation (leading to a Quality Factor, Q) and other modifying factors. The rem is the unit of dose equivalent which expresses on a common scale for all ionizing radiations, the irradiation incurred by exposed persons. It is the unit that is used for totalling doses from exposure to different types of radiation.

Dose in rems $=$ rads $\times$ Q $\times$ any other modifying factors.

Summary of quality factors based on I.C.R.P. recommendations

External exposure

X- and γ-radiation	1
β-particles and electrons	1
Neutrons	2 to 10·6 (see Table 5.6)
Heavy recoil nuclei	20

Internal exposure

X- and γ-radiation	1
β-particles	1
α-particles	10
Heavy recoil nuclei	20

These values mean, for example, that in the case of X-, γ-, and β-radiation, the numbers of rads and rems are numerically equal.

References

1. Putman, J. L. *Isotopes,* 2nd edn., Penguin Books, London, 1965.
2. International Commission on Radiological Protection No. 23. *Report of the Task Group on Reference Man.*, Pergamon Press, Oxford, 1975.

3. Report of the United Nations Scientific Committee on the Effects of Atomic Radiation. *Ionizing Radiation: Levels and Effects. Volume I: Levels.* New York, 1972.
4. Webb, G. A. M. Radiation exposure of the public—The current levels in the United Kingdom. *N.R.P.B.–R24,* 1974. Available from H.M.S.O., London.
5. The evaluation of risks from radiation. *I.C.R.P. Publication 8.* Pergamon Press, Oxford, 1966.
6. The hazards to man of nuclear and allied radiations: *First Report Cmd. 9780, Second Report Cmnd. 1225.* H.M.S.O., London, 1956, 1960.
7. Report of the United Nations Scientific Committee on the Effects of Atomic Radiation. *Ionizing Radiation: Levels and Effects. Vol. II: Effects.* New York, 1972.
8. National Radiological Protection Board. *Living with Radiation,* 1973. Available from H.M.S.O., London.
9. Recommendations of the International Commission on Radiological Protection (adopted 17 September 1965) *I.C.R.P. Publication 9.* Pergamon Press, Oxford, 1966.
10. Dunster, H. J. The application and interpretation of I.C.R.P. recommendations in the U.K.A.E.A. *A.H.S.B.(R.P.) R78.* 3rd edn., H.M.S.O., London, 1968.
11. Recommendations of the international commission on radiological protection. Implications of Commission's Recommendations that doses be kept as low as readily achievable. *I.C.R.P. Publication 22.* Pergamon Press, Oxford, 1973.
12. Department of Employment and Productivity. Ionizing radiations: precautions for industrial users. *Safety, Health and Welfare, New Series No. 13,* H.M.S.O., London, 1969.
13. Recommendations of the international commission on radiological protection. Report of Committee II on permissible dose for internal radiation. *I.C.R.P. Publication 2.* Pergamon Press, London, 1960.
14. Factories: The Ionizing Radiations (Sealed Sources) Regulations 1969.
15. Factories: The Ionizing Radiations (Unsealed Radioactive Substances) Regulations 1968.
16. Cross, W. G. Rough estimates of β-doses to the skin. *Health Physics,* 1965, **11**, (5), pp. 453–454.
17. *Handbook of Radiological Protection. Part 1: Data.* Prepared by a panel of the Radioactive Substances Advisory Committee, 1971. Available from H.M.S.O., London.
18. Bennellick, E. J. A review of the toxicology and potential hazards of natural, depleted and enriched uranium. *A.H.S.B.(R.P.) R58,* 1966. Available from H.M.S.O., London.
19. Radiation protection procedures. *Safety Series No. 38.* I.A.E.A., Vienna, 1973.
20. Recommendations of the International Commission on Radiological Protection. Protection against ionizing radiation from external sources. *I.C.R.P. Publication 15,* Pergamon Press, Oxford, 1970.
21. Recommendations of the International Commission on Radiological Protection. Data for protection against ionizing radiation from external sources: (Supplement to I.C.R.P. Publication 15). *I.C.R.P. Publication 21,* Pergamon Press, Oxford, 1973.
22. International Atomic Energy Agency. Respirators and protective clothing. *Safety Series No. 22,* Vienna, 1967. Available from H.M.S.O., London.
23. Walton, G. N. (ed.). *Glove Boxes and Shielded Cells for Handling Radioactive Materials.* Butterworths, London, 1958.
24. Dunster, H. J. Protection against radiation and contamination. *Laboratory Practice,* 1963, **12**, (3) pp. 243–248.

25. Department of Health and Social Security. *Code of Practice for the Protection of Persons against Ionizing Radiations arising from Medical and Dental use.* H.M.S.O., London, 1972.
26. Recommendations for data on shielding from ionizing radiation. Part 1, Shielding from gamma-radiation. *B.S. 4094: Part 1: 1966.* British Standards Institution, London, 1966.
27. Duncan, K. P. and H. J. Dunster. The place of environmental monitoring in the radiological protection of the worker. *ATOM.,* 1966, **115**.
28. International Commission on Radiological Protection. Report by Committee 4 on General principles of monitoring for radiation protection of workers. *I.C.R.P. Publication 12.* Pergamon Press, London, 1969.
29. Bullen, M. A. and J. A. Garrett. Radiation monitoring: methods and instruments. Part 1: Monitoring tasks and problems. *Biomedical Engineering,* 1973, **8**, p. 5.
30. Bullen, M. A. and J. A. Garrett. Radiation monitoring: methods and instruments. Part 2: Monitoring instruments. *Biomedical Engineering,* 1973, **8**, (6).
31. Recommendations of the International Commission on Radiological Protection (as amended 1959 and revised 1962). *I.C.R.P. Publication 6.* Pergamon Press, Oxford, 1964.
32. Dunster, H. J. Surface contamination measurements as an index of control of radioactive materials. *Health Physics,* 1962, **8**, (4), pp. 353–356.
33. Langmead, W. A. The standardization of pressurized suits. *A.H.S.B. (R.P.) R76,* 1967. Available from H.M.S.O., London.
34. Hounam, R. F. *et al.* The evaluation of protection provided by respirators. *Ann. Occup. Hyg.,* 1964, **7**, (4), pp. 353–363.
35. Stevens, D. C. and J. B. Ritchie. Evaluation of air-hoods and air-blouses used at *A.E.R.E. A.E.R.E.-R5184,* 1966. Available from H.M.S.O., London.
36. Butler, H. G. and R. W. Van Wyck. Integrity of vinyl plastic suits in tritium atmospheres. *Health Physics,* 1959, **2**, (2), pp. 195–198.
37. Monitoring of radioactive contamination on surfaces. *Technical Reports Series No. 120.* I.A.E.A., Vienna, 1970.
38. Dunster, H. J. *et al.* Accidental exposures to ionizing radiations. *A.H.S.B.(R.P.) R71,* 1966.
39. Recommendations of the International Commission on Radiological Protection. Report of Committee V on the handling and disposal of radioactive materials in hospitals and medical research establishments (1964). *I.C.R.P. Publication 5,* Pergamon Press, Oxford, 1965.
40. *Safe Handling of Radionuclides,* 1973 edn., Safety Series No. 1, I.A.E.A. Vienna.
41. Department of Employment and Productivity. *Code of Practice for the Protection of Persons Exposed to Ionizing Radiations in Research and Teaching.* London, H.M.S.O., 1968.
42. Symbols for ionizing radiation. *B.S. 3510: 1968.* British Standards Institution, London, 1968.
43. Shielding for high-energy electron accelerator installations. U.S. Department of Commerce. *National Bureau of Standards Handbook 97.* Washington, 1964.
44. Gibson, R. (ed.). *The Safe Transport of Radioactive Materials.* Pergamon Press, Oxford, 1966.
45. Regulations for the safe transport of radioactive materials. *Safety Series No. 6.* I.A.E.A., Vienna, 1973.

6. Entry into confined spaces

T. H. Wareing

Textile Product Manager,
Courtaulds Ltd

Accidents arising from entry into confined spaces are in a special category; they are all preceded by a deliberate and conscious act—the entry into the confined space. Many factory accidents occur under the most commonplace of circumstances: a slip on an apparently good surface, a missed step on a well lit stair. They also happen to people innocent of any unsafe practice: a tool dropped from a height onto a person working or passing below.

In the case of work associated with confined spaces we know where the accident, if there is one, will take place. We even know the persons most likely to be affected. When we couple this with the observation that confined space work is usually of a non-routine nature and therefore requires special consideration, it is fair to say that the possible causes of an accident should be foreseen, and the accidents are always avoidable.

A mishap in a confined space has, by the very nature of the hazards usually encountered, a high probability of being fatal and this emphasizes the need for special care. There are statutory regulations covering the circumstances under which entry into confined spaces is permitted, but these can only operate in retrospect. They provide the means to prosecute offenders, and to allow victims to secure damages. If the regulations are invoked in this manner, something has already gone wrong. Because of the wide field of diverse hazards covered, such regulations can be specific only on very few points—size of manholes for instance—and can give only a very general guide to safe procedure. If the work is to be carried out safely the interpretation of the regulations must rest with experienced and responsible people on the premises concerned. What is a desirable background atmosphere of safety regulations in which the problem of entry into a confined space can be properly approached?

(a) The overall safety policy in a factory should be a top management concern. Safe working procedures do not arise spontaneously, they have to be devised and enforced. This can only be done effectively if the senior factory executive not only wants safe procedures, but is seen to want them.

(b) In any large organization of several factories the total number of confined space entries can be quite large. The hazards and problems are likely to exhibit some basic similarities from one factory to another. For these reasons it is desirable that a company booklet is produced, interpreting the statutory and common law requirements in as much detail as the common ground between factories permits.

Of course such an arrangement need not be confined to the large group of factories. Any factory where tank entry is other than a very rare event, can benefit from the preparation of such a booklet. Tank entry should never become routine. Due to the obvious dangers of familiarity, if a particular type of tank entry is frequent, then the whole system should be carefully examined to reduce or eliminate the need for entry.

(c) There is a statutory obligation to certify that a confined space is free from dangerous fumes, lack of oxygen, etc., and another to certify for how long it is safe to enter. The spirit of these regulations calls for more than a signature on an odd scrap of paper. The need is for a properly designed form in two copies. A flimsy to be retained by the issuer, and a copy, preferably on stiff card to be displayed at the actual site of the operation. The certificate should have ample space for special conditions, but many of the more common items can be pre-printed for 'yes/no' entry.

It is important to realize that the prohibition notices that, under the Health and Safety at Work Act 1974, the Factory Inspectorate can issue can be applied to entry to a confined space. Instances of such applications of the Act are already on record. The power to issue prohibition notices is a welcome tool in the Factory Inspector's locker. It is far better to prevent the incident than to prosecute offenders after the inquest! It was gratifying to see on the notice I was shown that full use of the Act was made, so that the last part was virtually the basis for a satisfactory entry certificate. Receipt of such a notice should not cause annoyance; it may have saved you from a gaol sentence after an accident.

With the general background procedure established we can now consider the details of a safe tank entry system.

Isolation of the vessel

The closing of valves is generally not sufficient. The only completely satisfactory method of isolation of inlet and outlet pipes is to disconnect them or to use bolted blanks. Valves can be sealed for liquors and not be gas tight, and this can lead to hazardous conditions where volatile liquids or liquids containing dissolved gases are concerned. The isolation should be supplemented by the notice '*Danger: men working in tank*'.

The isolation of chutes and dump valves from hoppers should not be forgotten, as many solids can give rise to gassing or explosive hazards under suitable conditions.

Where the vessel contains moving parts such as stirrers or cutters, these also require adequate isolation. The removal of fuses is a useful supplement but no more than that. It is too easy for someone to come along, perhaps on a following shift, think they have been left out by mistake and replace them or fit others. The same arguments apply to pulley belts. Locking isolator switches in the off position is much safer, and multiple hasps which will take several padlocks are available.

Cleaning

Removal of the last part of the contents which may not have drained off, and the removal of any sludge should be carried out, as far as possible, from outside the tank. In some cases neutralization of the contents may be appropriate, but the advice of a suitably qualified and experienced person should always be sought, as new hazards can be introduced. For instance, a steel tank used for storage of caustic liquors could be neutralized with acid, but apart from the residual acid, hydrogen and arsine can be generated by the action of dilute acid on the steel of the tank. Other chemicals, whilst effectively neutralizing each other, can react with explosive violence!

Washing out with high-pressure hoses, followed by steaming, is the most generally applicable method. Where sludge is present, filling with water and agitation with paddles, compressed air, or a perforated steam pipe is usually effective. In the case of liquids with low boiling and flash points, such as carbon bisulphide, even steaming is considered hazardous. In this case, filling with cold water, which is then brought to the boil with steam, is a useful preliminary to steaming. Unfortunately, with such liquids the tanks are often without any bottom outlet because of the great risk which would arise from a leaking joint. This makes removal of the last of the contents almost impossible.

From an installation cost point of view, a few large storage tanks are often preferred to a lot of small tanks—a view engendered by a sometimes misplaced confidence in modern structural materials and construction methods. Where these large tanks have no bottom outlet and are used for storing highly hazardous materials, and the sludge formed is heavy, all problems are at maximum. Fortunately, this sort of situation has been faced and overcome. With the tank emptied down to the sludge and re-filled with water, manholes can be cut by a technique using a rotating cutter which floods with water when the tank is pierced, and completes the cutting under a water seal. Removal of the sludge, under a maintained water cover if necessary, is then relatively simple. The hole can later be fitted with a manhole lid or welded up if the absence of a low-level joint is still considered essential.

Few accidents arise from abstruse causes; it is usually the omission of an obvious precaution that causes an accident. For this reason I make no apologies for mentioning the obvious when discussing steaming operations. When the steam is shut off make sure the outlet is not also closed. The collapsing empty tank may not injure anyone, but it could cause fracture of connecting pipework, with the release of dangerous substances.

The use of organic solvents for tank cleaning is generally not a satisfactory method. Whilst they may be satisfactory diluents for the previous contents of the tank, they are less efficient at dissolving tarry or gummy residues. They also produce hazards of their own. Many, such as benzene and isopropyl alcohol, are highly inflammable and others, such as trichloroethylene and benzene again have toxic properties. Carbon tetrachloride, it should be noted, can give rise to phosgene when heated; a further reason for avoiding its use.

At the completion of cleaning, pyrophoric substances such as finely divided carbon or iron sulphide may be present. If their presence is suspected the inner walls should be kept wet until the substance is removed, and preferably kept wet throughout the entire operation.

Ventilation of vessel

Whether or not cleaning has been necessary, some form of ventilation should be provided to sweep out traces of dangerous fumes, or to clear out an atmosphere which may be deficient in oxygen. A tank with top and bottom outlets is easy to ventilate. If the vessel has been steamed, opening whilst still hot will make use of natural ventilation. In other cases a fan can be applied to the lower opening, ensuring that the air intake is from a clear atmosphere. Without bottom outlets, air should be piped from a fan to the bottom of the tank. A compressed air pipe may also be used for the same purpose.

An exhaust fan with an inlet extending to the bottom of the tank is sometimes used to draw fumes from the tank, fresh air entering in its place. This method should be used with caution as there can be risk in passing an inflammable or explosive mixture through a fan impeller. With particularly dangerous substances, such as carbon bisulphide, this procedure is most inadvisable. If an inert gas has been used to purge the tank, this gas should, of course, be replaced by a breathable atmosphere. One useful method of doing this is to refill the tank with water, emptying it when the inert gas has dispersed.

Ventilation should not be confined to purging the atmosphere before entry, but in most cases should be continued throughout the operation. Fumes can be generated and oxygen used up during brazing and welding. The fumes produced during rubber covering, lining with resin/fibreglass, and even painting can easily become a hazard and this calls for adequate ventilation during work and before re-entry for further spells of work.

In one recent fatal accident, workmen in a pit were using an internal combustion engine. This is a highly dangerous operation and should always be avoided. Where such an engine *must* be used, the twofold danger should be recognized. It is not sufficient to pipe out the exhaust as this type of engine rapidly uses the air in a tank and the induced draught may draw in fumes from outside the tank.

Inspection and testing

At some stage in the preparation for entry tests for dangerous gases, explosive hazards, and lack of oxygen will be necessary. At just what stage testing and inspection fit into the procedure cannot be stringently laid down. If there are likely to be difficulties or hazards associated with the blanking off or cleaning from outside, an inspection, and the issue of a separate permit to work certificate, may be necessary. For instance, opening a joint in a line to insert a blank may result in an escape of gas. In such cases protection is required for the workers, and it must also be provided where the fumes are likely to disperse. In one instance fumes escaping when a manhole was opened were detected in quantity 50 yards away in another room, carried there by a strong draught. Similarly, disturbing of sludge or deposit during cleaning from outside may release fumes, and this must be taken into account when stating the safety measures to be followed.

Testing may come quite early in the preparatory work. In some cases testing may be a first stage to see if there are any dangerous fumes to be removed. Atmospheric testing should be designed to determine:

(a) the presence of fumes injurious to health above the permitted limits,
(b) whether sufficient oxygen is present for breathing,
(c) whether an explosive mixture is present.

In testing for specific gases, the tests designed by the Department of Scientific and Industrial Research, and issued by H.M. Stationery Office, should be used if the gas is on the list. These have all the desired attributes of sensitivity, simple apparatus, and ease of measurement. In other cases standard textbooks will have to be consulted or outside advice sought. A knowledge of the past history of the tank is of obvious importance. In this context it should be specially noted that steel tanks which have contained strong sulphuric acid should always be tested for hydrogen sulphide and arsine. These can arise from the hydrogen generated by the diluted acid formed during washing out—acting on impurities in the acid. Neglect of testing for these gases has cost several lives.

Within my own experience, a knowledge of the past history of a tank led to correct precautions being taken. When a small factory was closed, I had charge of the safe disposal of steel tanks which had contained carbon bisulphide. Some of these were cut up for scrap but one was sold intact. A few months later the firm who had bought it and knew of its previous contents rang up and asked our advice as they wished to weld a pipe connection to the underside of the tank. The most appropriate course of action appeared to be to visit the firm, retest the tank, and review their safety precautions. This was done and the work was completed successfully.

The testing of tanks for specific gases will not ensure that the tank has a sufficiency of oxygen. In cases where a deficiency of oxygen is suspected or possible, analysis by absorption should be made. Lack of oxygen can arise from flame processes, an internal combustion engine, oxidation of the tank walls over a long period, purging with inert gas or neutralizing the contents with a carbonate or bicarbonate. In the latter case, generation of carbon dioxide will displace air at the bottom of the tank. Where the deficiency of oxygen can occur during work in a confined space fatal accidents have occurred not only to the workers, but also to would-be rescuers who have gone in without taking proper precautions.

No confined space should ever be lightly declared safe without careful thought to the possible hazards, by a qualified person with sufficient local experience. The following case quoted in the *Quarterly Safety Summary* for April–June, 1967, makes the point quite clear, in a chapter of accidents which would be ludicrous if it were not so tragic.

A worker entered a water cistern and immediately fell from the platform to the water 0·6 m below. A second man ran for assistance and brought back several people, two of whom entered and immediately collapsed. Another man had sounded the fire alarm and this brought the fire chief from the station across the road. While he was putting on his self-contained breathing apparatus, a man who said he could swim also entered, and was overcome. The fire chief entered, but removed his mask to call for rope or ladder and collapsed. His safety line slipped off when he was being pulled out. In all, five lives were lost in a matter of 20 min. The subsequent testing and enquiry revealed that the water coming from an underground source contained hydrogen sulphide and there was a lethal concentration of this gas above the surface of the water. A lethal gas concentration in an apparently harmless water tank!

The above abortive rescue is the most dramatic, but newspapers and safety bulletins seem to abound in variations on the same theme. Even as I write (July 1975), I read of three deaths in a manhole on the M25. A man was overcome at the bottom of a 6 m shaft. Two men went to his rescue and also died. A man may collapse from some ailment at any time, including when he is in a confined space, but the odds must always be that the collapse is because of fumes. If that is so, without proper rescue equipment he will die anyway. All an unequipped would-be rescuer can do is die with him. Far better to raise the alarm and get breathing apparatus. Then the rescuer will live and so may the rescued.

The presence of an explosive mixture, where the specific gases are not all known with certainty, or for which there is no easy test, can be detected using a Haldane or McLuckie apparatus. Electrical meters are also available commercially. It is a poor excuse at the inquest to report that the meter was faulty, so prudence dictates that these should always be tested each time they are used.

Issue of certificate

As already mentioned, the certificate should be properly designed and made up into book form. Many items common to all jobs can be included as headings for *yes/no* entry. This makes for ease and speed in issuing a certificate and also gives a reminder of many of the items to be considered and checked. There should be adequate space for detail of special circumstances and for the results of gas tests.

Design of the form should satisfy the following conditions:

(a) distinctive colour, 'Tank Entry' a different colour from 'Permit to Work without Entry';
(b) two copies required, one for reference, one for issue;
(c) the copy issued should be on stiff paper or card to withstand handling and display at site of job;
(d) as many of the precautions as possible to be preprinted with space for *yes/no* entries;
(e) space for special conditions;
(f) space for test results;
(g) space for signatures of issuers and users;
(h) space for dates between which the certificate is valid;
(i) the certificates should be numbered.

Some of these items need only appear on one copy of the certificate. The test results which may continue at intervals through the period can be conveniently entered on the reference copy. The space saved on the copy issued can be utilized for the signature of the foreman in charge of the work, and the 'job complete' signature.

A certificate which fulfils all these conditions is shown on pages 94–95. It is produced in book form and both copies are coloured orange. A similar book of green forms for 'Permit to Work without Entry' is also produced with altered layout to suit the different type of work.

The procedure for request, preparation, and issue of certificates should be clearly laid down in a memorandum to all persons concerned. In a factory of many departments there can be a very large hole in the system waiting for the unwary. A certificate is required and the department where the work is to be carried out think the engineering trade concerned has asked for a certificate. The tradesmen think the department has asked for it. At best there is time wasted, with people ready to do a job and plant stopped, and a certificate prepared in haste. At worst, misplaced enthusiasm can result in entry before the certificate detailing precautions is issued, with possible tragic results.

During the actual work, tests for hazardous gases should be continued. The frequency will depend on the background knowledge of the likelihood of changes in conditions. When the entry is extended over a number of working periods, atmospheric tests should be made at each re-entry. The results should always be recorded, and the most convenient place is in a space reserved on the file copy of the certificate itself.

If a fire occurs in a confined space, fresh hazards can be present after the fire has been put out. Obviously there may be a lack of oxygen due to the fire, and an increase in carbon dioxide. Many fire extinguishing materials emit harmful fumes. For these reasons the confined space must be purged and tested again. However, equally important, there may have been an oversight in the specifications for safe working, or a failure to carry out the conditions. The whole purpose of safe working conditions is to avoid such occurrences, and work should not restart until the cause of the outbreak has been investigated and working conditions altered to avoid a repetition.

Rescue

Rescue facilities must be available. The persons selected for this duty must be fit and capable of lifting out an injured man. They must also have had some training in rescue, first aid, and the use of breathing apparatus. Without such capabilities and training there could be a case to answer in Common Law in the event of an accident.

It must be possible to get a man out of the confined space. Manholes in old tanks may not be up to current statutory size. Enlargement of the manhole is one solution but the other solution should be considered; is it possible to avoid entry altogether, either by re-thinking the job, or using new techniques for cleaning or repair? It has been known for the conditions of entry to be so stringent that a disgruntled Departmental Manager returned the entry certificate with the comment that making entry safe would take so long that he found a way of doing the job without going inside.

For any entry into an unsafe atmosphere for rescue, or for work where fumes cannot satisfactorily be removed, breathing apparatus is the only solution. Canister respirators are not allowed as these do not give protection against very high concentrations of gas, any concentration of carbon monoxide, or a deficiency of oxygen.

Types of breathing apparatus fall into three categories:

(a) self-contained, where the supply of air is carried on the person;
(b) face mask with a supply of air supplied from cylinders outside the confined space;
(c) face mask with an open hose leading to the outside air.

Of these the face mask supplied from external cylinders is the most pleasant to use. The wearer has the minimum of encumbrance, the air supply is usually refreshingly cool, and no effort is required in breathing. The choice depends on individual circumstances, but a suitable arrangement is often compressed air masks for workers in the confined space and self-contained apparatus for rescuers. Where there is a danger of being overcome by fumes, availability of resuscitation apparatus is mandatory. Lightweight oxygen resuscitators are now available; these are simple to use and small enough to be taken into a confined space by a rescuer. Resuscitation and removal can then take place simultaneously.

Bibliography

1. *Factory Acts*. H.M.S.O., London.
2. *Memorandum on Explosion and Gassing: Risks in the Cleaning, Examination and Repair of Stills, Tanks, etc.* H.M.S.O., London.
3. *Safety pamphlet No. 18. Repair of Drums or Tanks: Explosion and Fire Risk.* H.M.S.O., London.
4. *Methods for the Detection of Toxic Substances in Air.* H.M.S.O., London.
5. *Quarterly Safety Summaries.* British Chemical Industry Safety Council.
6. F. L. Creber, *Safety for Industry.* Royal Society for the Prevention of Accidents.
7. *Health and Safety at Work, etc., Act 1974.*

Courtaulds Limited	British Celanese Limited

Works

CONFINED SPACES

EVERY ITEM MUST BE FILLED

SAFETY APPLIANCES AND EQUIPMENT TO BE USED

	No	Yes: Type	Use	Spare
Breathing apparatus				
Dust masks				
Lifebelts or harness				
Protective helmets				
Eye protection				
Protective clothing				
Spark-reducing tools				
Safety handlamps				
Lifting tackle				
Lifting tripod				
Special ventilation				

NOTES TO FOREMEN

1. The carrying out of these conditions is the responsibility of the Foreman in charge of the work. He must personally satisfy himself that all conditions are observed, that the work is performed safely at all times and that assistance is obtained, where necessary, of trained personnel who, if rescue watchers, must be physically fit and adequately equipped to extract a person from the confined space in emergency.

2. This Certificate must be explained to the men doing the work and then be displayed in the work area.

3. The Departmental Foreman must be notified of the intention to start work in the area, and the workmen must leave the area if he so instructs.

WORK STATE

(Foreman to cross out one or other item, and sign)

THE WORK WILL CONTINUE BEYOND THE TIME LIMIT
and a new certificate is required

(or)   THE WORK IS COMPLETED

Date .... Time ....

Signed ....

PERSONNEL CONTROLS

Max. number of persons allowed inside at one time	1.
Max. time each person allowed in at one stretch	2.
Length of rest pauses between stretches	3.
Number of watchers—Rescue	4.
Fire	5.
Engineering	6.
Process	7.

JOB CLEARANCE

(to be completed by the Departmental Manager or Foreman unless cancelled by the Certificate issuers)

THE WORK DONE UNDER THIS CERTIFICATE HAS BEEN CHECKED AND THIS ITEM OF PLANT IS ACCEPTED AS SAFE FOR PRODUCTION USE

Date .... Time ....

Signed ....

TIME LIMIT

THIS CERTIFICATE IS VALID

ATMOSPHERIC TESTS MUST

TEST RESULT				
	Date			
	Time			
	Safe			

NTRY CERTIFICATE

OR DELETED AS APPROPRIATE

AS/LIQUID/SOLID ISOLATION BEFORE TEST	Done	Return Check
....	....	....
....	....	....
....	....	....
....	....	....
....	....	....
....	....	....
....	....	....

DATE:

No. 46449

To: ....

....Dept.

THIS IS TO CERTIFY THAT the work described below may be performed in, on or adjacent to the undermentioned plant, confined space or area which is:

* (a) isolated, sealed and free from danger

or

* (b) fit and safe to work in subject to the conditions of this Certificate.

LECTRICAL/MECHANICAL ISOLATION BEFORE WORK

....	....	....
....	....	....
....	....	....

Welding/burning/naked lights allowed/not allowed
(delete one or other)

PLANT, CONFINED SPACE OR AREA:
(use reference number if possible)

PECIAL CONDITIONS

....	....	....
....	....	....
....	....	....
....	....	....
....	....	....
....	....	....

WORK AUTHORISED:

* Delete item which does not apply.

om   on   to   on

RE MADE AT INTERVALS OF:

....	....	....	....	....
....	....	....	....	....

SIGNATURES of persons authorised by the Works Manager:

Technical ....

Engineering ....

....

Form reproduced by permission of Courtaulds Ltd

7. Fire control: manual

P. Nash

Deputy Director,
Fire Insurers' Research & Testing Organisation,
Borehamwood, Herts.

Some statistics of fires in industry

During the past 10 years, the total number of fires in buildings attended by local authority fire brigades has increased from 90 000 to 130 000 each year, and the total number of all fires attended by brigades from 165 000 to 330 000.

Table 7.1   Selected types of fires attended by Local Authority Brigades

Year	Manufacturing industries buildings	Distributive trades buildings	Educational, medical and dental buildings
1966	8 000	5 000	2 000
1973	10 000	6 000	5 000
Total between 1966–1973 inclusive	73 000	46 000	25 000

The increases shown are partially due to increased commercial and industrial activity, but the situation is not one to be accepted without concern. The most prevalent sources of ignition of industrial fires attended by local authority fire brigades in 1966 and 1973 are as shown in Table 7.2.

Table 7.2   Most prevalent sources of ignition of industrial fires

Cause	Number		% of total	
	1966	1973	1966	1973
Electrical apparatus, misuse of, overloading, etc.	1 236	1 554	9·8	10·2
Electric wiring installations	404	410	3·2	2·7
Gas apparatus	339	438	2·7	2·9
Children playing with sources of fire	1 026	1 144	8·2	7·5
Malicious or intentional ignition	306	954	2·4	6·3
Mechanical heat or sparks	905	1 354	7·2	8·9
Burning of rubbish	1 289	1 894	10·3	12·5
Slow combustion stove	216	66	1·7	0·5
Disposal of smoking materials	1 173	1 026	9·3	6·8
Other known sources	3 695	4 252	29·4	28·0
Unknown	1 984	2 086	15·8	13·7
Total	12 573	15 178	100·0	100·0

The figures quoted in this section were obtained from the Annual Fire Statistics.[1]

Need for early warning of fire

About one-third of the total direct fire loss each year is due to some 600 large fires, each costing £20 000 or more, mostly in industrial or storage premises, and nearly all discovered from outside the building when they were well advanced. Analysis of brigade response times, in comparison with the time taken for a fire to develop under conditions of normal ventilation, shows the importance of an early warning system in detecting a fire early. This is necessary not only at night, but

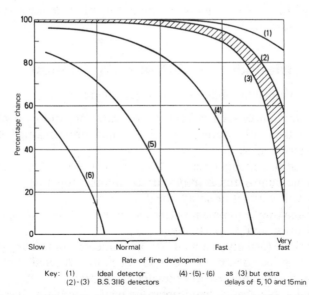

Key: (1)   Ideal detector   (4)-(5)-(6)   as (3) but extra
     (2)-(3)   B.S. 3116 detectors   delays of 5, 10 and 15min

Fig. 7.1   Chance of brigade arrival before flashover in compartment of fire origin (County Boroughs)

also during the day when certain areas may be visited by personnel at times separated by long intervals. A normal heat-sensitive alarm system connected to the brigade will generally give warning soon enough for the brigade to arrive before the fire has spread beyond the compartment of origin, but a delay of only 10 min in relaying the call can seriously reduce the chance of their doing so—almost to zero for a rapidly-developing fire. It is important that the alarm should be given at the central factory control direct to the industrial or local authority fire brigade.

Need for efficient first-aid fire-fighting

The Annual Fire Statistics for 1973[1] show that of the total number of fires in buildings attended by local authority fire brigades, some 25 per cent are extinguished before brigade arrival. A further 10 per cent are extinguished by the brigade using 'small means' such as beating out, removal of combustibles, fire extinguishers, and buckets of water, sand or earth. A further 40 per cent are extinguished by hose reel jets alone, using water on the appliance, while the remainder require the use of major fire-fighting equipment.

In view of the increasing number, cost, and severity of fires fought by brigades, there is need for more effective 'first-aid' fire-fighting in the early stages of a fire. A better understanding of the fire hazard, correct planning of emergency equipment and procedures, and adequate training of personnel, is needed. Research and development into new or improved fire-fighting agents, and better definition of the optimum conditions of their use against fires of various types, is essential.

Manual fire-fighting equipment

Equipment for manual fire-fighting can be classified as:

   (a) hand appliances such as buckets for water or sand; shovels for sand; implements for beating out fires; and fire-resisting blankets for smothering fires;
   (b) portable fire-extinguishers, with various agents for a range of risks; portable pumps, for example stirrup pumps, for water;
   (c) hose-reels with jet or spray nozzles;
   (d) foam-making equipment such as branch pipes, mechanical foam generators, and proportioners;
   (e) major hose (usually 38 mm or 70 mm diameter), and jet or spray nozzles;
   (f) foam, dry powder, carbon dioxide or halon wheeled appliances, e.g., 90 kg trolley-mounted, dry powder dispenser;
   (g) special appliances or hand-operated installations for special fire-risks, for example foam installation for heat-treatment tanks.

The best form of fire protection is to avoid fires altogether. The next-best solution is to prepare for emergencies by selecting the correct type of equipment for the risk, and the placing of this equipment in readily-available positions.

Classification of types of fire

The applicability of fire-fighting agents or equipment can be expressed by classifying the types of fire for which they are suitable. Two classifications are at present in use:

U.S. classification

Originated by the Underwriters Laboratories of America, and used in North America, Australia, South Africa, this classification has not been adopted officially in the United Kingdom, although referenced in one document of H.M. Factory Inspectorate of the Health and Safety Executive. It lists the following classes:

Class A: fires in ordinary combustible materials where the quenching and cooling effects of quantities of water, or solutions containing large percentages of water, are of first importance.

Class B: fires in flammable liquids, greases, etc., where a blanketing effect is essential.

Class C: fires in 'live' electrical equipment where the use of an electrically non-conductive extinguishing agent is of first importance.

European classification

The classification of fires adopted by the Working Group on Fire Extinguishers of the Comité Européen de Normalisation is as follows:

Class A: fires in ordinary cellulosic materials such as wood and paper.

Class B: fires in flammable liquids such as petrol, kerosene, and fuel oil.

Class C: fires in gases or liquefied gases such as propane and butane, discharging from a jet.

Class D: fires in combustible metals such as magnesium, lithium, sodium, and radioactive metals.

There is no classification for fires in electrical apparatus, which are considered to be covered by Classes A and B, but extinguishers are tested for their suitability for use against live electrical equipment. The International Standards Organisation has concurred with this view, subject to Class B being subdivided into (a) Fires in pools of flammable liquids such as petrol, kerosene and fuel oil and (b) Fires in running fuels and their vapours.

Fire-fighting agents

The fire-fighting agents used in hand operated portable extinguishers and equipment include:

(a) water, sometimes with additives to increase efficiency and/or prevent freezing;
(b) foams for use on non water-miscible flammable liquids, including air foams made from protein, fluoroprotein, fluorochemical (AFFF) or synthetic foaming liquids;
(c) foams for use on water-miscible flammable liquids, e.g., alcohols, made from 'all-purpose' foaming liquids;
(d) medium-expansion air foams made from synthetic foaming liquids;
(e) dry powders;
(f) inerting and inhibiting gases, e.g., carbon dioxide, halons 1211, 1301, and 2402.

Water

The main advantages of water on Class A (we are using the European classification) fires are its ability to absorb large quantities of heat and to evaporate to 1 700 times its liquid volume to give an inert atmosphere of steam. Its ready availability, cheapness, and freedom from toxicity ensure its pre-eminent place among agents. The heat extracted by water from a fire when heated from, say 10°C to 100°C (0·4 kJ/g) is far less than the heat absorbed in the change of state from water to steam at 100°C (2·3 kJ/g). For efficient fire-fighting, the applied water should therefore be vaporized—an objective rarely realized completely either in first-aid or in full-scale fire-fighting. The cooling of hot surfaces by water, some of which may be evaporated, but much of which flows off without evaporating, is the usual situation. Efficiency of heat transfer may sometimes be improved by the use of fine spray to cover surfaces more adequately, but in practice this results in a reduced throw of water from the extinguisher spray nozzle, which may reduce the advantage gained. British practice is to specify extinguishers giving a jet with a minimum throw of 6·4 m, and to depend on impact to break down the jet, and to spread it over the hot surfaces.

Water may be treated with various additives to increase its effectiveness on different types of fire. Wetting agents are sometimes useful to assist penetration and let the water reach the seat of the fire. This is effective for baled materials, particularly those that naturally resist wetting, such as thatch. An increase of viscosity can reduce the rate at which water will run off surfaces, thus giving longer for heat transfer to the water. This method is being used in forestry plantations in the United Kingdom to extinguish the many small outbreaks with the minimum of water. A combination of viscosity additive and a fire-retardant additive gives an additional advantage. Thus, the ammonium phosphates may be used in conjunction with methyl cellulose or Fuller's earth. In the United States, a 'surplus' mineral deposit which combines viscosity and fire-retardant effects is sodium calcium borate, or 'Firebrake'.

The so-called 'loaded stream' extinguisher, containing an aqueous solution of the salt of an alkali metal, usually potassium or sodium carbonate, is used in the United States to extinguish U.S. Class A and Class B fires. The solution also has a depressed freezing point. Chemicals used in this country to lower the freezing point are often based on sodium or calcium chloride, or ethylene glycol. The former increase the extinguishing power of the solution, but they lead to increased corrosion, and so require an inhibitor. Ethylene glycol, while reducing the freezing point of water, also reduces its extinguishing power.[5]

Water is effective in the extinction of some types of flammable liquid fires (Class B), and it is convenient to classify flammable liquids into three groups as follows.[6]

(a) Those that can be cooled below the fire point by direct heat transfer between the flammable liquid and the water droplets falling into it. The fire point of the liquid should be at least 25°C above the temperature of the water, so the method is only suitable for liquids such as kerosene, gas oil, transformer oil, and lubricating oils.

(b) Those that are miscible with water, so that the fire point is raised by dilution to 45°C or above. This method often requires large quantities of water. For example, ethyl alcohol and acetone need to be diluted by some 7 and 30 times their initial volume.

(c) Those that can only be extinguished by direct cooling of the vapours in the combustion zone by heat transfer to the water droplets—in petrol, for example.

In extinguishing flammable liquid fires with water spray applied from hand lines, the most important factors are the experience of the operator, the rate of application and the drop size of the water spray, and the fire point of the flammable liquid. The time for which the liquid has been burning before water application, and the depth of the flammable liquid, are also important.

Nozzles for use against flammable liquid fires in the first group should give mass median droplet sizes in the range 0·4–0·8 mm, the lower limit to ensure sufficient impetus of the spray to penetrate to the flammable liquid surface, and the upper limit to avoid a stabilized fire due to splashing of the flammable liquid by the spray. Since the smaller size of droplets gives more efficient heat transfer, it is best to use an 'applicator', i.e., a special nozzle mounted on a rigid flow tube, to enable the water to be applied in fine droplets close to the flammable liquid surface. When extinguishing flammable liquids which form a 'hot zone' at the surface layers after prolonged burning, there is considerable likelihood of steam formation as the water spray is injected into the liquid, with an attendant danger of 'frothing over' of the burning liquid.

Liquids of the second group can sometimes be extinguished by mixing only the surface layers with water. This requires gentle application of a very fine spray (drop size less than 0·4 mm), preferably by an applicator. The flames from these liquids are often non-luminous, and extinction is difficult to detect. Some liquids in this group are only partly miscible with water, and those which have a low fire point, for example ether and methyl ethyl ketone, are difficult to extinguish with water spray.

Liquids in the third group cannot generally be extinguished except with the finest droplets, of size 0·3 mm and less. Penetration of the flame zone can be increased by entrainment of air with the spray, or by the use of an applicator. As these liquids can safely be extinguished by foams and dry powders, it is preferable to use these agents rather than water sprays.

The ability to cool hot surfaces in the vicinity of the fire is an important advantage of water sprays, and they can often be used to 'hold down' a fire, or prevent its reflashing, while extinction is completed with another agent.

Water should *not* be used on electrical fires, particularly those of high voltage, unless there is no suitable alternative available. When it is used, the electrical supply must be disconnected before fire-fighting, and even so, care should be observed. In general,[7] it is safe to use water *sprays* on electrical equipment outside certain safe distances which depend on the potential of the equipment, but water *jets* should never be used within the range where they break up into individual drops. The streams from hand extinguishers containing water or water solutions vary greatly in their conductivity, but as they are likely to be used within dangerous ranges where the solution is ejected as a jet, it is best to prohibit their use on electrical fires.

Water spray can sometimes be used effectively on liquefied gas fires, although it is safer to delay fire-fighting until the source of leakage can be stopped. Water can rarely be used safely on flammable metal fires, and its use can be highly dangerous due to sputtering of the molten metal when the water is applied.

Foams

Fires in flammable liquids can be extinguished by covering the free liquid surface with a blanket of foam, a bubble structure made by aerating an aqueous solution of a suitable foaming liquid and 'working' the bubbles formed into a

Foam liquid	Description
Protein	Based upon a hydrolized solution of waste protein with additional stabilizers, bactericides and anti-freeze agents. Protein derived from hoof and horn meal, blood, fish scales, soya bean, chicken feathers, etc.
Fluoroprotein	As protein, with the addition of a fluorinated surfactant to give stability and improved flow properties.
Fluorochemical or 'AFFF'	An aqueous solution of fluorinated surfactants which has the ability to form a film on top of the fuel surface, when applied gently, due to its low surface tension.
Synthetic	Derivate of lauryl alcohol, usually ammonium lauryl ether sulphate with additional stabilizers. Usually used to make 'medium expansion' and 'high expansion' foams.

stable structure. The foam liquids suitable for use with non water-miscible foaming liquids of fire points less than 100°C include those listed at the foot of p. 102.

The foaming liquid is mixed with water to a concentration of some 4 to 6 per cent by volume, and air is then induced into it in a 'foam-making branch-pipe'

Fig. 7.2   Standard test for foam performance on flammable liquid.

which serves to mix the air and the solution, and to form a stable bubble structure. The foam is applied to the surface of the flammable liquid as a 'blanket' which, while not completely impervious to the flammable vapours, can protect the liquid surface sufficiently from the flame radiation to reduce the efflux of vapours below the rate necessary to sustain combustion. Foams for surface application to flammable liquids are usually of 'low' expansion, that is, the ratio of the volume of the foam to that of the solution from which it is made is of the order of 7–10 units.

The performance of foams can be measured in a standard test described in the draft U.K. Defence Standards for foam liquids of the protein, fluoroprotein, and fluorochemical types. In this test, foam is applied to a 0.25 m^2 area of fire, by means of a standardized laboratory branchpipe, at a rate of 3 l/m^2/min of foaming solution. The physical properties of the foam i.e. expansion, critical shear stress and rate of liquid drainage from the foam, are measured before application of the foam to the fire. The times to gain '90 per cent control', that is, to reduce the heat radiation from the flames to one-tenth of its initial value, and to extinguish the fire are measured. The time for the fire to burn back against the foam layer from a small reignition source is also measured. The acceptance levels of performance against the two principal aviation fuels, aviation kerosine (AVTUR) and aviation gasoline (AVGAS) are shown in Table 7.3. The drainage from the foam layer for a specified period after applica-

Table 7.3   Requirement for foam performance, Draft Defence Specifications. Figures in brackets show normal expectation

	A.V.T.U.R.	A.V.G.A.S.
Fluoroprotein		
90 per cent control—S	50 (35–45) max.	75 (50–60) max.
Extinction—S	80 (50–60) max.	110 (75–100) max.
Burn-back—min	15 (25–30) min.	10 (15–18) min.
Protein		
90 per cent control—S	100 (55–80) max.	140 (60–120) max.
Extinction—S	130 (75–115) max.	180 (80–165) max.
Burn-back—min	15 (20–30) min.	10 (15–20) min.
Fluorochemical		
90 per cent control—S	25 (15–20) max.	40 (27–33) max.
Extinction—S	45 (20–25) max.	60 (38–42) max.
Burn-back—min	10 (18–22) min.	5 (8–12) min.

The best of 2 tests out of 3 can be selected.
Minimum expansion = 7 for all foams.
Minimum 25% drainage time = $3\frac{1}{2}$ minutes for fluorochemical foam.
Minimum 25% drainage time = $6\frac{1}{2}$ minutes for protein and fluoro-
protein foams.

tion is also measured. An upper limit is placed on control and extinction times, and on foam drainage, to ensure adequate performance on practical fires.

The application of foam to flammable liquid fires is characterized by a 'critical rate' below which the fire cannot be controlled.[9] Fig. 7.3 shows the effect of rate of application of foaming solution (water and foaming liquid) on the time to control a fire, and it may be seen that the critical rate of application

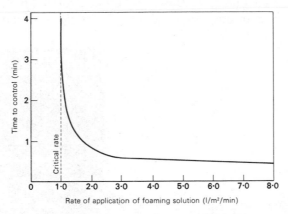

Fig. 7.3   Effect of rate of application of foaming solution on time to control flammable liquid fire.

is approximately $0\cdot07$–$1\cdot0$ l/min of foaming solution per m^2 of fire surface area. The control time drops rapidly at first, and then less rapidly, as the rate of application is increased. If the derived curve is plotted (Fig. 7.4) to show 'quantity of solution to control' against 'rate of application', an 'optimum'

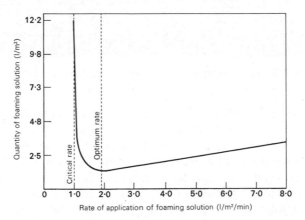

Fig. 7.4   Effect of rate of application of foaming solution on quantity of solution to control flammable liquid fire.

or 'most economical' rate at which the least quantity of foaming solution is used, often appears, and lies in the range twice to five times the critical rate. This rate of application should always be planned for if possible, provided that the fire is constrained within limits and does not endanger life, e.g., a fire in a dip-tank, or other flammable liquid storage. Where life is endangered or the fire may spread rapidly beyond its initial boundaries, a higher rate should, if possible, be used to obtain the most rapid fire control. A typical foam-making branch-pipe with a throughput of about 230 l/min of foaming solution at

Fig. 7.5   Foam being projected from a foam-making branch pipe.

7 kgf/cm^2 pressure, is shown in Fig. 7.5. Hand foam extinguishers are equipped with miniature foam-making nozzles at the end of the efflux tube, and are designed to deliver their charge of 9 litres of foaming solution in about 1 min.

Normal protein foam may be used to control fires in mixtures of water-miscible and non-water-miscible flammable liquids such as alcohol/petroleum mixtures, but a reduced performance may be expected in comparison with the use of protein foam in hydrocarbons. For example, an admixture of ethyl alcohol in petrol in the range 0 to 20 per cent increases the critical rate of application by a factor ranging from 0 to 3 for gently-applied foam, or a factor of

Table 7.4(i)   Critical rates of foam applied gently to various flammable liquids

Foam liquid \ Flammable liquid	Petrol	Industrial methylated spirit	Methanol	Isopropyl alcohol	Acetone	Methyl ethyl ketone	Physical properties of foam Expansion units	Physical properties of foam Critical shear stress N/m²
			Critical rate l/m²/min					
A	1·0 (3% sol.)	Very rapid breakdown, unsuitable for practical use					7·0	20
B_1	Not tested	1·5 (10% sol.)	Not tested	3·5 (12% sol.)	2·0 (8% sol.)	27·5 (15% sol.)	2·8	6
B_2	Not tested	Not tested	Not tested	Not tested	Not tested	3·5 (10% sol.)	3·0	4
B_3	1·75 (12% sol.)	Not tested	1·25 (10% sol.)	Not tested	Not tested	3·25 (12% sol.)	3·0	4
C_1	Not tested	1·5 (4% foam liquid 4% additive)	Not tested	2·5 (4% foam liquid 4% additive)	Not tested	Not tested	7·4	66
C_2	5·0 (5% foam liquid and additive)	Not tested	1·0 (5% foam liquid and additive)	Not tested	Not tested	27·5 (5% foam liquid and additive)	2·4	30
D_1	4·0 (4½% sol.)	2·75 (4½% sol.)	Not tested	3·25 (4½% sol.)	Not tested	2·5 (4½% sol.)	6·0	43
D_2	3·5 (3½% sol.)	2·25 (3½% sol.)	Not tested	2·25 (3½% sol.)	Not tested	2·5 (3½% sol.)	5·2	53

A—Normal protein foam liquid

B_1, B_2, B_3
C_1, C_2, D_1, D_2 } 'All purpose' foam liquid

Table 7.4(ii) Optimum rates and minimum quantities of foaming solution to control various flammable liquid fires (two-thirds control)

Foam liquid		Flammable liquid					
		Petrol	Industrial methylated spirit	Methanol	Isopropyl alcohol	Acetone	Methyl ethyl ketone
A	Optimum rate l m² min	1·5 to 2·0	Very rapid breakdown, unsuitable for practical use				
	Minimum quantity l m²	1·5 to 2·0					
B₁	Optimum rate l m² min		6·0		10·0	15·0	39·0
	Minimum quantity l m²		2·5		4·5	4·0	7·0
B₂	Optimum rate l m² min						12·0
	Minimum quantity l m²						4·0
B₃	Optimum rate l m² min	8·0		5·0			5·0
	Minimum quantity l m²	4·5		4·0			5·0
C₁	Optimum rate l m² min		11·5		11·0		
	Minimum quantity l m²		4·5		4·0		
C₂	Optimum rate l m² min	9·0		5·0			12·5
	Minimum quantity l m²	4·0		1·0			2·0
D₁	Optimum rate l m² min	5·5	6·0		10·0		3·5
	Minimum quantity l m²	5·0	3·0		6·5		3·5
D₂	Optimum rate l m² min	6·0	8·5		8·0		4·5
	Minimum quantity l m²	5·5	2·5		4·0		3·5

A—Normal protein foam liquid

B₁, B₂, B₃
C₁, C₂, D₁, D₂ } —'All purpose' foam liquid

0 to 8 for foam projected forcefully onto the flammable liquid.[10] Similar increases in the quantity required for control at higher rates of application may also be expected.

For use on water-miscible flammable liquids, special 'all-purpose' foam liquids have been developed, based on normal protein liquids but containing additional stabilizers which resist the extraction of the water from the bubble walls by the solvent. All-purpose foam liquids generally give a performance[11] comparable with that of normal protein foams on non water-miscible liquids, whereas the performance of normal protein foams on water-miscible liquids is quite unacceptable, even at much increased rates of application. On water-miscible solvents, the performance of all-purpose foams depends considerably on the actual solvent and whether the foam is applied gently or forcefully, but in all cases the performance can at least be regarded as adequate. The critical rates of application of solutions of all-purpose foam liquids to various solvents, and the quantities of foam liquid required to control fires at the most economical rates of application, are shown in Tables 7.4(i) and (ii).

Dry powders

Dry powders are being used to an increasing extent in the extinction of fires in solids, flammable liquids, gases and combustible metals, and in hand extinguishers and portable appliances they are tending to replace on an increasing scale the use of water and foam. They may be classified in terms of their usage, as follows.

For use in flammable liquid fires (Class B). Powders for this purpose are usually based on sodium or potassium bicarbonate, or mixtures of these salts. The powder is treated with additives to promote moisture repellency and free-flowing qualities, and its particle size range is chosen to give an efficient, non-packing powder. Dry powders are only effective while they are being applied, so it is necessary that the appliance used should be capable of extinguishing the whole of the fire zone. If it cannot, or if hot reignition sources remain, the fire is likely to reflash to its original intensity.

Powders are effective when the rate of discharge and quantity available ensure complete extinction, but their lack of post-extinction protection makes their use in windy conditions, or when hot reignition sources are present, more problematical. The Department of the Environment publishes a specification for Class B powders based on sodium bicarbonate.

For use on normal solid combustibles (Class A) and flammable liquids (Class B). Powders for this purpose have to provide a surface-inhibition property in order to check flaming *and* smouldering combustion. The powders usually selected are based on the ammonium phosphates, with appropriate flow and moisture-repellent additives. No specification for Class A/B powders is yet available in the United Kingdom.

For use on flammable liquids in conjunction with protein-based foams. Powders in this class are termed 'foam-compatible, Class B' and they are necessary because the bicarbonates, with the metallic stearates as flow additives, can destroy foam very rapidly. In situations where powder is used to knock down a fire rapidly, and foam is then applied to 'seal' the fire area, a foam-compatible powder is essential. Modern foam-compatible dry powders are based upon potassium chloride, or sodium bicarbonate, with a suitable flow additive of the siliconized type. The Ministry of Defence publishes a specification for foam-compatible dry powders.[12]

Powders for metal fires. Several powders have been developed for use on fires in combustible metals such as magnesium and its alloys, sodium, and the radio-active metals. These powders are generally available either in hand extinguishers, or in polythene film 'packs' for use on radioactive metal fires in glove-boxes, etc. Powders for metal fires have been based on common salt with a plasticizing agent, graphite, ternary eutectic chloride mixtures, mixtures of powdered PVC and sodium borate, or on boron trioxide.[13] The last two agents contain boron and are not suitable for use on radioactive metals required for further use in atomic processes. They are, however, suitable for magnesium fires.

These agents generally operate by forming a molten glassy layer on the surface of the metal, excluding air and inhibiting further combustion. Because of this type of action, it is best to apply them as gently as possible to the metal surface, and an extinguisher with a low discharge velocity, preferably with an applicator, should be used.

Halons and inerting gases

The 'halogenated hydrocarbons' or 'halons' are fire-inhibiting agents based upon hydrocarbons, in which some of the hydrogen atoms have been replaced by the halogens chlorine, bromine or fluorine. The hydrocarbon usually used is methane, though ethane is the root molecule in at least one agent. A number of these agents has been developed over the years, but those in current use are chlorobromomethane (CBM or HALON 1011), bromochlorodifluoromethane (BCF or HALON 1211), bromotrifluoromethane (BTM or HALON 1301) and dibromotetrafluoroethane (HALON 2402). In addition, 1.1.1 trichloroethane is sometimes added to CBM in the ratio of three parts to one part CBM. All the above agents are accepted by the Home Office as suitable for use in hand fire extinguishers. All halons give rise to acid gases on decomposition in a fire, but the more efficient the agent, the less decomposition is necessary to achieve

Table 7.4 Toxicity and extinction properties of vaporizing liquids, carbon dioxide and nitrogen

Agent	Dangerous concentration (DC)			Extinguishing concentration (EC)			$R = \dfrac{EC}{DC}$	$K = \dfrac{B}{A}$	Order of increasing toxicity	Order based on volume DC	Order based on weight DC
	% v/v	kg m³	Wt rel. to CO₂ (A)	% v/v	kg m³	Wt rel. to CO₂ (B)					
CTC	1·2	0·08	0·43	9·7[8]	0·66	1·24	8·1	2·9	7	7	7
MB	0·6	0·026	0·13	7·1[8]	0·3	0·57	11·8	4·4	8	8	8
CBM	2·6	0·15	0·77	6·35[8]	0·35	0·66	2·4	0·86	4	5	5
BCF	24·0	1·76	9·2	5·2[9]	0·38	0·72	0·22	0·08	2	3	2
BTM	50·0†	2·4†	12·5	4·9[8]	0·33	0·61	0·098	0·049	1	1 =	1
1.1.1. Trichloroethane	2·2	0·12	0·64	11·5[1]	0·6	1·14	5·75	1·78	6	6	6
CO₂	10·0*	0·19*	1·0	28·0[8]	0·54	1·0	2·8	1·0	5	4	4
N₂	50·0†	0·625†	3·25	42·0[10]	0·53	0·98	0·84	0·3	3	1 =	3

* Figure known to be dangerous to humans.
† Figure corresponding to an oxygen concentration of 10%.

extinction. Halons are at their best on fires which are not deep-seated. They are therefore most suitable for Class B fires, but as they do provide some cooling, they can also be used on Class A fires, provided the latter are not so deep-seated as to cause extensive breakdown of the agent, with greatly increased toxicity. It is best to use the halons in well-ventilated areas, or the ventilate thoroughly after use. A comparison of the toxicity and extinction properties of the halons, carbon dioxide, and nitrogen, is shown in Table 7.5, based on published results obtained with laboratory animals.

The gaseous halons (1211 and 1301) may be used effectively on the inerting of volumes or spaces in which fire may occur, or has occurred. They require a much smaller volumetric addition for this purpose than do the 'natural' inert gases carbon dioxide, nitrogen or steam.

Carbon dioxide is available in hand fire-extinguishers of various sizes. It disperses without residue on discharge, and is suitable for use on delicate and expensive electronic or electrical equipment, particularly if this is largely enclosed so as to maintain an inert atmosphere for a few minutes. It may give a 'cold shock' to its surroundings on discharge, and care must be taken to keep this to a minimum. While carbon dioxide can be used on all types of fire, it is most effective in relatively enclosed spaces which can be inerted for sufficiently long for the fire to be extinguished and cooled.

Liquid nitrogen, while not available as a regular extinguishing agent, is effective on fires in normal solids, flammable liquids, and electrical equipment. Its low temperature makes it necessary to avoid thermal shock to the equipment. Because of problems of storage, hand extinguishers containing liquid nitrogen are not likely to become generally available.

Medium and high expansion foams

These foams are used to control fires in both solid and liquid combustibles. They are made in hand-held or portable foam generators in which a large volume of air is induced either by the venturi-effect of the water/foam solution ejected from a nozzle into the body of the generator, or by a motor-driven fan which forces air through a gauze onto which the foam-making solution is sprayed. Medium expansion foam has expansions from about 30 to 250 units, and high-expansion foam from 250 to 1 000 units, the 'expansion' being the volume of the made foam divided by the volume of foaming solution from which it is made. The foaming liquids used to make these foams are usually synthetic and often based on ammonium lauryl ether sulphate with stabilizers. They can be used in 1–2 per cent concentration to produce relatively stable foams.

Medium expansion and high expansion foams depend for their performance on a combination of their cooling effect, due to the water carried to the fire, and on their smothering effect on the fire in reducing the availability of oxygen below that required for combustion. With persistent, deep-seated fires, it is probable that a cold inert gas filling of the bubbles would be desirable, so that smouldering can be quickly suppressed. No hand extinguishers dispensing high expansion foams are yet available in the U.K., although the principle has been patented. Manual appliances for medium- or high-expansion foams consist of portable generators in which the foaming solution is sprayed into an air-inducing venturi nozzle, and thence to a foam-forming gauze screen, giving rates of foam delivery up to 28 m^3/min. Portable or mobile high-expansion foam generators giving rates of foam delivery up to 378 m^3 are also available, usually operated by a ducted fan driven by a petrol, electric or hydraulic motor.

Standards and requirements for manual fire-fighting equipment

Many Western European countries promulgate standards for hand fire-extinguishers, and the appropriate standards for the different types of extinguishers in the U.K. are listed in Table 7.5. Recommendations for inspection and maintenance of fire extinguishers are given in a British Standard Code of Practice.[15]

Table 7.6 British Standards for hand fire extinguishers

1	Portable fire extinguishers of the water type (soda acid)	B.S. 138:1948
2	Portable fire extinguishers of the water type (gas pressure)	B.S. 1382:1948
3	Portable fire extinguishers of the water type (stored pressure)	B.S. 3709:1964
4	Portable plastics-bodied fire extinguishers of the water type (gas pressure)	draft
5	Portable fire extinguishers of the foam type (chemical) (including amendments up to March, 1961)	B.S. 740 Pt. 1 1948
6	Portable fire extinguishers of the foam type (gas pressure) (including amendments up to October, 1960)	B.S. 740 Pt. 2 1952
7	Portable carbon dioxide fire extinguishers	B.S. 3326:1960
8	Dry powder portable fire extinguishers	B.S. 3465:1962
9	CTC and CBM portable fire extinguishers	B.S. 1721:1960
10	Portable fire extinguishers of the halogenated hydrocarbon type	B.S. 1721: Feb. 1967 (draft revision)

Standards for fire extinguishers in the U.S. and Canada are promulgated by the Underwriters Laboratories, Inc., and by the Underwriters Laboratories of Canada, Inc., and are published by the National Fire Protection Association, 470, Atlantic Avenue, Boston, Massachusetts.

The Fire Offices' Committee, Aldermary House, Queen Street, London, E.C.4, a central technical body of the tariff fire insurance offices, maintains a list of extinguishers approved for insurance purposes,[16] and their rules contain requirements for the appropriate scale of installation of fire extinguishers in insured premises. European and international standards for fire extinguishers are also being developed.

Fire-fighting hose reels

Efficient protection of premises can often be arranged by the use of wall-mounted hose reels. Those commonly available for public use give a water flow rate of 23–46 l/min, according to the pressure available, and are equipped with 21 m of hose-reel hose capable of extending in either direction. British Standards Institution and the Ministry of Public Building and Works both publish specifications for wall-mounted hose-reels[17, 18] or hose-reel hose.

Hose reels for use by trained fire brigades generally give greater rates of flow: some 46–55 l/min for low-pressure hose reels, and as high as 159–182 l/min for special high-pressure equipment. The use of special hose-reel guns, giving either jet or spray deliveries at will, is increasing, and a specification for this type of equipment is in preparation by the fire authorities.[19]

Special applicators can be fitted to hose-reels for reaching inaccessible danger points, or for dealing with flammable liquid fires.

Mobile fire-fighting equipment

To extend the range of hand fire extinguishers, certain types of mobile equipment are available which provide larger supplies of agent. These include mobile foam engines with capacities up to about 136 l of solution, carbon dioxide dispensers with up to 55 kg of gas, and dry powder dispensers delivering up to 68 kg of powder. This equipment is usually trolley-mounted.

Training of personnel

A fire is most successfully fought where the equipment is suitable, readily available, and adequate for the task, and where the personnel involved have already had sufficient training in the use of the equipment and the likely risks to enable them to fight a fire in an efficient, unflurried manner.

To achieve this, the potential risks must be realized, and their likely magnitude assessed. Suitable equipment must be provided, and regular, realistic, training must be given against practice risks of similar severity. Every person who fights a fire for the first time finds difficulty in handling the equipment properly, but after a while—say five to ten practices—substantial advances in fire control invariably result. Suitable training and regular practice, therefore, are essential for efficient results.

This paper is Crown Copyright, reproduced by permission of the Controller, H.M. Stationery Office. It is contributed by permission of the Director of the Building Research Establishment. (The Fire Research Station is the JFRO of the Department of the Environment and Fire Officers' Committee.)

References

1. *United Kingdom Fire and Loss Statistics, B.R.E.,* Garston.
2. *N.F.P.A. Handbook of fire protection,* Current Edition.
3. *Ministry of Labour: Fire-fighting equipment, fire alarms, and fire drills in offices and shops.* H.M.S.O., London, 1966.
4. O'Dogherty, M. J. International standards for fire-extinguisher construction and performance. *Ministry of Technology and Fire Offices' Committee, Joint Fire Research Organization, Fire Research Note No. 682,* 1967.
5. Elkins, G. H. J. Use of anti-freeze in water extinguishers. *Fire,* 1965, **57** (716), pp. 448–456.
6. Rasbash, D. J. and Stark, G. W. V. Extinction of fires in burning liquids with water spray from hand lines. *Institution of Fire Engineers Quarterly,* 1960, **20** (37), pp. 55–64.
7. O'Dogherty, M. J. The shock hazard associated with the extinction of fires involving electrical equipment. *Joint Fire Research Organization Technical Paper No. 13.* H.M.S.O., London, 1965.
8. Foam liquid, fire extinguishing. Defence Standard 42–3, Issue 1, 1969. *Min. of Defence, Directorate of Standardisation.*
9. Nash, P. Recent research on foam in the United Kingdom. *Institution of Fire Engineers Quarterly,* 1961, **21** (41), pp. 14–33.
10. Fittes, D. W. and Richardson, D. D. The influence of alcohol concentration on the effectiveness of protein foams in extinguishing fires in gasoline alcohol mixtures. *Journal of the Institute of Petroleum,* 1963, **49**, pp. 150–152.
11. Nash, P. and French, R. J. Foam compounds for use against fires in water-miscible solvents. *Journal of the Institute of Petroleum,* 1961, **47**, pp. 219–222.
12. Dry powder, fire extinguishing, foam compatible. *Ministry of Defence Spec. No. DEF 1420.* H.M.S.O., London, 1963.
13. Elkins, H. J. Fire-fighting composition N.R.D.C. *U.K. Patent Spec. 1063207,* 1967.
14. Thorne, P. F. A review of the toxic properties of some vaporizing liquid fire extinguishing agents. *Ministry of Technology and Fire Offices' Committee, Joint Fire Research Organization. Fire Research Note No. 659,* 1967.
15. Fire-fighting installations and equipment, Part 3—Portable fire extinguishers for buildings and plant. *British Standard Code of Practice CP 402.*
16. *List of approved portable fire-extinguishing appliances.* (Latest issue with amendments.) Fire Offices' Committee, 1965.
17. Wall-mounted hose-reels for fire-fighting. *British Standard 5274: 1976.*
18. Rubber reel hose for fire-fighting purposes. *British Standard 3169.* British Standards Institution, London, 1959.
19. Nash, P. Some factors in the design of hose-reel equipment. *Institution of Fire Engineers Quarterly.* 1962, **22** (45), pp. 22–39.
20. *Offices, shops and railway premises act, 1963.* H.M.S.O., London (see also ref. 4).

21. Nash, P. Dry powder extinguishing agent for magnesium fires. *Fire (International)*, 1963, **1** (1), pp. 16–17. *Institution of Fire Engineers Quarterly*, 1963, **23** (51), pp. 275–276, **23** (51), pp. 275–276.
22. Nash, P. Fire extinguishers. *Proceedings of the 2nd National Fire Protection Conference*. Fire Protection Association, London, 1964.
23. Nash, P. Fire protection in factories. *The Plant Engineer*, 1966, **10** (2), pp. 36–42.
24. Fittes, D. W. and Nash, P. Use of foam on fires involving water-miscible solvents. *Institution of Fire Engineers Quarterly*.
25. *Home Office (Fire Service Department) Manual of firemanship*. H.M.S.O., London.
26. *Fire Protection Review: Fire protection directory*, current edition, Benn Bros., Ltd, London E.C.4.

8. Fire control: automatic

J. Newall

Fire and Security Group,
Mather and Platt Ltd

Automatic fire protection equipment has played, and continues to play, a great part in saving life and property from fire hazards in industry. The provision of fixed pipework systems employing water as the extinguishing medium is acknowledged to be an efficient means of protecting buildings and many other classes of risk against extensive damage resulting from an outbreak of fire. Such systems can be divided into three main classes—automatic sprinklers, drenchers, and waterspray systems.

Automatic sprinklers are installed inside a building and operate automatically when a fire occurs. Drenchers, which may be automatically or manually controlled, are fitted outside a building in order to protect it against spread of fire from a neighbouring property. Waterspray systems are a specialized development of automatic fire protection, designed for extinguishment of flammable liquid fires, or as a means of safeguarding plant, structures, and machinery against fires involving highly flammable liquids, gases, and solids.

Automatic sprinkler systems

Automatic sprinkler and fire alarm systems use water to extinguish or control fire in its incipient stages, before it has time to develop. Pipework, fitted with sprinklers spaced at regular intervals, is installed throughout the protected building, and is connected to a reliable water supply. When fire occurs, only the sprinklers in the immediate vicinity automatically operate and discharge water to control it. The operation of any one sprinkler causes an audible alarm to sound.

Automatic operation of the alarm gong is caused by the action of water flowing through the pipes leading to the opened sprinklers. Usually the discharge of water from the sprinklers is sufficient to extinguish a fire completely, but if any obstruction prevents the water falling directly onto burning materials, the fire will be held in check until the arrival of the fire brigade.

Automatic sprinklers are universally accepted as the most important of all fire protection systems; it is known that many thousands of fires have been extinguished by sprinklers during the last 60 years and in addition there must have been many more minor fires where the loss has been considered too small to be worth reporting.

The value of the automatic sprinkler system in the eyes of the fire insurance companies is clearly illustrated by the high annual rebates on fire insurance premiums which are allowed to the owners of sprinklered premises.

Principle of operation of automatic sprinklers

In an automatic sprinkler system water is carried to every part of the protected building in pipes graded in size and generally suspended from the ceiling or roof. Water is held back in the pipes, until a fire occurs, by automatic valves which form an integral part of the sprinkler. When the temperature rises above a predetermined value the heat-sensitive element, incorporated in the sprinkler, ruptures or fuses and allows the valve to open under the influence of the water pressure. The modern design of sprinkler incorporates an operating element which, unlike the earlier designs of sprinkler held closed by metal struts held together by fusible solder, is immune from the effects of the corrosion.

A typical modern sprinkler is illustrated in Fig. 8.1 and incorporates a quartzoid bulb heat-sensitive element. The quartzoid bulb is made from a transparent corrosion-free material strong enough to withstand any water

Fig. 8.1   The Grinnell type 'F' quartzoid bulb sprinkler.

pressure likely to occur in the system. The bulb contains a highly expansible liquid which exerts considerable disruptive force when heated to the temperature of operation, but because the liquid shrinks as it cools the bulb is unaffected by the most intense cold and no internal pressure will develop.

During manufacture a small amount of vapour is trapped when the bulb is hermetically sealed. When the liquid expands under the influence of heat, the vapour is gradually absorbed until the liquid completely fills the bulb. Any further increase in temperature is accompanied by a rapid rise of internal pressure, which is sufficient to shatter the bulb into small pieces, so ensuring prompt and free opening of the sprinkler.

The water passage of the sprinkler illustrated in Fig. 8.1 is closed by a valve assembly of flexible construction ensuring permanent compression of a soft metal gasket against the valve seating. The assembly is held in position by the quartzoid bulb. Flat facings on the bulb and the self-centring cap and cone ensure accurate alignment. The whole assembly is extremely stable and will withstand pressure surges or external vibration without displacement.

When the bulb shatters the strut and valve assembly are thrown clear of the sprinkler yoke to leave a clear orifice discharging water onto the deflector, which distributes it onto the fire area.

The quartzoid bulbs are produced in a range of ratings to cover the temperatures generally met within industry. For normal situations in temperate countries the 68°C bulb is recommended. The temperature rating of a sprinkler is usually indicated on the deflector, and the liquid filling of the quartzoid bulb is coloured according to its temperature rating. The quartzoid bulb sprinkler of the type illustrated can be mounted either in the upright or pendent position.

Where there is any risk of the sprinkler being damaged, for example when erected under a very low ceiling, specially designed guards can be fitted. Where it is necessary for the pipework to be concealed above a false ceiling, for example in offices, shops, and department stores, it is also necessary to provide sprinklers which are both aesthetically pleasing and unobtrusive. Another sprinkler has been specially developed for this application, in which a masking ring is provided to cover the opening in the ceiling through which the sprinkler is connected to the pipework.

To secure official recognition and to qualify for the rebates allowed by insurance companies, sprinklers must be of a pattern approved by the Fire Offices' Committee.

Types of installation

In order to simplify the control of the sprinkler equipment in a protected building it is usual to divide it into sections, generally referred to as installations, of which there may be several in premises covering an extensive area. Each installation is equipped with a set of automatic installation controlling valves and alarm gong.

Water must never be allowed to freeze in a sprinkler installation. For this reason there are in general three types of sprinkler systems:

    (a) the 'wet' system,
    (b) the 'dry' system,
    (c) the alternate 'wet' and 'dry' system.

The 'wet' system is the description given to an installation where the sprinkler pipes are constantly charged with water under pressure. The system is only employed where there is no danger of frost, i.e., in climates where freezing is unknown, or where buildings are continuously heated during winter months. Sprinklers fitted on a 'wet' installation are usually fitted pendent below the pipework.

The 'dry' system is applied in premises where the water in the sprinkler pipes is liable to freeze. All the installation pipework is charged with air at a moderate pressure and the water is held back by a differential valve which, when a sprinkler opens and releases the air in the pipes, lifts and allows water to enter the pipework. The 'dry' system has contributed very largely to the wide adoption of automatic sprinklers in such cold climates as those of Norway, Finland, Sweden, Switzerland, Canada, and parts of the United States. It offers the same degree of reliability as the 'wet' system.

The alternate 'wet' and 'dry' system is used to protect buildings in which freezing is likely to occur only during part of the year. These systems are set on the 'wet' system during the summer when there is no danger of frost, but are changed over to the 'dry' system by drawing off all the water and charging the sprinkler pipework with air under pressure before frosts are likely to occur.

The installations remain on the 'dry' system during the winter and are changed back to the 'wet' system in the spring when the danger of frost has passed. Sprinklers on 'dry' and 'alternate' systems are usually fitted upright above the pipework to facilitate complete drainage of the system.

If a choice of system is possible a 'wet' installation is always preferable because water is distributed over the fire immediately a sprinkler operates, whereas the air pressure in the pipework of a 'dry' or 'alternate' installation has to be reduced through the open sprinklers before the differential valve will lift and allow water to enter the pipes. In order to reduce the time taken for water to reach the sprinklers which have opened by fire, an accelerator should be fitted to 'dry' or 'alternate' systems.

When an installation is on the 'dry' system, or the alternate 'wet' and 'dry' system, the main valves and all exposed piping below valves which contain water are housed in a room which is kept at a temperature above freezing point, or in a frost-insulated cupboard. Drain valves at all low parts of an installation on the 'dry' system are to be opened occasionally in order to drain the pipes of any water or condensation which may have accumulated.

Installation controlling valves—'wet' system. The controlling valves of a typical 'wet' installation are shown in Fig. 8.2 and include stop valve 1, alarm valve 2, combined drain and test valve 3, test valve 4, pressure gauges 5 and 6, alarm motor isolating valve 7, strainer 8, alarm motor and gong 9 and drip plug 10.

Stop valve 1 is generally of the wedge type and is normally strapped and padlocked in the open position. The stop valve is provided as a means of cutting

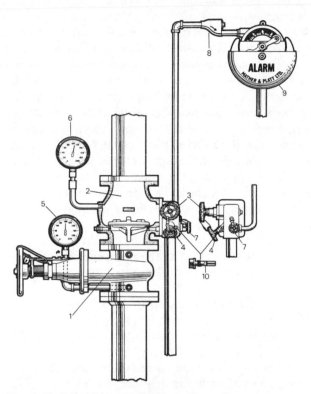

Fig. 8.2   Grinnell installation controlling valves—wet system.

off the water supply to the sprinkler installation after the extinguishment of a fire; it must not be closed until the person responsible has established definitely that the fire is out. It should not be closed at any other time, except when it is necessary to carry out alterations or extensions to the sprinkler installation.

Alarm valve 2 is fitted with a seating on which rests a clack consisting of a flat circular plate, faced with a special composition disk and having a cylindrical stem which is free to slide vertically in the seating guide. In the face of the seating is an annular groove through which water can pass to the alarm motor and gong 9 when the clack is raised. Whilst the alarm valve clack is resting on its seating, which is the normal position when there is no water flowing through the installation pipework, the annular groove in the seating is closed by the composition facing of the clack. When fire causes a sprinkler to operate water will flow from the supply through open stop valve 1, alarm valve 2, and the installation pipework to the open sprinkler.

In passing through alarm valve 2 the water raises the clack, which permits some water to flow through the annular groove in the seating to alarm motor and gong 9 to give an audible alarm. The clack remains in the raised position so long as there is any flow through alarm valve 2, i.e., until stop valve 1 is closed. When sprinklers are in operation during a fire the alarm motor and gong may be isolated by closing valve 7, which should be strapped open at all other times.

If required, an electrical pressure operated switch may also be connected to the annular groove to give an alarm in either the local or works fire brigade station, but in order to comply with the requirements of the fire insurance companies it must be in direct communication with the annular groove, and no means of isolation should be provided such as a valve or cock.

In order to avoid false alarms when the pressure of the water supply increases gradually above the normal installation pressure, for example, when a town's main pressure is increasing at night, a small compensating check valve is fitted in the centre of the alarm valve clack. This permits the flow of water at a limited rate from the water supply into the sprinkler installation pipework without raising the alarm valve clack.

Test valve 4 is used at regular intervals to test the alarm system. When the valve is opened it has the same effect on the alarm equipment as the opening of a sprinkler, but the water drawn from the installation is conducted to a drain connection.

Test valve 3 is provided to drain the sprinkler installation of water after a fire has been extinguished and before replacing sprinklers that have operated. It is also used periodically to test the efficiency of the water supplies under operating conditions.

There is a permanent connection to the drain from the alarm motor pipe connection through a small orifice in drip plug 10. This is provided in order to allow water to drain from the alarm motor pipe connection and the annular groove in the seating. The orifice should be kept clear and should therefore be examined regularly.

Installation controlling valves—'dry' and 'alternate' systems. The sprinkler pipework of a dry installation or an alternate 'wet' and 'dry' installation is arranged with a gradual fall to permit the draining of any water which has been carried to open sprinklers during a fire, or to facilitate complete drainage of the pipework of an 'alternate' installation when it is being changed from the 'wet' to the 'dry' system.

The controlling valves of a typical alternate 'wet' and 'dry' system are shown in Fig. 8.3. The controlling valves are shown set on the 'dry' system.

They include stop valve 1, alarm valve 2, and differential air valve 3.

Stop valve 1 and alarm valve 2 are identical to those described for the 'wet' system. Differential air valve 3 is provided with two combined circular clacks, 4 and 5, mounted on a common spindle, resting on seatings 6 and 7 fitted in the

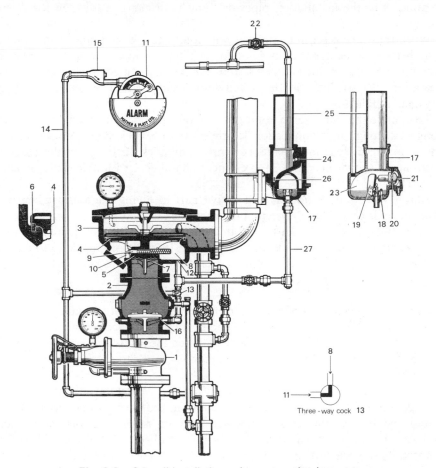

Fig. 8.3   Grinnell installation valves set on the dry system.

valve body. Upper clack 4 is fitted with a composition disk whereas lower clack 5 makes metal to metal contact with seating 7. The area of upper clack 4 is considerably larger than the area of lower clack 5, and therefore the installation pipework can be charged with air at a pressure much less than the pressure of the water supply acting on the underside of lower clack 5 and maintain the valve closed.

It is common practice to charge the pipework with compressed air to about 2·8 bar.

Immediately a sprinkler opens air can escape from the installation pipework and the pressure on the top side of upper clack 4 falls. When this pressure is reduced to a value whereby it can no longer hold the combined clack, the clack lifts and water flows from the supply through differential valve atmospheric chamber 8 to the installation pipework and is discharged through the open sprinkler.

When the valve is set in its normal position, ready for automatic operation, chamber 8 between upper clack 4 and lower clack 5 contains air at atmospheric pressure. Once combined clacks 4 and 5 have been lifted during automatic operation they are held in the raised and open position by spring catch 9 engaging notched upright 10.

To operate the alarm some water passes from chamber 8 to alarm motor and gong 11 through pipe connection 12, three-way cock 13, pipe connection 14, and strainer 15. It is important that three-way cock 13 be set in the correct position, as shown in Fig. 8.3, to connect chamber 8 with alarm motor and gong 11.

Alarm valve 2 does not play an active part in giving the alarm when the installation is on the 'dry' system, but when the installation is set on the 'wet' system three-way cock 13 must be set in the appropriate position to allow water from alarm valve 2 to pass *via* annular groove 16 and alarm cock 13 to alarm motor and gong 11; also the combined clacks 4 and 5 are held in the raised position by the spring catch.

Air valve accelerators. Accelerators are fitted on all except the smallest 'alternate' or 'dry' sprinkler installations. The purpose of an accelerator is to speed up the lifting of the differential valve clack when the installation is on the 'dry' system and, therefore, reduce the time taken for water to reach the seat of the fire after a sprinkler has opened.

A typical accelerator 17 is illustrated in Fig. 8.3. When the accelerator is set for operation with the sprinkler installation on the 'dry' system, valve 18 is closed and bob-weight 19 is in a raised position resting against the end of plunger 20. The movement of plunger 20 is controlled by diaphragm 21 which is normally subjected to equal air pressure on both sides. During the process of setting the valves on the 'dry' system and whilst the installation is being charged with compressed air, some air from the sprinkler pipework passes through open valve 22 into accelerator chamber 23, through restricted orifice 24 into air vessel 25 and through portway 26 to the chamber behind diaphragm 21.

After charging the sprinkler installation to the required air pressure, the pressure will be uniform throughout the sprinkler pipework, accelerator chamber 23, air vessel 25, and the chamber behind diaphragm 21, therefore diaphragm 21 is in balance. Accelerator valve 18 is closed and water-sealed and therefore air cannot leak from chamber 23.

When a fire causes a sprinkler to open there is an immediate fall of air pressure in the sprinkler pipework and accelerator chamber 23. The air pressure in air vessel 25 and behind diaphragm 21 is maintained at practically its initial value, as the rate of escape of air from air vessel 25 into chamber 23 is limited by orifice 24. Thus diaphragm 21 is deflected by the higher air pressure in air vessel 25 and moves plunger 20. Bob-weight 19 falls to a horizontal position and, in falling, raises and retains accelerator valve 18 in an open position. Compressed air from the sprinkler installation can now pass through open valve 18 and pipe connection 27 into the differential air valve intermediate chamber. The air pressure in the intermediate chamber below the upper clack thus becomes equal to the air pressure in the sprinkler pipework above the upper clack. The pressure of the water from the supply immediately raises the combined clack assembly of the differential air valve to allow water to flow rapidly into the sprinkler pipework.

On an installation fitted with an accelerator the differential valve clack lifts a few seconds after the operation of a sprinkler. If no accelerator is fitted it takes considerably longer for the differential valve to operate, it being necessary for the air pressure in the sprinkler installation to fall to a value whereby the differential of the combined clack is overcome by the pressure of the water supply.

Water supplies for sprinkler installations

An adequate supply of water must be available for a sprinkler installation at all times, and must be thoroughly reliable. In the British Isles, sprinkler installations recognized by insurance companies are placed in one of six classes according to the nature of the water supplies which serve them. The insurance premium rebate for which the risk then qualifies depends upon this classification. To qualify for the highest rebate, an installation must have two independent water supplies, one of which must be virtually unlimited, such as a connection from a public water supply. In addition a minimum running pressure of 1·8 bars should be available at the highest sprinkler in the installation at all times.

The water supply for the lowest class of installation acceptable must be capable of maintaining a minimum running pressure of 0·4 bar at the highest sprinkler.

Approved water supplies for sprinkler installations are:

(a) public mains,
(b) elevated private reservoirs,
(c) automatic pumps boosting the supply from public mains, or drawing water from reservoirs or other static source,
(d) pressure tanks,
(e) elevated tanks.

Public mains. This is the most common supply and is usually extremely reliable unless the risk is at a considerable elevation above the ground level of the area, when it may be difficult to obtain sufficient pressure. The insurance companies require a direct connection to be made from the mains to the sprinkler installation valves.

Elevated private reservoirs. This supply is relatively uncommon, but if a suitable reservoir can be provided at a reasonable cost, it is probably the most reliable of all supplies, as it is immune from any demands which may be made on public water supplies at the time of a fire.

Automatic pumps. The water pressure in the public supply mains is sometimes inadequate for a sprinkler installation, although there is an ample supply of water available. Under these conditions it is common practice to employ a centrifugal pump with its suction branch coupled direct to the town's main, or through a break tank, the pump delivery being connected to the sprinkler trunk main. The pump must be automatic in its operation and its rating is determined by the number of sprinklers which may be expected to operate in the particular premises.

An adequate supply of water is sometimes available from such sources as rivers, canals, reservoirs, or static tanks. Automatic pumps may be used to bring water to installations from these sources if the water is reasonably clean.

Centrifugal pumps may be driven by a.c. or d.c. electric motors, or diesel engines. When driven by electric motors the electrical supply must be taken direct from a public supply and must always be available.

Pressure tanks. A pressure tank is frequently used in combination with one or other of the water supplies described previously. It is a cylindrical steel shell of riveted or welded construction, with a total capacity up to 46 000 litres.

The tank is charged with the required amount of water, and then with compressed air until the required working pressure is reached. The height of the highest sprinkler in the installation above the base of the tank determines the air/water ratio and the required working pressure.

The pressure tank must be housed in a building which is heated during the winter months or protected in some other way against frost.

Elevated tanks. Sufficient water pressure to supply a sprinkler installation can be provided from a tank placed at the requisite height above the highest sprinkler. In this case the tank is constructed from cast iron plates provided with internal flanges to facilitate erection and jointing. The necessary accessories include a top cover to prevent foreign matter from falling into the water, and a water level indicator. Access ladders and manholes can be fitted during construction.

Precautions should be taken to ensure that the water contained in the tank is prevented from freezing during the winter months.

The quantity of water which an elevated tank must contain depends upon the hazard classification.

The correct water level in an elevated tank is maintained by a ball-float valve supplied with water through a connection not less than 25 mm diameter from the town's main or by means of a pump.

The drencher system

Drencher systems are designed to provide a discharge of water over windows, doors, or other openings in a building in order to prevent the transmission of fire from adjacent premises. Drencher systems have proved their value on numerous occasions by preventing the spread of fire across narrow streets such as are to be found in the congested areas of many large cities.

Some drencher systems are designed to come into operation automatically under the influence of heat from any nearby building which may be on fire. Under these conditions sealed drenchers are employed and, as will be noted from the illustration in Fig. 8.4, closely resemble sprinklers except that they are fitted with a special type of deflector.

Fig. 8.4   Grinnell automatic sealed type drencher used for window openings.

These sealed drenchers are mounted at intervals on a pipework system which is fixed to the extension of the protected building and is charged with air under pressure. Each sealed drencher contains a valve which is held in the closed position by a quartzoid bulb until heat from a fire causes the bulb to shatter into small pieces. On collapse of the bulb the valve opens and air escapes from the pipework system. The resultant drop in air pressure brings about the opening of a special control valve through which water is free to flow and ultimately to discharge from any drenchers which have opened due to the fire. An automatic warning of fire will be given immediately the control valve operates.

In the event of fire, only the automatic drenchers which are in the zone affected by the heat will operate. Water is only admitted to the exposed pipework when there is a fire in the vicinity, hence there is no risk of damage from frost at other times.

Drencher systems are also supplied suitable for manual control by means of a manually operated valve situated inside the building. The drenchers employed in such systems have open waterways and, where the protection required is extensive, may be divided into groups, each group of drenchers being controlled by a separate valve.

Water spray systems

Some of the worst fires which have occurred in the history of fire-fighting have been those involving oils and similar flammable liquids. Unless promptly tackled at the source, such fires rapidly grow to catastrophic proportions, often with irreparable damage to valuable plant.

Water spray systems have been developed to deal with risks of this type and are installed in a wide range of plant throughout the world. For example in power stations protection of indoor and outdoor transformers, turbo-alternators, switchgear and other oil-filled equipment is afforded by water spray systems. Further examples include mixing machines and varnish kettles in paint works, oil cellars in steel mills, and oil-fired boiler rooms and oil-quenching tanks in general industry.

In this type of protection water is applied in the form of a conical spray consisting of droplets of water travelling at high velocity. Three principles of extinguishment are employed in the system—emulsification, cooling, and smothering. The result of application of these three principles is to extinguish the fire within a few seconds.

The high velocity discharge of water in droplets breaks up the surface of the oil or similar flammable liquid, to form an emulsion which will not support combustion. The rate of burning of a flammable liquid depends upon the rate at which vapour is given off from the surface of the liquid, and the supply of air or oxygen to support combustion. When a flammable liquid burns, the rate of vaporization increases until the fire reaches a maximum rate of burning, with the surface of the liquid near to the boiling point. The system described above when forming an emulsion intersperses cold water with the flammable liquid, thus cooling it and preventing the further escape of the flammable vapours. Whilst the water droplets are passing through the flame zone some water is formed into steam, diluting the air or oxygen feeding the fire and creating a smothering effect.

Projectors for such systems are designed in various sizes to give an even distribution of water over the area to be covered. The various sizes give a combination of differing flow rates and discharge angles.

Automatic operation in the event of fire is an essential feature of this type of protection, but there are some applications in which manual control is desirable, for example, in the protection of lubricating oil systems and governor gear systems of turbo-alternators in power stations. In general, automatic operation is employed and typical arrangements of these systems are shown at Figs. 8.5 and 8.6.

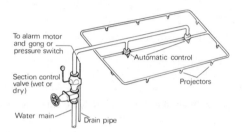

Fig. 8.5   Arrangement of projectors and Grinnell quartzoid bulb automatic controls.

Fig. 8.5 illustrates the type of protection installed where fires in their incipient stages are likely to be small. The open projectors are arranged in small groups to cover the risk and are connected by pipework to quartzoid bulb controls. When fire causes a control to operate water is discharged simultaneously through all the projectors in the group. When this type of equipment is installed in situations where there is a danger of freezing, the pipework between the controlling valves and the quartzoid bulb controls is charged with compressed air and the equipment functions in a similar manner to a 'dry' sprinkler installation.

Where fires are likely to be larger, or to spread rapidly over an extended area, a larger number of projectors must be arranged to operate simultaneously. A typical application of this method of control is illustrated in Fig. 8.6. In this case projectors are mounted on empty pipework commanding the entire risk including the floor area. Quartzoid bulb detectors, mounted in an independent system charged with compressed air are so positioned that, wherever a fire may originate, one at least will operate, to allow the compressed air in the pipework to escape. At a predetermined air pressure the automatic deluge valve will open and water will immediately discharge from all the projectors to extinguish the fire.

Water spray systems have been developed to protect plant against fires which may involve highly flammable liquids, gases and solids. Commonsense often dictates that when a fire involving a highly flammable liquid or gas breaks out, the vapours should be permitted to burn at the point of escape until the flow of fuel can be stopped. Efforts to minimize fire damage and to reduce the risk

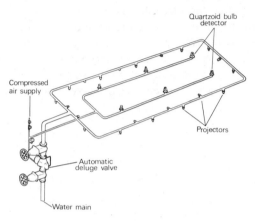

Fig. 8.6   Arrangement of Grinnell quartzoid bulb detectors, projectors and automatic deluge valve.

of explosion are then concentrated on protecting adjacent equipment such as tanks, pipes, and supporting structures by means of the cooling effect of water spray. This method avoids the danger of spreading the heavier-than-air explosive vapour which might collect in cellars and low levels thereby creating hazardous flash explosion conditions.

This type of system makes use of specially designed projectors applying water in a spray formed of finely divided droplets discharged at medium velocity, and protects the plant by cooling, by controlled burning of flammable liquids, and by the dilution of explosive gases.

When a system of this type is brought into operation over an enclosed storage tank involved in a fire, and containing flammable liquid or gas, the protective film of water flows over the external surface of the exposed vessel and the supporting structure, to reduce the temperature and prevent a dangerous absorption of heat.

The rate of burning of a flammable liquid depends upon the rate at which vapour is given off from the surface of the liquid, and the supply of air or oxygen to support combustion. This medium velocity water spray system controls these conditions in three ways: first, by direct cooling of the surface of the liquid thus reducing the rate of vaporization; second, by producing steam which dilutes the air and vapour feeding the fire; third, by introducing droplets of water into the flame, thus cooling it, and reducing the transfer of heat back to the liquid surface. Fires involving highly flammable liquids can thus be controlled to a point where personnel can approach the fire with little danger to themselves.

When highly flammable gases leak from their containers to form a mixture with air, a source of ignition may cause an explosion or fire. If such a water spray system is operated when a leakage has been detected and before a fire has started, it provides air turbulence which promotes quicker dilution of the gases by air and water vapour, thus often producing a mixture which is too weak to support combustion. If a fire did occur such a system could prevent a dangerous increase of pressure inside the containing vessel and protect the supporting structure against failure.

These systems are equally suitable for indoor and outdoor situations and can be designed so that none of the installation pipework contains water until a fire is detected, thus avoiding risk of water freezing in the installation.

Different forms of automatic operation can be used and the one selected will depend upon a number of factors. In some circumstances quartzoid bulb sealed sprayers—similar to sprinklers—may be located in the hazardous area. More frequently, however, it is advisable to make provision for a group of sprayers to operate simultaneously and an automatic control which permits the admission of water into a section of the system is used. If the risk demands a large number of sprayers to be brought into operation simultaneously an automatic deluge valve is employed to control the water supply. In such installations automatic quartzoid bulb detectors are mounted on an independent range of pipework charged with compressed air and, in case of fire, operate to cause the opening of the automatic deluge valve. Alternatively, a device may be employed which detects a rate of rise in temperature.

9. Static electricity

Dr N. Gibson

Senior Research Physicist,
I.C.I. Organics Division

Nature of the fire hazard from static electricity

Static electricity is recognized as a possible fire hazard in the petroleum, chemical, and explosives industries, during the manufacture of plastic and rubber sheeting, in hospital operating theatres, and generally where flammable gases, vapours, and dust clouds are processed.

The generation of static electricity is a surface phenomenon associated with the contact and separation of dissimilar surfaces.[13] For the recognition of the type of industrial processes in which static electricity will be generated it is only necessary to recognize that contact between dissimilar bodies causes a disturbance of the electric charge along the interface. Positive charge is localized on one body and negative charge on the other. Separation of the bodies can leave excess positive charge on one body and excess negative charge on the other. If both surfaces are electrically conducting, e.g., metals, the charge can move relatively easily and both bodies are virtually restored to their original electrically neutral state by the movement of charge through the last point of contact as they separate. When metals separate little, if any, excess charge is retained on each body after separation.

If, however, one or both of the bodies are poor conductors of electricity, e.g., plastics or liquids of high electrical resistivity, the movement of charge is impeded and excess charge can be retained on each body after separation.

When the bodies are in contact the potential difference between them is low, about 1 V. In separating the bodies, however, work has to be done against the electric field between the negative and positive charges. This increases the electric potential of the system. The potential of a charged body may be increased from 1 V to over 1 000 V by separation.

Surface contact, separation, and movement involving poorly conducting materials are an intrinsic feature of many industrial processes, for example, the flow of high resistivity liquids, e.g., petroleum products, the filtration of liquids, the movement of powders in grinding, blending and sieving operations, the pneumatic transfer of powders, the movement of people or vehicles across insulating floors, and the movement of transmission belting or other forms of sheet materials over pulleys and rollers. In these and similar processes the generation of static electricity can and does occur.

An idea of the order of magnitude of the electric charge, current, and potential involved in electrostatic phenomena may be gained from the following examples. High resistivity liquids flowing in pipelines generate charge at the rate of 10^{-6}–10^{-9} C/s (i.e., ampères). Powders emerging from grinding units into metal containers carry charge into the container at the rate of 10^{-4}–10^{-8} A. A charging rate of 10^{-6} A can cause the potential of the container to rise at the rate of 1 000 V/s if the container is insulated from earth. Potentials of 10 000–40 000 V can readily be obtained in this manner. People moving across insulating flooring, e.g., carpets made from synthetic fibres, in a dry atmosphere can frequently accumulate electrostatic charge and attain potentials greater than 10 000 V; this may cause them to feel an electric shock when they touch an earthed object, for example, a door knob.

Static electricity is essentially a low current, high voltage phenomenon. The fire hazard occurs when the generated charge accumulates on an object and increases its potential to a value greater than the breakdown potential of the surrounding atmosphere and causes an electric spark. Sparks produced in this way can have sufficient energy to ignite many different types of flammable material.

The ability of the spark to cause ignition—its incendivity—depends not only upon the amount of charge stored on the object and the sensitivity of the flammable material but also on the electrical characteristics of the object on which the charge accumulates. Sparks from electrically conducting objects tend to be more incendive than sparks from electrically non-conducting objects.

Incendivity of sparks from conductors

The energy (E) associated with an electrically conducting object carrying an electrostatic charge (Q) is

$$E = \frac{Q^2/C}{2} = \frac{CV^2}{2} \tag{1}$$

where E = energy (J)
Q = electric charge (C)
C = electrical capacitance of the object (F)
V = electric potential of the object (V)

When electric charge is released in a spark from a conducting object then charge may flow freely through and across the conducting surface and virtually the whole of the energy is released in the spark.

Although some of the energy may be dissipated resistively and radiated as electromagnetic waves and although the incendivity of the spark is determined not solely by its total energy but also by the form of the discharge (e.g., the distribution of current in the spark with respect to time), from the practical point of view the value of E calculated using eq. (1) provides a useful guide as to whether or not the spark has sufficient energy to ignite a particular flammable atmosphere.

One variable in eq. (1) is the electrical capacitance of the object. This can be measured by the usual techniques of current electricity.[10] Typical values of the electrical capacitance of objects on which charge can accumulate are given in Table 9.1, and these can be used in the calculation of E.

Table 9.1 Electrical capacitance of conducting items

Object	Electrical capacitance ($\mu\mu F$)
Small metal items (scoops, hose nozzle)	10–20
Small containers (bucket, 45 l drum)	10–100
Medium container (200–400 l)	50–300
General plant items (reaction vessels) immediately surrounded by earthed items	100–1000
Human being	200–300

$$1 \ \mu\mu F = 10^{-12} \ F$$

Table 9.2 Energy of sparks from conductors

Potential (V)	1 000				10 000			
Capacitance ($\mu\mu F$)	10	100	300	1 000	10	100	300	1 000
Energy (mJ)	5.0×10^{-3}	5.0×10^{-2}	1.5×10^{-1}	5.0×10^{-1}	5.0×10^{-1}	5.0	1.5×10^{1}	5.0×10^{1}

Potential (V)	20 000				50 000			
Capacitance ($\mu\mu F$)	10	100	300	1 000	10	100	300	1 000
Energy (mJ)	2.0	2.0×10^{1}	6.0×10^{1}	2.0×10^{2}	1.3×10^{1}	1.3×10^{2}	3.9×10^{2}	1.3×10^{3}

Conducting bodies insulated from earth can fairly quickly attain potentials of 10 000–20 000 V and potentials up to and greater than 50 000 V are not unknown. The possible energies of sparks from such bodies calculated from eq. (1) are shown in Table 9.2.

The development of the highest level of voltage on bodies of large capacitance requires not only a high charging rate but also that the body be well insulated from earth. Nevertheless energies in the range 1–100 mJ can be sufficiently easily attained for them to be a real possibility in processes where electrostatic charge accumulation is permitted to go unchecked.

Incendivity of sparks from non-conductors

The assessment of the incendivity and hazard of sparks from non-conducting surfaces is more difficult. The high surface and volume resistivities of such materials impede the flow of electric charge through them and across their surface. Furthermore, unlike a conductor, a non-conductor is not an intrinsic equipotential surface and the density of electric charge can vary across it. The energy of the spark cannot therefore be calculated from eq. (1). It depends, amongst other things, on the nature of the object to which the discharge is released (e.g., a pointed or round object), the nature of the non-conducting surface (e.g., its electrical resistivity, smoothness) and the uniformity of electric charge over the surface.

A limited amount of data has been published on the incendivity of sparks from the more common forms of non-conducting surfaces found in industry. It has been shown[11] that under certain atmospheric conditions, the discharge of 300 cm^2 of electrostatically charged polyethylene sheet to an earthed metal electrode can produce sparks with equivalent electrical energies up to 1 mJ and that these discharges can ignite solvent vapour–air atmospheres. Electric discharges have been reported[6] varying in energy from below 0·2 mJ up to about 100 mJ from the surface of high resistivity aviation fuel electrostatically charged by filtration. In laboratory tests Bruinzeel ignited pentane–air atmospheres with these discharges but he states that in the course of millions of fuelling operations of civil aircraft no explosions have yet occurred due to the discharge of static electricity generated during the fuelling operation. The non-incendivity of discharges from high resistivity liquid is considered[20] to be probably due to the fact that they take the form of a corona discharge in which the energy is released relatively slowly over a period and at any one time is insufficient to ignite the vapour–air mixture. However, it must be emphasized that the conditions under which the corona discharge may turn into an incendive spark discharge are not known. In powder handling systems electric discharges have been reported from powders stored in large containers but in the majority of cases it is not clear whether the spark originated from a charged metal object in or near the powder or from the powder itself. No comprehensive quantitative data have been published on the energy or incendivity of sparks from electrostatically charged powders. Experience with large-scale units involving the grinding and transfer of powders suggests, however, that under certain conditions the possibility of an incendive spark from the powder cannot be completely eliminated.[23]

Sensitivity of materials

The sensitivity of a material to ignition by an electric spark—its minimum ignition energy—can be determined by passing sparks of known energy through the material and noting the minimum energy required for ignition.[26] The minimum ignition energy measured for a flammable material depends to some extent on the test method used to determine it. For practical assessment of the electrostatic hazard, however, data published by national testing stations (e.g., Fire Research Station) or measured using the techniques adopted by these bodies, are sufficiently realistic for them to be used in assessing whether or not a spark from an electrostatically charged body could ignite the material.

Typical values of minimum spark ignition energies for the different forms of flammable materials are given in Table 9.3.

Table 9.3 Typical minimum spark ignition energy values

Flammable material	Typical minimum spark ignition energy values (mJ)
Sensitive detonator explosives	0·001–0·1
Vapour–oxygen mixtures	0·002–0·1
Vapour–air mixtures	0·1 –1·0
Chemical dust clouds	5 –5 000

Conclusions

Comparison of Tables 9.2 and 9.3 show that the accumulation of electrostatic charge on conducting bodies can lead to electric sparks of sufficient energy to ignite a wide variety of flammable materials. The ability of an electric conductor to release the maximum available energy in a short duration, concentrated spark makes the accumulation of electrostatic charge on a conductor the major electrostatic hazard in the majority of industrial operations.

Published data[6, 11, 23] show that under certain circumstances discharges from nonconducting bodies can ignite explosive materials, gas–air and vapour–air atmospheres and the possibility of ignition of the more sensitive chemical dust clouds cannot be excluded.

In any process the fire hazard from static electricity can be assessed by comparing the charge generated in the process and the possible spark energies with the sensitivity of the material to ignition by an electric spark. With certain approximations the necessary electrostatic calculations may be made relatively simply.[10]

Safety precautions against electrostatic hazards

Static electricity can cause a fire or explosion only under the following conditions:

(a) an inflammable or explosive atmosphere must be present;
(b) an electric charge must have been generated, must have accumulated on plant, product, or operator, and produced an electric field greater than the breakdown field strength of the surrounding atmosphere;
(c) the resultant spark must have an energy greater than the minimum spark ignition energy of the surrounding atmosphere.

An explosion or fire cannot occur if one of the above conditions is not satisfied.

In any situation the safety precautions must aim at the elimination of one or more of the above factors with the minimum interference with the operation of the plant and process. The general form of the precautions against electrostatic hazard may be itemized as follows:

(a) control of charge generation by suitable design of plant, and operating procedures;
(b) control of charge accumulation on plant, product, and personnel;
(c) elimination of flammable atmospheres;
(d) design of operating procedures to minimize the possibility of a spark.

We will first consider safety precautions that have virtually universal application and then others which have limited application and are best described with reference to specific industrial operations.

General precautions against electrostatic hazard

Earthing and bonding

The accumulation of dangerous electrostatic charge on an electrical conductor can be avoided by bonding, i.e., by providing an electrical connection of relatively low resistance between the point of charge generation and the point of charge accumulation. In practice, however, it is normally more convenient to prevent charge accumulation by earthing, i.e., by providing each conductor with a low resistance path to earth.

Earthing and bonding systems can also be required to give protection against electrical systems and equipment, and against lightning, as well as against static electricity. Here we are concerned only with the earthing required to give protection against static electricity.

Electrostatic charging rates rarely, if ever, exceed 10^{-4} A and therefore an electrical conductor has only to have a resistance to earth of less than about 1 $M\Omega$ (10^6 Ω) for dangerous accumulations of electric charges to be prevented.

Exceptions to this may occur in the explosives industry owing to the very low potential, e.g., 100 V required to initiate an incendive spark.

If, however, a conductor has a resistance of the order of 10^6 Ω then certain elements in the path to earth, e.g., grease, oil, paint, rust, or corrosion, may have an electrical resistance that varies with time and could exceed the safe value during the operation of the plant. It is therefore better to set the safe upper limit for the earth resistance at a level which will ensure a stable leakage path for electric charge. If the resistance to earth is only a few ohms then good contact is assured across all joints in the earth line.

A satisfactory earthing system is one that ensures:

    (a) the resistance to earth of all conductors will remain less than 10 Ω throughout the life of the plant;

    (b) the earthing system is recognizable as such and is replaced and checked after plant maintenance or modification;

    (c) the complete earthing system is checked at prescribed times, e.g., annually.

A satisfactory low resistance to earth may be achieved either through the normal construction of the plant and associated electrical equipment or by the provision of special metal earthing wires, and bonds across joints. Large plant items are normally earthed satisfactorily by the virtue of their mechanical connections to the steel framework of the building. A separate earthing system is more likely to be required for items subject to corrosion and for portable items.[24]

Prevention of charge accumulation on personnel

A man, insulated from earth by non-conducting footwear and/or floors, can, quite unknown to himself, become electrostatically charged and attain potentials of 10 000–20 000 V. He may become charged by generating static electricity on himself, e.g., by movement relative to his clothing, by the process of charge sharing, e.g., whilst sampling an electrostatically charged product, or by electrostatic induction, e.g., by standing near a charged object.

Man is essentially an electrical conductor and can be prevented from accumulating dangerous levels of charge by the use of low resistance footwear and floors. Specifications for such items are given in the relevant British Standards.[1, 2, 3, 4] Two types of product are available—anti-static and conductive. Anti-static footwear and floors aim to prevent the accumulation of charge on the man and to give protection against electric mains failure. Conductive footwear and floors, which are primarily designed for use in explosive handling areas, prevent the accumulation of static charge but do not give protection against electric mains failure. It is recommended that they should only be used in situations where special precautions are taken in the design, installation and maintenance of the electrical installation and equipment virtually to eliminate the possibility of electric shock and fire. Anti-static footwear made with controlling insoles may not, under certain conditions of use, give protection against electric shock.[4]

Many anti-static and conductive products are made from polymeric materials, e.g., rubber, and the resistance of such materials can change markedly during use. It is therefore necessary to check that the resistance of the material remains within the safe limits during use. The frequency of retesting is to some extent dependent on the type of product and the use to which it is put; with flooring an interval of 6 months is reasonable. Ideally footwear should be monitored at the beginning of each working period. The British Standards Institution test is unsatisfactory for the daily testing of footwear but a simple monitoring instrument operated by plant personnel can be used.[12] Such instruments are commercially available in certain countries, e.g., the U.S.A.

Ionization of the atmosphere

Electrostatic charge may be removed from a body by increasing the conductivity of the surrounding atmosphere by ionization. Air may be ionized by means of an electric discharge or by the use of radioactive isotopes.

Various forms of electrostatic eliminators based on electric ionization have been successfully used in the textile and paper industries. Before they can be used in inflammable atmospheres, however, it is necessary to be certain that the ionizing sparks and associated electrical equipment cannot themselves cause ignition of the inflammable atmosphere.

Radioactive ionization has been most successfully applied in processes where the rate of charge generation is low; the relatively low rate of ion formation limits its usefulness in situations where electrostatic charge is being generated quickly. Furthermore it can only be used where the necessary precautions against radioactive hazard can be taken.

Humidification

The possibility of dangerous accumulations of static electricity can be minimized by operating in atmospheres of relative humidity greater than 60–70 per cent r.h. High relative humidity aids the formation of conducting layers over the surface of objects which provide low resistance paths to earth.

Use of non-metallic materials

Many non-metallic materials, e.g., plastics, rubber, fibre-glass, are poor conductors of electricity: they readily become electrostatically charged and retain charge for some time. Normal earthing procedures do not prevent the accumulation of charge on insulating materials.

The general safeguard is to use non-metallic materials that have been made electrically conducting by the incorporation of additives. Specifications for footwear, flooring, transmission belts, conveyor belts, hoses, sheeting, tyres, and wheels made from low resistance polymeric materials have been defined by the British Standards Institution.[5] Normal earthing methods can be used to prevent the accumulation of static electricity on conducting versions of non-metallic materials. As with footwear and floors, checks on the resistance of the items are necessary at periodic intervals throughout their life. Test methods are described in the British Standard.[5]

Electrostatic hazards associated with particular operations

Flammable liquids

Processes generating static electricity.[6, 7, 8, 16, 17, 19, 20] (a) Pipeline flow: Charge separation occurs at the liquid/pipe wall interface and the charging current is approximately proportional to the second power of the flow velocity. Typical values are in the range 10^{-7}–10^{-10} A.

  (b) Filters, valves: Constrictions in pipe lines provide localized areas of high charge generation. Charge generation in filters can be as high as 10^{-5}–10^{-6} A.

  (c) Immiscible mixtures: The presence of an immiscible component, e.g., free water, in a liquid can produce up to a 50 times increase in the rate of charge generation.

  (d) Water settling: The settling of immiscible liquid droplets can produce high field strengths in storage tanks even after liquid has ceased to flow into the tank. Charge can also be generated by the release of gas through a liquid.

  (e) Free-fall of liquids: The free-fall of liquids tends to increase turbulence and emulsification—factors that stimulate charge generation.

  (f) Mixing and stirring.

Sensitivity of materials. The minimum spark ignition energies of all but the most sensitive flammable solvent vapour–air mixtures are in the range 0·1–1·0 mJ.

Safety precautions. (a) Earthing: All metal plant items should be earthed as described on page 137. For mobile containers, e.g., rail and road tankers, drums, the earthing connection is made before the mechanical connection of the hose and before liquid transfer starts. The earth connection is broken only after transfer has ceased and the hose has been disconnected. In systems employing cathodic protection, e.g., ships, care has to be taken that the earthing arrangements do not short circuit the insulating flange and introduce the possibility of a spark as lines are disconnected. Items on the shore side of the flange can be insulated *via* the shore earthing system, and items on the seaward side earth *via* the vessel.[18] Clips on the end of earth lines should be spring loaded and capable of breaking through rust or paint layers.[24]

(b) Use of ionic agents: If the electrical resistivity of a liquid is greater than 10^{11}–10^{12} Ωcm then it may retain charge for some time even when it is in direct stationary contact with its earthed container. Charge accumulation on high resistivity liquids can be prevented by the use of ionic agents to decrease the liquid resistivity. As little as 2–10 p.p.m. of Shell ASA-3 additive is sufficient to reduce the resistivity of hydrocarbon liquids to below 10^{11} Ωcm.[20, 27]

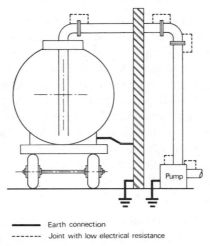

——— Earth connection

------- Joint with low electrical resistance

Fig. 9.1 Schematic diagram of typical earthing system for a road or rail tanker.

(c) Elimination of flammable atmospheres: A flammable atmosphere may be avoided by operating with the vapour concentration outside the flammable range, by the use of an inert gas, or, in the case of large storages, by the use of floating roof tanks. The flammability limits of vapour–air mixtures are published by the U.S. Bureau of Mines.[9] For reasons given, carbon dioxide is not recommended as an inert gas.

(d) Control of flow velocity: During the transfer of high resistivity liquids containing no immiscible components only low rates of charge generation are likely to occur if the flow velocity is below about 6 m/s. If an immiscible component is present flow velocities greater than 1 m/s are not recommended.[7,8,20]

(e) Vessel inlet and outlet points: Bottom inlet and outlet points or the use of diplegs with submerged ends are recommended to avoid charge generation by free-fall.

(f) Filters, valves: Localized areas of high charge generation can be minimized by the use of valves, etc., of maximum bore. The effect of high charge generation at filters, etc., can be minimized by locating them as far as possible from vapour–air atmospheres, i.e., remote from the entrance to storage tanks.[20]

(g) Level indicators: Float operated level indicators require robust earthing connections. If the dipping of storage tanks is necessary, this is best carried out in a dip pipe extending from the roof of the tank to near the bottom. Dipping with earthed metal tapes, etc., outside a dip pipe should not be carried out until after pumping has ceased and sufficient time has elapsed for any charge on the liquid to dissipate to earth.[20]

(h) Personnel: People handling flammable liquids can be earthed in the manner described.

(i) Use of non-metallic materials: As far as possible only non-metallic materials of low resistance should be used with flammable liquids. Heidelberg and Schon[15] have suggested the following size limitations on non-conducting plastic containers: flammable liquids with low ignition energy, e.g., carbon disulphide—maximum size of container 1 litre; other flammable liquids with flash points below 21°C—maximum container size 2–5 litres: liquids with flash points 21°C–100°C— maximum container size about 60 litres. New forms of non-metallic materials for use with liquids are continuously being developed and tested; the possible electrostatic hazards associated with these materials will have to be established when they become available commercially.

Powders and dusts

Many powders that are not classified as explosives can form dust clouds that burn with explosive violence when ignited in a confined space.

Processes generating static electricity. The majority of powders, including many containing metal elements, are poor conductors of electricity and generate static electricity when in contact with plant or other products. Processes generating static electricity include mixing, pouring, sieving, grinding, micronizing, pneumatic transfer and the movement of powders with scroll feeds, etc. The charge generated in a particular operation can be measured by collecting the powder as it leaves the system in a metal container insulated from earth. If the potential (V) developed on the container is measured with an electrostatic voltmeter then the charge (Q) on the powder can be calculated from the equation $Q = VC$ where C is the electrical capacitance of the measuring system. Levels of charge generation measured in industrial operations are given in Table 9.4.

Table 9.4   Electrification of organic powder in plant operations

Operation	Charge per kg ($C. kg^{-1}$)
Sieving	2×10^{-9}–2×10^{-11}
Pouring	2×10^{-7}–2×10^{-9}
Scroll feed transfer	2×10^{-6}–2×10^{-8}
Grinding	2×10^{-6}–2×10^{-7}
Micronizing	2×10^{-4}–2×10^{-7}

Sensitivity of materials. The minimum spark ignition energies of dust clouds (excluding those formed from explosives) vary from about 5 mJ to greater than 5000 mJ. Data on particular products have been given by Hartmann,[14] Maisey[22] and in numerous publications of the U.S. Bureau of Mines. It must be emphasized, however, that the minimum ignition energy of a dust cloud depends not only upon its chemical constitution but also upon its physical characteristics, e.g., particle size. Published data can only be used to assess the sensitivity of a material if it has been obtained from tests on samples of similar physical form.

Safety precautions. (a) Earthing: The earthing of metal plant as outlined on page 137 will prevent the accumulation of charge on the plant. Care is needed to provide earth continuity across joints where a powder layer may introduce a higher resistance element into the earth path. A typical earthing system for a grinding unit is shown in Fig. 9.2.

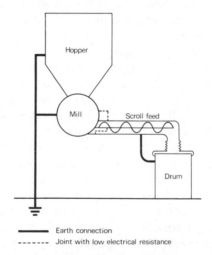

Fig. 9.2   Schematic diagram of typical earthing system for powder grinding unit.

(b) Removal of charge: There is no generally applicable conducting additive for use with powders to facilitate the dissipation of charge from the powder via the earthed plant. Humidification and air ionization can be used but they are subject to the limitations given on page 139.

(c) Elimination of flammable atmospheres: For every substance there is a dust concentration in air below which an electric spark cannot cause ignition. Typical values for this minimum explosive concentration are 10–15 mg/litre. Processes operating with dust concentrations below the minimum explosive concentrations are safe provided non-uniformity of dust dispersion cannot cause localized areas of high concentration. Inert gas can be used to eliminate the flammable atmosphere. This is most successfully applied in closed systems: in systems open at some point to the atmosphere considerable quantities of air can be entrained and carried into the system with the powder. An alternative inerting technique is to mix an inert dust, e.g., limestone, with the sensitive material. This can reduce the risk of an ignition and of a propagating explosion.

(d) Personnel: In situations where the accumulation of charge on a plant operative can result in a spark capable of igniting the dust cloud this hazard can be eliminated by the use of anti-static or conducting footwear and floors (see page 137). Care has to be taken that spillages of high resistivity powder do not effectively insulate an operative from earth.

(e) Use of non-metallic materials: There are few quantitative data on the electrostatic hazard associated with the use of non-conducting materials in the presence of dust clouds. It is unlikely that the use of non-conducting containers up to about $0.2 \, \text{m}^3$ capacity will introduce a hazard into a system provided that all the conducting items are earthed. It is good practice, however, to use low resistance non-metallic materials for very sensitive dust clouds, i.e., minimum ignition energy less than 10–20 mJ. Large electrical discharges are known to have occurred during the large-scale transfer of powders into non-conducting bulk storage systems. No quantitative data have been published on the fire hazard associated with these discharges. In any situation an assessment of this hazard can be made by comparing the electrical energy introduced into the storage system by the charge powder with the energy required to ignite the powder.

(f) Protection of equipment against explosion damage: in large-scale powder handling systems it is often necessary to base safety not on the complete elimination of all sources of ignition but on the provision of explosion venting or explosion suppression equipment. The design of explosion vents for powder systems is a complicated procedure[21, 22] but a useful practical guide has been given by the U.S. National Fire Protection Association.[25] The area of vent required to protect a system depends on the volume and form of the dust enclosure and upon the explosive characteristics of the dust. Typical values are $0.1 \, \text{m}^3$ of vent area for every 0.2–$3.0 \, \text{m}^3$ of enclosure. Explosion suppression devices rapidly inject an inert liquid or gas into the system at the start of an explosion, suppress it, and prevent the development of dangerous explosion pressures. The most appropriate explosion suppression equipment for a plant is normally decided by the supplier of the equipment.

Explosives

Explosives are recognized as sensitive materials and precautions are taken against ignition by any means during their manufacture and storage. The minimum ignition energy of sensitive explosives can be as low as 0·001 mJ and stringent precautions are necessary therefore to avoid the accumulation of static electricity in their presence. This is normally achieved by the efficient earthing of the plant and personnel and by placing restrictions on the use of non-conducting materials that could retain electrostatic charge.

Gases

Pure gases charge less readily than either liquids or solids. If a gas contains solid or liquid particles, however, it will generate static electricity on being forced through an orifice. Electric currents of the order of 10^{-5}–10^{-7} A can be generated by releasing wet steam, carbon dioxide, or propane from a high pressure enclosure through a narrow crack or nozzle. Wet steam contains water droplets. Both carbon dioxide and propane cool when expanding rapidly; this causes solid particles to be formed in carbon dioxide jets and liquid droplets to be formed in a propane jet.

Electrostatic charge generated in this way is a known cause of fires. The fire is normally caused by a spark from an insulated conductor in the path of or near the electrostatically charged gas stream.

An example of this is the fire hazard caused by steam leaks in flammable areas; very often the wire enclosures protecting asbestos lagging are not efficiently earthed, they accumulate charge from the steam jet and become the source of an electric spark.

The primary safeguard against this form of hazard is the earthing of all the electrical conductors in or near the gas stream, e.g., the metal nozzle, any metal plant, and personnel.

It is the generation of electrostatic charge during the expansion of carbon dioxide that makes this gas unsuitable for the rapid inerting of flammable areas.[20]

Conclusions

Static electricity is generated in many industrial processes. In the majority of situations the major fire hazard arises from the accumulation of charge on electrical conductors and the primary safeguard is therefore the earthing of all metal items of plant. The magnitude of any hazard from other sources such as the accumulation of charge on personnel, products, non-conducting materials, etc., can be assessed by comparing the possible spark energies with the minimum spark ignition energies of the flammable materials in the plant. In any particular plant the hazard may be eliminated by selecting the combination of safety precautions that is most appropriate to the plant.

References and Bibliography

1. British Standards Institution. *Electrically Conducting Rubber Flooring*. B.S. 3187: 1959.
2. British Standards Institution. *Anti-static Rubber Flooring*. B.S. 3398:1961.

3. British Standards Institution. *Electrically Conducting Rubber Footwear*. B.S.3825: 1964.
4. British Standards Institution. *Anti-static Rubber Footwear*. B.S. 2506:1964.
5. British Standards Institution. *Electrical Resistance of Conductive and Anti-static Products Made from Flexible Polymeric Material*. B.S. 2050:1961.
6. Bruinzeel, C. Electric Discharges During Simulated Aircraft Fuelling. *J. Instit. Petroleum*. **49**, No. 473, May, 1963.
7. Bustin, W. M., Culbertson, T. L., Schleckser, Jr., C. E. General Considerations of Static Electricity in Petroleum Products. *Proc. Amer. Petrol. Inst.* **37**, Part III, 1957.
8. Bustin, W. M. Static Electricity Safety Studies—Marine Operations. *Proc. Amer. Petrol. Inst.* **43**, Part III, 1963.
9. Coward, H. F., Jones, G. W. Limits of Flammability of Gases and Vapours. *U.S. Dept. of Interior, Bureau of Mines Bulletin*, 503, 1952.
10. Eichel, F. G. Electrostatics. *Chemical Engineering*, **74**, 6, 1967.
11. Gibson, N., Lloyd, F. C. Incendivity of Discharge from Electrostatically Charged Plastics. *Brit. J. Appl. Phys.* **16**, 1965.
12. Gibson, N., Lloyd, F. C. The Testing of Anti-static and Conductive Footwear. *The Brit. J. Industrial Safety*, **7**, Autumn, 1966.
13. Harper, W. R. *Contact and Frictional Electrification*. Oxford University Press, 1967.
14. Hartman, I. *Industrial Hygiene and Toxicology*, Vol. 1. Explosion and Fire Hazards of Combustible Dusts. Interscience Pub. Inc., New York, 1958.
15. Heidelberg, E., Schon, G. *Die Berufgenossenschaft Betriebssicherheit*. Explosion Hazards due to the Electrostatic Charging of Plastic Containers. July, 1960.
16. Herzog, R. E., Ballard, E. C., Hartung, H. A. Evaluating Electrostatic Hazard During the Loading of Tank Trucks. *Proc. Amer. Petrol. Inst.* **41**, Part VI, 1961.
17. Holdsworth, M. P., van der Minne, J. L., Vellenga, S. J. Electrostatic Charging During the White Oil Loading of Tankers. *Proc. Amer. Petrol. Inst.* **41**, Part VI, 1961.
18. The Institute of Petroleum: *Electrical Safety Code*. The Institute of Petroleum, London, 1965.
19. Klinkenberg, A. *Advances in Petroleum Chemistry and Refining*, Vol. 8. Theoretical Aspects and Practical Implications of Static Electricity in the Petroleum Industry. Interscience Publishers, 1964.
20. Klinkenberg, A., van der Minne, L. J. *Electrostatics in the Petroleum Industry*. Elsevier Publ. Co. Inc., Amsterdam, 1958.
21. Maisey, H. R. Gaseous and Dust Explosion Venting Part 1. *Chemical and Process Engineering*, September, 1965.
22. Maisey, H. R. Gaseous and Dust Explosion Venting Part 2. *Chemical and Process Engineering*, December, 1965.
23. Muller-Hillebrand, D. Electrostatically Initiated Explosion of Sugar Dust in Air. Investigations at Kopingebro in 1961, *Socker Handlingar II*, **18**, No. 3, 1963.
24. National Fire Protection Association. *Static Electricity*. National Fire Protection Association, Boston, U.S.A. *NFPA No. 77M*. 1961.
25. National Fire Protection Association. *National Fire Codes*, Vol. II. Combustible Solids, Dusts, Chemicals and Explosives. National Fire Protection Association, Boston, U.S.A., 1962–63.
26. Raftery, M. M. *Explosibility Tests for Industrial Dusts*. Fire Research Technical Paper, No. 21. H.M. Stationery Office, 1968.
27. Shell Chemical Co. Ltd. *Safety in Fuel Handling: ASA-3*. Shell Chemical Co., Ltd., London, 1963.

10. Grinding

K. B. Southwell

Technical Marketing Manager
Universal Grinding Wheel Co. Ltd

The past decade has seen considerable changes in the application of grinding wheels throughout industry. These changes have brought about increases in wheel speeds in some application areas, whilst in others the development of new machines and grinding techniques has seen the grinding wheel at last demonstrate its potential to compete with the lathe and milling machine in obtaining high metal-removal rates.

These increased demands on the grinding wheel have been met by the manufacturers without sacrificing their high safety standards. Where failures do occur a very high percentage of the cases are caused by incorrect use, incorrect mounting, bad handling or wrong storage.

In the following pages it is intended to highlight the major causes of wheel failures and how they can be prevented.

Identification and selection of abrasive wheels

To understand what contributes to a wheel failure and to appreciate the need for restrictions on the maximum operating speed and for other safety precautions, it is essential to know the general composition of grinding wheels and the system by which they are identified.

Grinding wheels fall into the category of bonded abrasive products, and as such consist of abrasive particles held together by a suitable bonding medium. The abrasive grain does the actual cutting and clearly the type of abrasive chosen must be the one most suitable for the particular application concerned. However, the performance of the abrasive in grinding is also governed by the bonding characteristics of the product, which must therefore be designed to suit the job in hand.

Abrasives

Aluminium oxide. This is always denoted by the letter 'A'. The 'regular' variety of this abrasive is produced by the fusion and controlled reduction of the naturally occurring raw material bauxite. This abrasive is of a tough nature and is principally used for grinding and cutting materials of high tensile strength. Regular aluminium oxide is normally brown in colour, but can turn blue under certain treatments or conditions of wheel manufacture.

Other forms of aluminium oxide abrasives are produced by the fusion of chemically refined bauxite, and these vary in colour, the majority being white or pink. The manufacturers use a further symbol prefixing the latter 'A', i.e. WA, 32A, PA etc. to denote a difference from the regular variety. These abrasives have different chemical and physical characteristics which result in a more friable, harder abrasive, which is particularly suitable for surface grinding tools and cutters, and for most applications where the arc of contact between wheel and work is large, and they are specially effective on heat-sensitive hardened steels.

Silicon carbide. This is always denoted by the letter 'C'. This abrasive is made in an electric resistance furnace from high-grade silica sand and petroleum coke. Silicon carbide is a harder abrasive than the aluminium oxide types and is more friable. It is most suitable for grinding hard materials and those of low tensile strength, such as tungsten carbide, copper, brass, glass, porcelain.

Grit sizes

The sizes of the grits used are classified according to the sieves on which they are retained. Sieve numbers represent the number of apertures per linear inch, and therefore small grit numbers correspond to coarse grits and vice versa. The standard range is 8–1 200.

Coarse grits ranging from 8 to 30 are predominantly used on abrasive machining operations, whilst for precision grinding applications the call is for 36–100, and for form and thread grinding 100–220 grit sizes are mainly used. Finer grits are used for superfinishing operations.

Grade

The grades of grinding wheels range from E (soft) to Z (hard). This designation refers to the tenacity with which the bonding material holds the abrasive particles in place and not to the hardness of the abrasive itself.

Structure

This is denoted by a number which ranges from 1, which represents a dense structure with a high grit concentration, to 16– open structure with a low grit concentration.

Bonds

Bonds are the media by which the abrasive is held together in a carefully controlled matrix to give the requisite strength and grade applicable for the many different applications. There are four main bond types, as follows.

Vitrified. This is identified by the letter 'V'. Vitrified bonds are glasses or vitreous porcelains, and are formed in situ during the firing stage of the wheel manufacturing process. Vitrified bonded wheels excel in precision grinding operations where form and dimensional accuracy of the finished component are of paramount importance.

Resinoid. This is identified by the letter 'B'. Resinoid bonds are blends of synthetic resins and fillers. The majority of resinoid wheels are used for heavy duty work such as abrasive machining and are employed on machines of high power and at relatively high speeds.

Rubber. This is identified by the letter 'R'. Natural and synthetic rubbers are blended to produce the bonding media for this range of products which are mainly confined to cut-off, centreless control wheels, camshaft finishing and other operations where good finishes, strength of product, and self-dressing properties are required.

Shellac. This is denoted by the letter 'E'. This is a natural resinous substance which is used as a thermoplastic moulding material. Its use is mainly restricted to special finishing operations and parting off wheels. Its chief characteristic is coolness of cut and ability to impart high finishes.

Handling and storage

Grinding wheels must always be handled carefully. Wheels are fragile to a degree which varies according to the specification, and great care should be exercised in handling and storage to prevent damage.

Often wheel breakages can be attributed to careless handling and storage. Wheels damaged from these causes may appear perfect yet represent a danger to operatives. Manufacturers inspect and test wheels prior to packing, ensuring that they are despatched in perfect condition.

Handling

The following rules are based upon experience and should always be observed.

(a) Handle wheels carefully and prevent dropping or bumping. If a wheel is dropped or bumped do not use. It is better to scrap or return to the manufacturer for testing than risk injuring an operator.

(b) Do not roll wheels (hoop fashion).

(c) Use trucks or suitable conveyors which will provide support for transportation of wheels which cannot be carried by hand.

Inspection

On receipt wheels should be inspected carefully to determine that they have not been damaged in transit. As an added precaution wheels should be subjected to the 'ring test', which is carried out in the following manner.

Ensure that the wheel is dry and free from loose packing material, e.g. sawdust. Small wheels may be held by slipping one or more fingers through the hole; the wheel should then be tapped gently, using the wooden handle of a small tool such as a file or screwdriver, 45° either side of the vertical centre line, 25 to 50 mm from the periphery. Rotate the wheel 45° and repeat test. Heavy wheels should be allowed to rest on edge in a vertical position on a clean hard floor and gently tapped with a wooden mallet. A clear bell-like tone is produced by sound wheels, but a dull sound instead of a clear ring is produced by cracked ones. If wheels are struck along the vertical centre line, the 'ring', even in a sound wheel, is sometimes muffled and may give the false impression that the wheel is cracked.

It should be noted that organic bonded wheels do not emit the same clear ring as do vitrified wheels.

Storage

All abrasive wheels should be stored under responsible supervision in suitable racks, bins or drawers in a dry area not subjected to extreme temperature changes, since organic bonded wheels are affected by excessive humidity.

Care should be taken to use wheels on a correct rotational basis, to minimize the length of time a wheel is stored as the shelf life of organic bonded wheels may be affected by oxidation if stored for a long period. Where there is any doubt, or when these types of wheels have been stored for over two years, the manufacturer should be consulted before use.

The type and size of wheels to be stored should determine the design of the racks. Generally the following suggestions should be observed.

(a) Thin organic bonded wheels should be stored away from excessive heat on a perfectly flat horizontal surface. Washers should not be interposed between stacked thin wheels.

(b) Straight or tapered wheels of appreciable thickness are best supported on edge in racks. Such racks should be constructed to provide two-point cradle support to prevent rolling, and with frequent partitions to allow wheel selection with the minimum of handling or tipping of the wheels.

(c) Cylinder wheels and large straight cup wheels may be stacked on flat sides with cushioning material between them; or may be stored as wheels in section (b).

(d) Tapered cups should be stored in stacks, preferably no more than six high with adjacent faces matching.

(e) Other small wheels may be stored in boxes, bins, or drawers.

Mounting of abrasive wheels

A high percentage of wheel failure can be directly or indirectly attributed to failure to carry out the correct mounting procedure.

It is an offence for any person to mount a wheel unless he has undergone the required training and is deemed competent by the management of his place of employment to carry out that duty in respect of the class or description of abrasive product to which that wheel belongs.

It is essential that the operator/wheel mounter should be capable of interpreting clearly the specification marked on the wheel and establishing its suitability for that grinding operation, always observing the manufacturer's recommendations. Abrasive wheels are marked with their maximum permissible operating speed in revolutions per minute, and this should be checked with the spindle speed of the machine on which it is to be mounted. The size of some wheels prevents marking and in such cases a notice must be displayed which clearly states the speed for this class and size of wheel.

The design, manufacture and maintenance of the flange plates is of paramount importance in preventing accidents. For plain-sided wheels they should be of good quality mild steel, being of a minimum size of one third the diameter of the wheel. Inner flange plates should be fixed to the spindle and the clamping surface should run true to the spindle. Inner and outer plates must be of equal size and equally recessed or undercut, except for the single type flanges used with threaded hole wheels.

A summary of the correct mounting procedure and precautions that should be noted is as follows.

(a) Wheels should be mounted only by fully trained, competent and certificated persons.

(b) A wheel should only be mounted on correctly designed and built machines, and the manufacturer's specified limits on sizes should be followed.

(c) Every wheel before mounting should be closely inspected and the ring test carried out.

(d) The specification and maximum operating speed on the wheel should be checked with the grinding operation and the spindle speed of the machine.

(e) Examination of the flange plates should be carried out as previously detailed and a check made for damage, distortion or foreign matter.

(f) If a bush is used, a check should be made that it does not project beyond the wheel width.

(g) The machine should be isolated electrically before fitting the wheel.

(h) The wheel should be mounted on the spindle or wheel arbor, with blotters of compressible material, between the wheel sides and the clamping flanges. The blotters should be slightly larger in diameter than the wheel flanges.

(i) Protection flanges designed for use with taper-sided wheels should have the same degree of taper on wheel and flange. Blotters should not be fitted to these wheels, nor to mounted points and wheels, abrasive discs, plate-mounted wheels, chuck-mounted cylinders, cup or segmental wheels, depressed centre wheels, thin rubber-bonded slitting wheels 0·5 mm and thinner and diamond wheels, except certain vitrified bonded types.

(j) The wheel should fit freely but not loosely on the spindle or the wheel arbor.

(k) The wheel flange should be tightened only sufficiently to grip and drive the wheel without slippage. When there are a number of screws, they should be tightened uniformly in diametric sequence as for a car wheel or cylinder head.

(l) Screws for inserted nut discs, cylinders and cones should be of sufficient length to secure the product, but should be restricted in length to prevent them fouling the bottom of the insert.

(m) Where applicable the correct balancing procedure should be carried out.

(n) In all cases the wheel guard should be secured and adjusted, and also where applicable the work rest.

(o) On precision and off-hand machines the wheel should be dressed in the appropriate manner until it is running true and the surface is in the required condition for grinding.

(p) When mounting or using mounted wheels or points, it must be checked that the overhang is correct for the speed selected for use with that size of wheel and diameter of spindle.

Operational hazards

The hazards involved in the use of abrasive wheels are mainly concerned with wheel breakages. One of the major problems that contributes to accidents from grinding wheels is the grinder or machine operator not adhering to his training and to general engineering safety procedures. It should be remembered that all grinding wheels will break, and that even at the lowest speed used on precision machines, the wheel is attaining a peripheral speed of 30 m/s, or to put it in more recognizable terms, the top speed at which a car can travel on a motor-way, 70 mi/h (108 km/h). The most common cause of wheel breakages can be associated with the following.

    (a) Damage during storage.
    (b) Incorrect mounting.
    (c) Incorrect speed of rotation.
    (d) Incorrect use.
    (e) Accidents.

Storage and mounting

A high percentage of wheel failures occur on the initial run up to operating speed, and this points to damage during transportation, incorrect storage or incorrect mounting. The section on correct storage and mounting has already been covered, but it should be remembered by all persons responsible for storage of abrasive products that if there is any doubt about a wheel, then it should not be issued. *Remember a faulty wheel can kill or seriously injure a grinder or operator.*

The condition to be avoided in mounting a wheel is one which puts a wheel under additional stress to that set up from normal usage.

Speed of rotation

As the speed of a wheel increases, the stress in the wheel increases at a rate proportional to the square of the velocity. The important velocity is that measured at the periphery. Manufacturers of abrasive wheels carry out speed tests of wheels at higher speeds than the maximum operating speed quoted on the wheel, but it is essential that this quoted figure should not be exceeded when the wheel is new and of maximum diameter. The number of revolutions per minute may be increased proportionally as the wheel diameter is reduced by wear, so long as the initial peripheral speed is not exceeded.

Incorrect use

Normal work pressure, applied as intended in an approved grinding set-up, does not produce serious mechanical stress in a wheel, but excessive work pressure could prove dangerous, and work should not be forced against a 'cold' wheel, but should be applied gradually allowing the pressure to build up slowly. On vitrified wheels heavy continuous pressure must be avoided when dry grinding. With mounted points or wheels, excessive pressure can be a source of danger, through bending or fracture of the spindle. If excessive pressure is required, then a change to a freer cutting wheel should be made.

The cause of stress due to heat results from variation in temperature within the wheel structure. During grinding, the mass of the periphery of the wheel is of a higher temperature than the mass adjacent to the bore. This results in compression in the outer zone and tension in the inner zone. If the temperature gradient is excessively steep, a radial crack may start from the bore, this type of failure applying mainly to dry grinding operations such as snagging or off-hand grinding.

On portable grinding machines the operator is in a vulnerable position, and strict attention to safety factors already covered is especially important. There is a tendency to treat these machines roughly, and although the majority of them use reinforced organic bonded wheels, they can still be cracked or damaged if dropped just a short distance onto a hard floor. All portable machines should be handled with respect and not thrown on the floor at the end of an operation.

As with the larger machine tools, wheel speed is a critical factor. On compressed-air type machines regular checks must be made to ensure that the speed quoted on the machine is not being exceeded, and planned maintenance should include special attention to the speed governor.

Accidents

Accidents to operators' eyes from grinding detritus can be avoided by wearing safety spectacles. The regulations on Health and Safety require that safety spectacles of the approved B.S.I. standard must be worn by operators on all dry grinding and cutting-off applications when using fixed or portable machines. This also applies when dressing an abrasive wheel or product.

The floor surrounding every fixed machine on which abrasive wheels are mounted must be maintained in a good and even condition, and as far as is practicable be kept clear of loose material and prevented from becoming slippery. For portable machines all reasonable practicable steps must be undertaken to comply with the above regulation.

On precision machines accidents account for a fair proportion of failures and injuries. These are the result of neglect and/or disregard of safety and good grinding practice.

Before starting a machine, ensure that the table traverse stops are accurately set, the workpiece is securely located, and the wheel and coolant pipes are clear of the workpiece. Infeed, downfeed and crossfeed rates should also be carefully checked.

When a grinding fluid is used, the supply should be turned on only after the wheel has reached operation speed, and should be turned off before switching off the wheel spindle. Grinding fluid running on a stationary wheel will result in an out of balance condition.

In off-hand grinding the work-rests must be adjusted close to the wheel and securely clamped after each adjustment. A maximum distance of 3 mm between wheel and rest is recommended, which will prevent the work being dragged between them. Never hold a workpiece in a piece of cloth or rag, and if some protection of the hands is required, use protective gloves.

It is most important to avoid any personal contact, or to allow clothing to come into contact with a moving wheel or workpiece. A loose necktie, shirt sleeve, overall cuff or long hair caught in either the wheel or workpiece can result in serious injury or fatality.

All machines must be provided with the correct wheel guards, protection flanges and work rests. It is an offence for any person to wilfully misuse or remove any of these appliances. Efficient controls for starting and stopping a machine must be positioned so as to enable the operator to readily cut off the power to the machine.

Bibliography

1. *Abrasive Wheel Regulations.* S.I. No. 535 Department of Employment. H.M.S.O., London, 1970.
2. *Exemptions Certificates.* Issued by H.M. Chief Inspector of Factories under Regulation No. 5 (Abrasive Wheel Regulations 1970 No. 535). Obtainable from H.M.S.O., London.
3. *Training Advisory Leaflet No. 1: Abrasive Wheel Regulations 1970.* S.H.W. 11 (Advice on mounting straight-sided abrasive wheels not exceeding 250 mm diameter). Department of Employment. H.M. Factory Inspectorate.
4. *Training Advisory Leaflet No. 2 Abrasive Wheel Regulations 1970.* S.H.W. 12 (Advice on mounting straight sided abrasive wheels exceeding 250 mm in diameter, tapered wheels, bonded abrasive discs, cylinder wheels, cup wheels, cone wheels, depressed centre wheels and cutting-off wheels). Department of Employment. H.M. Factory Inspectorate.
5. *Health and Safety at Work.* H.S.W.4—A (Safety in the Use of Abrasive Wheels). Department of Employment.
6. *Bonded Abrasive Products.* B.S. 4481:1969: Part 1 (metric units). General features of grinding wheels, blocks and segments. British Standards Institution.
7. *Bonded Abrasive Products.* B.S. 4481:1972: Part 2 (metric units). Dimensions of grinding wheels and grinding segments. British Standards Institution.
8. *Dimension of Hub Flanges.* B.S. 4581:1970 (metric units). Mounting of plain grinding wheels. British Standards Institution.
9. *European Safety Code F.E.P.A.* No. 12–GB–74. For the use, care and protection of abrasive wheels. The Abrasives Industries Association.
10. Protection of Eyes Regulations 1974 S.I. 1681. H.M.S.O., London.
11. *Grinding Wheel Application Catalogue.* Universal Grinding Wheel Co. Ltd, in the course of preparation.

11. Working with machinery

R. Bramley-Harker

Formerly Group Safety Adviser,
Wiggins Teape Ltd

Man is distinguished from other animals in that he alone of the living world uses artificial aids to his work and living; in the twentieth century the ingenuity of invention has given us a wide choice of tools and machines to be used by unskilled workers. Handwork or craftsmanship is looked upon as a superior skill, and machine work as a substitute for the best—a little inferior in some minds. Yet but for mechanical invention, living standards would not have moved very far away from those of the cave where all work (and production) was manual. We live by machines and machinery; we work with machinery.

There is no doubt about our capacity to work and live with machinery; yet there is a doubt—a lack of knowledge—of the special problems of the use of it. The many personal injuries, some leading to death, confirm that we do not fully understand the inventions for which we are responsible, and that we are not in complete control of the machinery we use. By its nature and use there is conflict between the person employed with machinery and the machine itself. Every piece of progress made down the centuries in machine design, invention, and use has led to the injury, and sometimes the death, of designers, makers, or users. Working with machinery is dangerous; control of danger is essential if we are to live with and profit by the use of machinery.

Guarding of machinery

In the U.K. statute law has been a major factor in the control of danger; for more than 100 years the Factories Acts have successively prescribed standards of protection from the dangerous parts and legal interpretation from the Courts has established a rigid field of compliance. The simple term 'secure fencing of the dangerous part of any machinery' has become an involved argument which finishes always in a simple conclusion: *Unless machinery is safe or made safe its use is prohibited.* This is despite the conduct of the person who is at risk; the statutory requirement is that the machinery is safe without regard to the qualities of man. This is a sound moral approach, for there is no quality control of man and it would be invidious to select any particular strata as worthy of protection against danger and to leave others unprotected. Should we protect the knowledgeable or the ignorant, the obedient or the disobedient, the careless or the prudent, the intellectual or the innocent? The law relieves us of such decisions and gives us a standard applicable 'to every person employed or working'.

Hence, what has become a standard term of reference—'guarding of machinery'—is used to indicate that which is required to prevent injury to those employed in connection with it. Each machine designer strives to produce machinery which will perform its desired function without mishap to the machine or to people; there must be 100 per cent accomplishment of this aspect if accidental personal injury is to be avoided. The designer should not give scope for those who use his machine to have to apply *their* experience and skill in using it safely; the designers' duty is not carried out if there is need to guard dangerous parts of the machinery. The designer knows what he intends his machine to do; in general he knows little of what is done and how it is done. The user has the advantage of deciding how, when, and for what purpose he will use the machine; he knows of the danger of the machine–person relationship. This is the danger which must be controlled. If a machine is so segregated from the person that contact of any sort is impossible, there is little danger to that person. This is why those engaged in accident prevention emphasize over and over again that the complete enclosure of the source of danger so as to exclude effectively the person who may be at risk, is the best method of controlling danger. Secure fixed fencing of the dangerous parts—inrunning gear wheels, revolving shafts, parts of non-operative equipment such as driving motors and their belts—is the only satisfactory method, and is the legal method. Even this method of preventing accidents must be designed and fitted having in mind subsequent lubricating, repair, and maintenance work, and the simple mechanical stress to which it is exposed. Fixing must be secure—to be removed only by using tools—its structure strong enough to withstand ordinary mishaps and misuses.

Fixed guarding of dangerous parts which does not enclose completely the dangerous parts may be equally effective in the prevention of injury. Care must be taken to ensure that there is no access through any opening which will permit the worker to gain contact with dangerous parts of the machine. This method is used where it is necessary to feed a component or material to a die and tool, or in a work area like inrunning intakes of calender machines. The space left unprotected must be big enough to permit access of the material, yet effectively exclude the hand. The power press is a typical example of the tool and die machine; an enclosing guard fixed in position permits piece parts to be fed either by hand or by mechanical appliance to the danger area, but effectively excludes the hand. A set of inrunning rolls such as a calender on a paper-making machine excludes the fingers of the hand by the adjustment of the round bar or angle bar in the intake, yet permits the passage of paper. These devices can be described as making the dangerous parts safe by position: they are not securely fenced, but safety is accomplished by the positioning of fencing or other structures to exclude the person, or his hand (see also pages 253–5).

Safeguarding by other than fencing

If all dangerous parts were required to be made safe by secure fencing it is certain that the commercial use of much useful machinery would be prohibited. So there are available a large number of devices and expedients which are equally satisfactory in the prevention of accidents. Feeding devices such as slides, chutes, magazines, and rollers guarantee the remoteness of the operator from the danger point.

Machinery and production are so diverse that even these methods are not sufficient to control the dangers. It happens sometimes that a worker is in such a position that in the event of personal error, misjudgement, or machine mis-operation (as by repeat stroke in the case of a power press), he is required to be 'rescued'. There are occasions when the use of a machine precludes any secure fencing and the more usual methods of ensuring operator safety. It is in these fields that other methods of safeguarding have been developed.

Interlocking devices

These devices are used to ensure that whenever access to the danger point is possible, the machine (its dangerous parts) cannot be in motion, and that until the secure fencing has been replaced in an effective position to exclude an operator, the machine cannot be set into motion; once the machine is in motion, the fencing cannot be removed until the machine has come to rest. These devices are often complicated, require skilled and knowledgeable design, and skilled and constant maintenance.

Trip devices

These devices are used in connection with dangerous parts of the machinery where practical, and where the fixing of secure fencing is precluded by the very processes for which the machine was designed. A typical example is the hand-fed printing machine, in which a sheet of paper is placed by hand upon a platen which is then mechanically driven towards a type-holding forme; on the completion of the printing the platen reverses, the printed paper is removed by hand and the next sheet of paper put on it. For accuracy of register, little or no protection can be provided and it is in effect a race between the trapping parts of the machine and the skill of the operator in removing the hand. If the race is won by the machine, it can still be safe if a tripping device is in place which, by contact of the hand, withdraws the power from the machine and applies a brake. Such devices are often used in connection with partial fencing (distinguished from secure fencing) and with distance fencing which permits operation of the process far enough away to ensure safety, and controls inadvertent or other approach to the danger point. The devices are sometimes wholly mechanical, but there has been developed a use of photoelectric cells to control the operation of the machine, and also proximity capacity sensitive devices in a similar design. In these applications there is one feature which differentiates them from the fixed guard or wholly mechanical designs: there is no outward and visible sign of the effectiveness of the device. A fixed barrier can be seen and its effectiveness assessed. A light beam can be seen but its effectiveness cannot. A capacity device cannot even be seen. Nevertheless this defect has not prevented their effective development; devices are available which automatically monitor by control circuits to ensure that the device will answer an emergency call. If this response is positive, the machine controls come into operation; otherwise there is 'failure to safety'. Here again the maintenance and check systems used in connection with their use are a paramount feature of their success.

Automatic guards

Processes on some machines involve complete exposure of the dangerous parts so that fixed guards, interlocking, and trip devices are not practicable. It is then that the rescue apparatus or so-called automatic guards are used. Their function is simple: a person (or his hand) is in danger—they remove him forcibly. Such a device requires careful design. Its operation must not be so fierce as to inflict injury itself; it must be able to work within a field in which there is adequate time and space for it effectively to perform its function; and its design must incorporate a safety factor so that it is always effective within the field of application. Here again, maintenance is a vital feature of the success of its use.

Maintenance codes of practice

With all machinery there is the problem of maintenance; with guards, safeguards, and guarding systems it is a major problem. In the U.K., lack of effective maintenance is viewed under Statute Law as a breach of the prime requirement: that of provision. Whenever any safety device is used, be it a simple fixed or more complex guard, it is essential that maintenance be provided, organized, and supervised.

With non-complex guards, like enclosures or encasements, purposeful careful inspection by supervisory staff is sufficient, but it should not be left at that. Management should lay down and enforce regulations regarding the period of inspection, the method of reporting and the correction of faults, and a record system which will ensure complete knowledge of the position by everyone concerned: worker, supervisor, and management.

The more complex the safeguarding system, the more onerous the duty of maintenance and repair becomes, the more experienced and skilled must be the person employed to inspect, examine, and repair. It is essential that codes of practice detailing the duties to be performed and the records to be kept should be enforced. The Power Presses Regulations 1965, detailing duties to be performed in connection with the guards for the tools and dies of power presses goes further: it is prescribed that each person engaged in the setting of tools on the press must have been satisfactorily trained to the syllabus of training prescribed, and that not only must maintenance of the guards be ensured but that those parts of the machine, such as clutch and brake which have a vital connection with the safe use of the machine, must be periodically examined by a competent person, and adequate maintenance ensured. It is in this field that there will be considerable future development; I believe it is imminent.

Every paper-cutting guillotine is protected either by an automatic guard or by a photoelectric cell. The automatic guard must have a minimum standard of performance; it becomes less effective in use due to wear and tear. It must be checked periodically and at once maintained to be effective. The photoelectric cell device and its use on guarding systems requires even more attention. Despite the automatic self-monitoring circuits which are (or should be) incorporated in all such designs, there is great dependence for effective safe control on the clutch and brake of the machine itself. Often different manufacturers have been involved: the guard maker is not fully conversant with the clutch and brake design, and certainly not conversant with the rate of deterioration or the performance in production use. The use of a code of practice is essential to ensure competent periodical examinations, with written records of what is discovered and what is necessary to ensure effective safe working. Record keeping is an essential part of this code of practice and great care should be taken to ensure that the person selected for the task is competent. Electronic circuits are not clearly understood by all maintenance engineers and the neglect of things not understood is quite a common cause of accident. Even simple electrical faults are often not understood.

Education and training of the worker

Workers are no longer expected to understand the complexity, and consequent habits, of machinery without education and training. This training must include purposeful education in industrial accident prevention. The advancement in the design and development of mechanical and electrical devices has left a large amount of acceptance of performance in the mind of the worker without any idea of the reasons, the methods, the failures and faults, and the consequences. No worker should be put to work upon a machine unless he knows the full intent of the safety devices provided to control dangers and, more than that, has sufficient knowledge to recognize the failures which lead to inefficiencies in the safety devices.

Probably the king-pin in this field of accident prevention is the first line supervisor. Many workers perform tasks only of very moderate skill and they are required only to supervise that a machine is provided with the material upon which it is working and that the product conforms to the specification prescribed. The interest in their own safety is marginal; their capacity to understand what is happening is not very well based; and their interest probably less well versed. The supervisor becomes the accident prevention officer; what the operator lacks he must have, and his training must include not only supervisory duties, but full implications of the risks of industry and the safe practices required. He should be helped by codified conduct which entails record keeping and report writing so that the safety officer and other management staff can assess whether he is performing his duties satisfactorily.

The machinery, its purchase, its performance

We do not select our workers to an exact specification. At best we can select only within a wide field and then train and educate to the use for which we wish to put them. On the other hand, we have almost full control of the machine; we buy it because it fits our requirements; we often quote exact performance standards. We forget, neglect, or are not very precise, however, about our safety standards. There is indeed a serious divergence among machine designers and buyers of what is a desirable standard and how it should be attained. The same care in specification of safety standards must be taken as in production standards; perhaps a higher care. This involves preknowledge of how machinery is to be used and where. Is the layout in the factory decided—has it been considered in relation to the machine, to its raw material manufacture, and the products of manufacture? The cost of the machine is often known, the cost of its housing, its installation, its servicing, and its operation is less precise but equally important from the accident prevention aspect.

Is every aspect of machinery use considered early enough? Engineers design equipment to take account of natural laws such as gravity. All too often they neglect the natural behaviour of the operative—his curiosity, his indolence, his comfort, and his impatience. This is a much neglected aspect of accident prevention; it will pay dividends to consider it.

12. Portable electric tools: safe working

J. G. V. Bland

Product Manager,
Wolf Electric Tools Ltd

Introduction

It is an established fact that the majority of accidents which involve power tools are caused by incorrect handling or inadequate maintenance. Contributory factors are ignorance of the correct information, carelessness, familiarity, inattention, apathy and lack of constructive training.

The principles of electrical engineering are not the subject of this article; however a brief description of the protective electrical safety systems for power tools should be outlined.

Electricity has been harnessed and controlled to power the widely varying appliances developed to speed industrial production and eliminate drudgery in everyday life. Amongst these appliances are electric tools. They make a substantial contribution to both industrial productive capacity, the maintenance of domestic and industrial services and the quality of life in profitable leisure activity around the home.

Electricity in itself is not dangerous when properly used with equipment which is correctly designed, manufactured, installed, operated and maintained. In all situations where electricity is used, however, this may not be the case, and then there is always the potential risk of an electric shock. Most of us have probably experienced this disconcerting occurrence on at least one occasion— a familiar, normally acceptable piece of electrical equipment sometimes 'behaves' in a manner which is very far from acceptable.

The normal and established electrical 'behaviour' that must be accepted, however, is that an electrical current breaking through, or caused to by-pass the controlling functional insulation of a power tool, or any other electrical appliance, will always seek out and follow the path of lowest electrical resistance and 'return to earth' to complete the natural electrical circuit. To guard against the possible consequences of this basic electrical fact, which range from discomfort, through injury or damage, to fatality, three principal systems of protection have gradually evolved. These are as follows:

(a) the use of an earth conductor to give a clear low-resistance path to earth;

(b) the use of a low voltage power supply to reduce the potential shock voltage; and

(c) the use of double-insulated construction, which, in the event of an electrical fault developing within the power tool or wiring circuit, effectively prevents the electricity from returning to earth through the operator.

The earth wire

In common with other members of the European community, Britain has adopted colour coding to identify the conductors (in a single-phase circuit) which make up the modern flexible current-carrying cable (or flexible cord as it is known) fitted to portable electric power tools. The insulation of a line or 'live' conductor is *brown*, the neutral is *blue*, and the earth conductor is *green/yellow* striped.

In a conventional and generally metal bodied single-insulated portable electric tool (classified Class I to B.S. 2769:1964), one end of this green/yellow striped 'lead' is connected to a terminal on the inside of the metal casing (usually in the switch handle), whilst the other end is connected to the earth terminal of a suitable three-pin plug top (or extension lead connector). Thus when the plug top is plugged into a socket of the power supply a direct path should then be made to earth.

In the event of a fault developing within the tool which allows current to leak to the outer casing it should then be conducted immediately to earth by this path. Simultaneously, the sudden surge of excess current bypassing the normal electrical resistance of the tool windings overheats and ruptures ('blows') the controlling fuse in the live conductor and stops a further supply of power entering the circuit. The operator is caused no harm or discomfort.

From this electrical principle, first introduced in 1906, resulted the Electricity (Factories Act) Special Regulations 1908 and 1944 to make the fitting of an earth wire mandatory whenever conventional single-insulated portable electric tools (and other electrical appliances of similar construction) are used at places of work. This requirement is still in force today.

The 'earth wire' safety precaution ensures protection against electric shock providing it is tested, checked and *known* that all the connections to the tool from the power supply are neither loose, damaged, disconnected or mis-connected in *any* way to break the correct continuity of the earth circuit. This assurance however cannot be guaranteed, and very real hazards exist within the system, especially when one realizes live and earth wire terminals being mis-connected have caused fatalities.

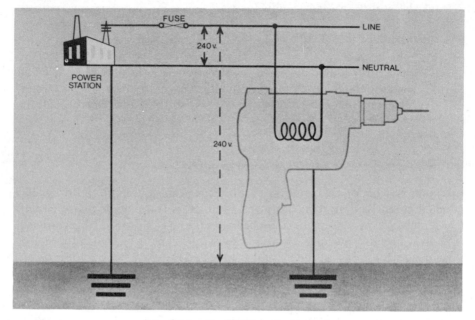

Fig. 12.1　The 240 V earthed system.

Low voltage protection

To reduce the voltage potential of an electric shock, recommendations are made by safety authorities, and in some situations it is mandatory, to use portable electric tools (classified Class I and II 110V and Class III 50V to B.S. 2769:1964) constructed for operation on a reduced voltage supply.

Unlike 50 V tools, which are restricted in power and performance, the 110 V tools retain the same power and performance characteristics as their 240 V counterparts, and unless there is a mandatory directive to the contrary, are recommended for the majority of low-voltage requirement.

Low-voltage machines and equipment should be fitted with discriminating plugs and sockets to ensure machines cannot be accidentally plugged directly into a mains voltage circuit.

Low-voltage power is obtained through step-down transformers (portable or fixed installation) which, as the name implies, will step down or reduce the standard grid voltage supply to the selected voltage. The 110 V transformer should be of an insulated construction and designed with a tapping to the earth screen taken from the midpoint of the secondary winding within which the 110 V voltage is induced. This 'centre tapping' as it is known ensures that, in the event of an electrical fault developing within a 110 V single-insulated tool, an operator will not receive a shock whilst the earth circuit is correctly connected. In the event of the earth circuit being broken, however, the maximum potential shock voltage that can be experienced is reduced to 55 V (half the output of the secondary winding). This voltage has not been known to have had any lethal effects on a normally healthy person.

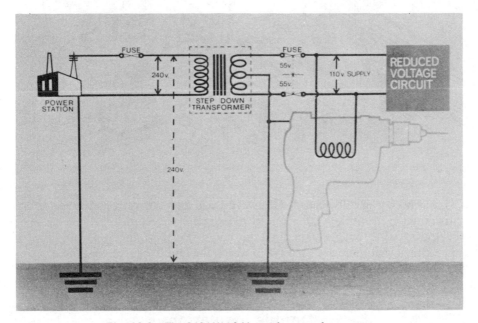

Fig. 12.2   The 240 V/110 V transformer safety system.

The protection against shock from both the foregoing systems is therefore dependent on circumstances beyond the control of the operator. He relies on authorized, competent personnel to properly connect and maintain all the wiring within the power supply circuit and especially the safety earth wire.

An electric shock from the 240 V supply can be fatal, depending upon the circumstances in which it is sustained—wooden floors, rubber boots, etc., can mitigate the severity of the shock.

Low-voltage electric shocks, in themselves not fatal, have indirectly been the cause of injury and loss of life. Operators have fallen from ladders and scaffolding or dislodged equipment from them onto people below.

Fig. 12.3

It has been tragically recorded, however, that an incorrectly connected safety earth wire, even within a 110 V electrical system, has become the direct cause of electric shock—some with a fatal outcome. Furthermore it is not always appreciated that there are six ways in which the familiar three pin plug top can be wired up. Only *one* way is correct, whilst *two* of the incorrect ways are lethal at 240 V even before switching an appliance on!

Double insulation

Double-insulated portable electric tools (classified Class II to B.S. 2769:1964) are constructed to an internationally accepted principle and standard for operator safety. Two distinctly separate insulating systems, or barriers, are 'built in' to the tool. The need for the earth (safety) wire is eliminated in the complete knowledge that the operator cannot receive a shock from a tool so constructed.

The first barrier, common to all electric motors, is a *functional* insulation which isolates the current carrying components from each other and any external metal components of the tool. The second barrier is a purely *protective* insulation. The design and construction of motor frames, switch handles, auxiliary handles, and the isolation of transmission components, is carried out with insulating materials of a dielectric strength which cannot be penetrated by an electric current in the event of an electrical fault developing within the tool, plug, socket, or wiring.

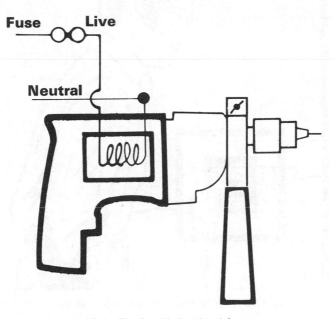

Fig. 12.4   The Double Insulated System.

Double-insulated tools are identified (usually on the nameplate) with the International symbol for double insulation of a square within a square (Fig. 12.5). The electrical safety of an operator is no longer dependent on the connecting up of safety systems beyond his control—protective safety is virtually in his own hands.

Official recognition for this superbly safe construction was obtained in the U.K. on 14th October 1968. The Electricity Regulations 1908 (Portable Apparatus Exemption) Order 1968 amended the mandatory requirements of The Factories Act 1961, The Electricity Regulations 1908 and approved the use of portable electric tools fitted with two-core cables in British Industry and, providing they are tested and approved by an authorized testing authority, no earth wire is required.

With the removal of the earth wire and the use of insulated materials, the risk of electric shock from accidentally misconnected wiring making the body of the machine live is also removed.

At this time the British Standard Institute is the recognized testing authority for portable electric tools in the U.K. (see Certificate of Approval No. 1 to the above Regulations) and the British Standard Kite Mark seal of approval to

Fig. 12.5 The International sign of Double Insulation.

B.S. 2769:1964 is the certification recognized by H.M. Health and Safety Executive with regard to the mandatory requirements of the industrial electricity regulations and portable electric tools of double-insulated construction.

It is worth noting that the British Standard Institution testing programme for double-insulated construction involves over 30 rigorous tests covering all aspects of electrical and mechanical design, construction, and reliability.

Fig. 12.6 The British Standard Kite Mark for portable electric tools.

Within this framework of testing, the secondary barrier of protective insulation is subjected to a 4 000 V test (16 times the normal 240 V grid voltage) to guarantee protective safety against electric shock.

Double-insulated machines are available wound for operation with a specified 240 V or 110 V power supply. The introduction of double-insulated construction has contributed to a marked reduction in the number of accidents involving electric tools. No reports of an accident involving an electric shock to an operator have been recorded with certified tools of double-insulated construction. The British Standard Kite Mark (Fig. 12.6) coupled with international sign of double insulation (Fig. 12.5) is an independent guarantee of safety to both operator and employer. This does not mean, however, that the potential risk of an electric shock or injury involving the use of a portable electric tool can be complacently discounted. Complacency is a luxury which can never be afforded.

The continuing efforts of research into safe design and construction by reputable electric tool manufacturers and careful legislation can only be complementary to a responsible and continuing programme of training in safe working habits. This point is clearly made in the Robens Committee Report and The Health and Safety at Work etc. Act 1974 which came into effect on 1 April 1975. From this well-detailed report and the new enabling act (discussed in depth elsewhere), firm guidelines have stemmed to place the responsibility for safety and health in any place of work (with defined exceptions) firmly on the employer, employee, and manufacturer of equipment.

It will be appreciated however that no matter how well designed or constructed a portable electric tool may be, or how specific the legislation, it is ultimately the user who determines the safe usage of it. Thus the user must be trained to be fully aware of the potential hazards from the power supply, the correct safe way to handle the machine and the value of routine preventive maintenance.

Routine electric tool checking is safety

The installation of the electrical circuits, the fitting of plug tops and connectors for extension cables must of course be carried out by authorized personnel. The majority of electric tools are subjected to rough usage as a normal procedure. Any operator, however, should be trained to check and question the points given in Tables 12.1–3 before using a tool. These questions asked and checked out whenever any portable electric tool is to be used will form a foundation to the safe-working habit.

Correct fusing is safety

It is important that before a portable electric tool is put into service a protective fuse of the correct rating is fitted in the power circuit or plug top by an authorized person.

There are a number of sophisticated circuit protection systems available for use with earthed portable appliances to ensure protection against overload and electrical faults; however they are beyond the scope of this article. In the main the portable electric tool situation involves a plug top and power socket, or connectors for extension cables, with fused protection.

Fused plug tops are readily available to appropriate British Standard specifications and it is recommended that these types of plug top are used with portable electric tools. This permits individual 'local' fusing to a correct rating for each tool in use with a standard range of plug tops. In the event of the fuse having to be renewed, it is important that one of the correct rating is refitted by

Table 12.1 An electrical check list

Points to check and action	*—and why?*
1. Is the voltage rating quoted on the nameplate the same as that of the power supply? (Replace damaged or defaced nameplates.)	Overspeeding with grinding machines which have been connected to the incorrect voltage supply has resulted in at least 2 per cent of recorded wheel burst accidents. A 110 V machine connected to a 240 V supply will rotate at approximately twice its rated speed. The ultimate cause of a wheel burst, centrifugal force, increases as the square of the speed. Under these circumstances there is a fourfold tendency for a wheel to burst.
2. Is the plug top in good condition and of the correct type? (Replace an incorrect or damaged plug top with a correct serviceable one.)	Damaged plug tops can leave conductors exposed; taped up plug tops can obscure very real hazards of damp and dirt forming tracks for short circuiting; loose connector pins may be guided into live sockets with 'shocking' results.
3. Is the cable properly secured with a cable cleat (cord clamp) over the tough outer covering? (Refit cleats to ensure the cable is correctly and securely positioned.)	The cable entry points of both tool and plug are subject to maximum flexing stresses and subsequent wear. The tough outer covering of the cable and cable cleat are designed to withstand this stress and retain the robust protection for the conductor insulation. A cable cleat left loose or incorrectly positioned places stress on the conductor connections or on the flexible but less robust insulation of the conductors. In either case it can result in broken or bared conductor wires and a potentially lethal situation.
4. Is the cable free from external damage and taped up repairs? (Replace damaged cables.)	Damage to the tough outer cable covering will leave the less robust conductor insulation exposed and vulnerable. Bared conductor wiring will result. Non waterproof 'insulating' tape can also obscure damaged insulation. Damp conditions, contact with metal work pieces or just lifting the cable over an obstacle may result in a potentially lethal situation.
5. Are both tool and power socket switched off *before* plugging the tool into the power supply?	The surprise resulting from a machine starting up in the hand or fiercely rotating unattended on a bench could result in damage not only to the surrounding area but, more seriously, in injury to the operator or other people before it could be brought under control.
6. Is the cable positioned clear of the workhead *before* switching on? (In workshop situations it is recommended that an overhead power-point system is installed whenever possible.)	As working situations vary, the only hard and fast rule that can be made for the position of a trailing cable is that it must be *clear* of the work head and *behind* the operator.

Table 12.2 A mechanical check list

Points to check and action	—and why?
1. Is the machine plug top disconnected from the power supply *before* making any adjustment to guards or fitting bits, blades, abrasives etc., as the case may be?	Circumstances may cause a machine to be accidentally switched on. A plug cannot be accidentally put into a socket.
2. Is the machine free from external damage and loose screws? (Replace damaged parts and missing screws.)	The resulting mechanical weakness is obvious when damaged gear boxes, motor frames and switch handles are held together with tape. There are less obvious dangers, such as pressure die cast aluminium parts being bent or cracked, thus causing damage to internal insulation and becoming live; severe damage to motor frames and switch handles of tough moulded insulating material cannot become live, however cracks which have been allowed to build up with dust (especially metallic dust) and damp are a very real and potentially lethal hazard.
3. Are guards, blades, abrasives, drill bits and adjustable components secure?	Insecure hub nuts, guards and depth controls are particularly dangerous—especially where revolving blades, cutters or abrasives are to be used.
4. Will the work head revolve freely by hand? (Always check it before plugging into the power supply.)	Damaged, loose, incorrectly fitted or adjusted guards can be fouled by blades and abrasive wheels. Breakages to them or other accessory components can result in injuries.
5. Is the workpiece secured? (Time spent making a workpiece secure is never wasted.)	Normal working pressure with a portable tool can dislodge an insecure workpiece. Attempts to hold the item with one hand and control the tool one handed, or wedge the work with foot, knee, elbow etc., will create a potentially dangerous situation where balance can be lost, resulting in injury.

Table 12.3 A personal check list

Points to check	—and why?
1. Loose clothing, sleeves, torn pockets, improvised bandage, loose bandage, trailing finger stall tie, necktie, a long unguarded hair style, etc.	Revolving chucks, bits, blades, abrasives and especially wire brushes will entangle at speeds of a few hundred to a few thousand revolutions per minute to result in severe injury in seconds.
2. Are footwear laces loose, broken or trailing?	Loose laces or insecure footwear can result in an off balance situation.

an authorized person. It is not unknown for fuses to have been replaced with an item of larger capacity rating or, worse still, with a solid conductor (small bolt, nail, heavy wire around the blown fuse cartridge) to 'make the machine work'. It cannot be overstressed that this latter practice is especially dangerous. Under these conditions the local protection is removed and the nearest protective fuse could be the comparatively heavily rated circuit wiring fuse. An operator or other personnel will be exposed to the high potential risk of electric shock, burns and even fire in the event of a fault developing within the tool or its cable.

Correct handling is safety

Each type of portable electric tool has specific safe-working techniques which must be made known to the operator and become part of his safe working habit.

In Tables 12.4–6, the safe working habits for the more popular machines such as a drill, saw, disc sander, or portable grinder are purposely given in considerable detail, to emphasize the hazard potential which good safe work habits will avoid. It should be pointed out that in general these precautions apply to planes, routers, and jigsaws, where rotating or reciprocating cutters and blades are used.

Table 12.4   Portable drilling machines

The safe working habit	*Why develop it?*
1. Select the correct bit for the material to be drilled and keep it correctly sharpened.	Excessive pressure can be exerted to 'force' an incorrect or blunt drill bit, resulting in breakage of the bit, with possible injury to the operator and other people.
2. The position of a hole to be drilled should always be clearly marked. When drilling metal a centre 'pop' mark should be made to locate the hole position. If such a mark cannot be made it is advisable to start the hole with a series of switch movements with the drill bit in light, firmly controlled contact with the workpiece to produce a locating mark.	Drill bits can skid across metal surfaces, especially curved ones, and cause an operator to overbalance. A locating mark will allow a full firm pressure to be safely applied.
3. Chuck keys and chucks must be in good condition. (Replace if the serrated teeth are damaged or broken.)	A damaged chuck scroll or key can result in small but painful injury to fingers and knuckles.
4. Drill bits in excess of the rated capacity of the machine should never be used.	It may be possible to make such a bit penetrate the material concerned. This usually involves a heavy pressure to overcome excessive machine speeds. This results in poor control. Under these conditions the bit is liable to 'bite' or 'snag' on breakthrough, and result in a twisting action against the machine rotation to cause injury to hands, wrist or arms— and possibly to other people, especially if the machine is powerful and is forced out of the operator's controlling grip.
5. Drilling pressure should always be eased as the drill bit starts to break through the material.	Easing the pressure prevents the bit from snagging with a twisting action on the operators hands, wrists and arms. This is more noticeable with metal drilling, especially thin sheet material, but can also occur on timber drilling with large-diameter bits. On timber it is advisable to reverse the work and complete the hole from the other side where possible.
6. Pilot holes in metal should be approximately the diameter of the chisel point of the large diameter bit for the hole eventually required.	An oversize pilot hole will generally make control of the bit and machine difficult and usually result in snagging. Similar injury potential to 4 and 5.
7. When using core drills into masonry greater control (and accuracy) is obtained by drilling through with a pilot bit of sufficient diameter for a mandrill to be used to 'guide' the core drill.	Large-diameter core drilling requires large-capacity powerful drilling machines which will react against snagged or binding core drills unless carefully controlled. Similar injury potential as in 4–6.

The safe working habit	*Why develop it?*
8. When deep drilling in timber or masonry, withdraw the drill bit from the hole frequently to clear the swarf.	Swarf will pack drill flutes and bind in the hole. Similar injury potential as outlined in 4–7.
9. Always secure work pieces with clamps when portable drilling work is to be done. .	If the drill bit does bite, an unclamped spinning workpiece, whether metal or timber, can result in serious injury.
10. When using a portable drill in a bench drill stand always secure the work piece to the stand base plate with a machine vice or appropriate clamps. Ensure drill is secure in stand and stand is firmly fixed to worktop.	As for Item 9 with the added hazard of an insecure machine to start with.
11. Telescopic guarding should be used with a portable drill mounted in a bench drill stand.	To comply with the mandatory requirements of Factories Act. To ensure operator protection from a revolving chuck, morse taper spindle or drill bit, becoming caught up in clothing etc., with the operating switch locked in the on position—or on run down as the drilling work is completed.

Table 12.5   Portable (rotary) sawing machines

The safe working habit	*Why develop it?*
1. The correct blade should be selected for the work and kept sharp.	Incorrect or blunt blades lead to excessive pressure being applied and increase the risk of a kick back, especially with a powerful machine. A blunt blade can still inflict serious injury to the person.
2. Check timber materials for nails before cutting.	Damaged blades can result in consequences similar to those outlined in Item 1.
3. The retractable lower guard should be checked frequently for full freedom of movement and *never* locked open with improvised wedges or other means.	Expediency offers no insurance or protection from an exposed rotating blade.
4. When required, the retractable lower guard should only be opened with the manual retracting lever (usually fitted as standard equipment).	Both hands must remain above the soleplate when it is necessary to expose the blade. The retracting lever can be quickly released to allow the guard to function automatically whilst both hands remain above the soleplate.
5. Always allow the motor to gain full *no load* speed before starting the cut into a workpiece.	The machine, especially the more powerful units, will kick back with possible loss of control if the blade is placed in contact with the workpiece before gaining a working cutting speed.
6. Allow the machine to do the work and never attempt to force the pace of cut into any material—especially green or wet timber.	A forced blade may bind in the work and it may cause a kick back with possible loss of control.
7. When abrasive discs are used to cut masonry or metal always fit them to the machine with flanges of at least one third the diameter of the disc.	To meet the mandatory requirements of the Abrasive Wheels Regulations 1970. To ensure adequate support and control of the disc.
8. Always wear goggles when cutting splinter-prone materials.	To meet the mandatory requirements of the Protection of Eyes Regulations. Eye injuries can be particularly grave as sight is irreplaceable.
9. Keep machine soleplate in contact with workpiece and allow blade or disc to stop revolving before withdrawing the blade from a partial cut.	A virtually free blade or disc, without control support from the machine soleplate, can come into contact with the workpiece and kick the machine towards the operator with a high injury risk—especially to wrists and hands.

Table 12.5 *continued*

The safe working habit	Why develop it?
10. Only mount a portable saw as a bench unit if the recommended support table, correct guard and switch control arrangements are available. Never improvise this type of support.	To meet the mandatory requirements of the Woodworking Regulations with regard to the necessary guarding, controls and riving knife.
11. Precautions must be taken to protect operators and other personnel when cutting masonry, stone, lead and asbestos products which produce toxic or offensive dusts.	To meet the mandatory requirements of the Factories Act regarding toxic and/or offensive dusts which are a health hazard or nuisance.
12. Always ensure that the working floor area is free from underfoot obstruction, various offcuts, etc.	To avoid a trapped supply cable or an overbalance situation resulting in injury from a revolving blade or disc. To meet the mandatory requirements of the woodworking and abrasive wheel regulations.

Table 12.6   Portable grinding and disc sanding machines

The safe working habit	Why develop it?
1. Select the correct grit, grade and type of abrasive wheel or disc. (Obtain abrasive manufacturers advice.)	Excessive pressure, or worse, a bouncing action, to force a wheel to 'cut' has been the cause of some 27 per cent of the wheel-bursting accidents and injuries recorded.
2. Check wheel for damage and cracks before mounting. (Ring test as outlined in the recommendations to the Abrasive Wheels Regulations 1970.)	Damaged wheels can burst as machine gains full no-load speed or on contact with workpiece. To comply with the Abrasive Wheels Regulations 1970.
3. Check that the speed marked on the wheel is *not* exceeded by the speed marked on the nameplate of the tool.	Centrifugal forces exerted on an abrasive wheel increase as the square of the speed, and the risk of a wheel bursting through overspeeding is a very real danger. To comply with Abrasive Wheels Regulations 1970.
4. Check that abrasive wheel has the correct specified Imperial or metric wheel bore. (Approximate conversions of specified wheel bore sizes are dangerous. The wheel must be a free, easy, but not loose fit on the mounting spindle.)	Too large a bore causes out-of-balance running. Too small a bore forced on a spindle places severe stress on the wheel. Opening up a small bore to fit usually results in out-of-balance running. All these conditions are dangerous and contribute to a very real risk of a wheel burst or accident.
5. Always fit the compressible 'blotter' washers between the wheel and the mounting flanges.	These reduce to a minimum the risk of a wheel burst due to uneven stresses being placed on the wheel when the flange assembly is tightened. To comply with the Abrasive Wheels Regulations 1970.
6. Mount the wheel or disc with the *correct* flange assembly.	To prevent injury from a wheel bursting unless the correct, adequate flanges to support and drive the abrasive wheel or disc are fitted. To comply with the Abrasive Wheels Regulations 1970.
7. The spindle hub nut to secure an abrasive wheel must be tightened only sufficiently to hold the wheel firmly.	Excessive pressure tends to 'crush' a wheel at the flange contact area. Should the pressure overcome the protective function of the resilient 'blotters' a strain could be set up within the wheel leading to a wheel burst.

The safe working habit	*Why develop it?*
8. Position guards correctly and make them secure *before* plugging machine into power supply.	Guards are designed to prevent accidental contact with wheel and protect the operator or other people from correct-sized wheel in the event of a burst occurring. To prevent the fitting of an oversize wheel and risks from unguarded overspeeding that would result. (See 3). A loose guard coming into contact with a wheel revolving at high speed could result in both damage and injury.
9. Before putting a new wheel into use allow it to run freely on trial for at least one minute on the machine at no-load speed before bringing it into contact with the workpiece.	To check that the wheel is undamaged, balanced and running smoothly. The machine should be held in such a manner and position to ensure that no one is in line with the open side of the guard in the event of a burst occurring due to an unbalanced condition or other unsuspected wheel damage.
10. Always allow the tool to attain full no-load speed before bringing the abrasive into contact with the work.	The wheel must be fed smoothly at a full 'no load' speed onto the workpiece to minimize impact 'snagging', which could result in a damaged or broken wheel, especially when cold.
11. Always allow the machine to do the work. (Keep the abrasive wheel in contact with the workpiece with a sufficiently firm pressure to allow it to cut freely.)	Excessive pressure does not increase 'stock' removal—it does however tend to 'load' or 'glaze' the wheel to *reduce* the effective stock removal rate. This in turn leads to greater pressure being exerted by the operator to 'make it work', thus aggravating the situation, with an increasing risk of damage to the wheel and subsequent injury to personnel.
12. Always keep rotary wheels clean and well dressed.	To minimize the risk of damage to the wheel from an unbalanced wheel 'bouncing' and potential injury from wheel breakage.
13. Always use the correct wheel dressing tool. The machine must be correctly supported whilst the wheel dressing is being carried out.	Pieces of abrasive wheel and other 'unofficial' dressing tools can become trapped in the wheel guard. They can be ejected as missiles to cause injury to others, or severe personal injury from wheel burst, an uncontrolled machine or trapped fingers. One hand cannot adequately control the machine whilst using the dressing tool in the other.
14. Vary the approach angle of the depressed centre grinding disc edge to prevent a thin taper being formed on the grinding edge.	Wafer thin edges created on depressed centre discs can break away and be projected as dangerous missiles.
15. Avoid using the face of a depressed centre disc for grinding or smoothing work.	These depressed centre discs are of reinforced laminated construction. The face of the disc will be damaged if brought into contact with work, especially an edge. Weakened reinforcing can cause a wheel breakage or burst.
16. A 178 mm flexible backing pad and flexible fibre disc assembly must never be fitted to an angle grinder with a speed higher than 6 500 r.p.m., and 230 mm discs should only be fitted on machines specified by the tool manufacturer. Sanding discs and flexible backing pad assemblies should not exceed the machine manufacturers recommended diameter.	The use of 178 mm reinforced flexible fibre backed disc at speeds in excess of 6 500 r.p.m. (or rotating a 230 mm disc at an excessive speed) is hazardous —the periphery of such discs are virtually rotating knife edges and, should they come into contact with a work edge, can snag and break away to become dangerous missiles.
17. Suitable goggles and protective screens must be used when grinding, sanding or wire brushing.	Particles of abrasive wire or abraded material are projected at very high speed and, though small, can inflict permanent eye damage. To meet the mandatory requirements of the Protection of Eyes Regulations.

Table 12.6 *continued*

The safe working habit	Why develop it?
18. Abrasive wheels and discs must not be stopped after use by applying them to a work surface under pressure—allow wheels to run down freely to a stop.	Applying force to stop a wheel places it in a stressed condition—stress in a grinding wheel can lead to damage, breakage, and possibly injury.
19. Fitting wire brushes to Grinders or Disc Sanders should only be done within the recommendations of the portable tool manufacturer.	Wire brushes are rarely balanced. The inherent action of wire brushing is to flex the wire as it contacts and leaves a surface. Excessive speed or an oversize wire brush usually results in a heavy control pressure being applied. This can lead to an early fracture of overstressed wires. These broken wires can be projected as high-speed and very penetrating missiles which have been known to pierce clothing.
20. Precautions must be taken to protect operators and other personnel when grinding, sanding or wire brushing masonry, stone, lead, asbestos products and other products which produce toxic or offensive dusts.	To meet the mandatory requirements of the Factories Act regarding toxic and/or offensive dusts which can be a hazard to health. (See Ref. 13 for detailed guidance).

Eye protectors should *always* be worn when using hammers and hammer drills, rotating and reciprocating cutters. The possibility of eye damage from small splinters, chippings and dust cannot be over-emphasized. The use and security of the correct guards for each type of machine cannot be over-emphasized—improvization in this area for less than 'just a minute' has resulted in serious injury.

Stress has been laid on abrasive wheel bursting because, although only about 16 per cent of the accidents reported are due to this happening, injuries from this type of accident are particularly severe.

This chapter has been compiled with the accent on safe working habits to avoid conditions and situations which could result in physical injury. Consider the bare facts of the overall annual accident rate in industry—1 000 killed, 500 000 injured and the productive loss of 23 000 000 working days. Much of the tragedy behind these figures could be avoided if simple precautions are observed and developed into safe working habits.

Apart from the intelligent avoidance of possibly fatal injury, however, good safe working habits—the use of sharp bits, cutters, and blades and well-balanced, correctly fitted abrasives—all contribute to the satisfaction of work being carried out professionally, efficiently, and profitably.

Bibliography

1. *Portable Electric Motor Operated Tools.* B.S. 2769:1964. British Standards Institution.
2. The Electricity (Factories Act Special Regulations) 1908 and 1944.
3. Grinding of Metals Special Regulations 1925 and 1950.

4. Factories Act 1961. H.M.S.O., London.
5. The Electricity Regulations 1908 (Portable Apparatus Exemption) Order 1968 (S.I. 1968 No. 1575) and Certificate of Approval No. 1 to these regulations.
6. The Employers Liability (Defective Equipment) Act 1969 (see also Chapter 37). H.M.S.O., London.
7. The Asbestos Regulations 1969 (S.I. 1969 No. 690).
8. The Abrasive Wheels Regulations 1970 (S.I. 1970 No. 535).
9. The Health and Safety at Work, etc. Act 1974 (see also Chapter 37). H.M.S.O., London.
10. The Woodworking Machines Regulations 1974 (S.I. 1974 No. 903).
11. The Protection of Eyes Regulations 1974 (S.I. 1974 No. 1681).
12. Health and Safety at Work Booklet No. 4: *Safety in the Use of Abrasive Wheels*, H.M.S.O., London, 1971.
13. Code of Practice: *Control of Dust from Portable Power Operated Grinding Machines*.

13. Welding operations

P. G. Sanderson

Manager, Commercial and Marketing Services,
Philips Electric Arc Welding Ltd

Manually held electrode holders, semi-automatic torches, or fully-automatic machines are all used in electric arc welding. During welding, the electrode and parts of the holder or torch are electrically alive and hot. The arc gives a visible light of high intensity: ultra-violet radiation and infra-red radiation being given off. Some fumes are given off, but these are not normally detrimental to health. However, arc welding equipment and processes can be used with complete safety providing they are used correctly, and with sufficient care; by taking precautions which, in general, are simply matters of common sense.

It is the duty of the occupier of a factory, the supervisors, and workmen alike, to comply with the provisos of any Factories Act, electricity regulations, and other local requirements which may apply.

Precautions against electrical risks

Welding equipment

Electric arc welding is normally carried out by transformers or motor generator sets which take their supply from the electrical distribution system in the factory, or alternatively, by engine-driven generators. Welding transformers, which may or may not incorporate rectifiers, are either oil-cooled or air-cooled, the latter sometimes being forced air-cooled by fan. Care should be taken to ensure sufficient ventilation around the transformer to avoid exceeding the rated temperature rise and thereby causing dangerous overheating of the equipment. It should be noted that B.S. 638 permits an oil temperature rise of 50°C above an ambient of 35°C: a total of 85°C. This means that the outer casing of the transformer can be up to this temperature.

In the case of engine-driven generators used indoors, care should also be taken to ensure adequate ventilation to dispose of the exhaust fumes.

Electrical circuit

In the majority of cases the electrical circuit is relatively simple, but it is important to understand the three essential connections for every welding circuit. These are (a) the welding lead, (b) welding return, (c) welding earth.

The welding lead. As the welding lead is normally connected to the electrode holder, this cable has to be capable of carrying the full welding current without overheating, and must be sufficiently flexible and robust to withstand everyday use.

The welding return or return current cable is often the most neglected part of the whole welding system. It is *not* the earth lead—it is the cable by which all the welding current returns to the welding set. Once this fact is realized, the importance of its efficiency becomes obvious. It must be at least of equal cross-sectional area to that of the welding cable, but it need not be as flexible. In some large installations, permanently installed copper busbars may be a better proposition. In such an installation it is advantageous to use multioperator equipment incorporating a three-phase transformer and 3, 6, 9, or 12 separate regulators. Simplification of the return current system is one of the advantages of multi-operator welding equipment which lends itself to the installation of a permanent return system. Normally the neutral terminal of the transformer can be permanently connected to the bed plate, work bench, ship, or structure to be welded

System layout

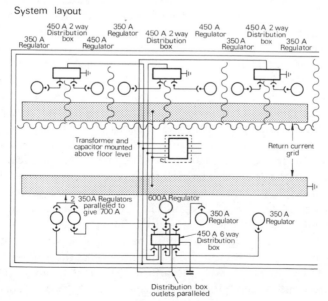

Fig. 13.1 Layout of typical welding shop employing individual welding booths and 'open' welding area.

by means of a copper conductor of adequate cross section, thereby eliminating trailing flexible return current leads. In some circumstances, a common return system can be installed by means of a copper conductor of suitable cross section passing through the welding bay. The conductor should be connected to the neutral terminal of the transformer secondary winding and will serve as a return current path for individual work pieces.

The welding earth is essential to ensure that the welding circuit is adequately earthed. The cross section of the welding earth should be capable of carrying the full welding current. In the majority of cases the earth should be as near to the work as possible, but the choice of earthing at the source of welding current supply or at the work depends on the consideration of the welding return circuit. However, with the welding earth applied at the source of supply a condition can arise in the event of a break in the welding return whereby some or all of the welding current returns to the supply by other paths while welding is continued. This can affect portable tools, conduits, and pipes in the vicinity. With the welding earth applied to the work there is no alternative route for the return welding current in the event of a break in the welding return cable, and the arc cannot be maintained and so welding stops.

Electrical joints

It must be emphasized that the welding circuit is frequently carrying currents of several hundred amperes. In addition to ensuring adequate capacity cables for the welding lead, welding return, and earth return, it is important to ensure that poor connections or poor joints are avoided. All joints, whether in the welding cable, the return cable, or the earth cable, must be efficient at the equipment, the electrode holder, and at the work. If a cable is joined throughout its length, coupling should be effected by plug and socket cable connectors or efficient cable joints, and not by nuts and bolts which often work loose, become corroded, and result in an inefficient circuit with danger to personnel and equipment.

On multioperator equipments the use of distribution boxes is recommended. These ensure good electrical connections and avoid hazards to personnel. The return and earth clamps are also an important part of the electrical circuit, but again are very often neglected. The preferable type of clamp is one which works on the vice principle which can be screwed down to ensure good contact with the work or bed-plate. The cable should be efficiently joined to the clamp and the clamp firmly fixed to the work, which should have been cleaned to ensure a good contact. The use of angle iron and the bare end of the cable weighted down on the job are to be avoided at all costs. Remember that the welding current can be several hundred amperes and has to return to the set through the return cable. Loose connections or temporary arrangements are obvious hazards which must be avoided.

Ideally, cables should not be dragged over floors and rough surfaces, but in practice, as this does happen it is important to carry out frequent examination of the cables for insulation defects. Cables with damaged insulation should not be used as they cause accidents or fire. Care should also be taken to ensure that cables do not become the cause of persons being tripped up.

Electrode holders

Electrode holders should be of sufficient capacity to hold the largest electrode to be used, and carry the maximum welding current without overheating. The holder must be provided with an insulating handle, as reliance must not be placed on the welder's gloves for insulating the operator's hand from live parts. A fully insulated type of holder is a further advantage as this helps to prevent inadvertent contact, or short circuit between the live holder and earthed parts. Care is necessary in leaving an electrode holder when welding is temporarily suspended. The safest procedure is to make the holder dead either by means of a plug and socket near the holder or at the transformer or regulator; alternatively, by disconnecting the complete equipment.

Fig. 13.2   Portable semi-automatic CO_2 welding equipment.

Safety device

The open circuit voltage of welding systems used in Britain ranges up to 100 V which is considered safe for many applications. However, under certain circumstances, such as working in confined metal tanks, working at height, or under damp conditions, this voltage could be a source of danger. In such circumstances, it is wise to fit a low-voltage device to the output side of the welding set. This reduces the open circuit voltage to approximately half the full normal open circuit voltage. When the electrode is touched to the work the higher voltage is automatically restored to enable the arc to be struck, whereupon the voltage across the arc drops to a much lower figure in the normal way.

General

Many welding sets are portable, and much recent equipment of the semi-automatic gas shielded equipments is more sophisticated, incorporating plastic tubes for gas and water supply. Do *not* pull equipment around by the cables, hoses, etc. It is also important not to risk damage by contact of cables or hoses, with hot weld material. Always have on display in prominent places instructions for the treatment of electric shock, and do not let familiarity breed contempt.

Precautions against other risks

The welding arc is a source of great heat and this, together with hot spatter and the heat of the plates being welded, can easily cause fires. Inflammable materials such as oils, rags, paint, shavings, etc., must be removed from the vicinity of welding operations. It is necessary to clear both sides of the work to be welded, and cover all openings to ensure that spatter does not fall through.

Personnel must be protected during welding against the heat and the possibility of burns. A leather apron or other suitable device is essential. In certain positions leather or asbestos sleeves and leather spats are necessary in addition to an apron. Overalls of flame-proof cloth without turnups at cuffs or trousers are a wise precaution. Leather or asbestos gloves must be used; rubber gloves are not suitable.

Helmets and goggles

For all welding work either a helmet or handshield is essential to protect the welder's head from radiations, spatter, and hot slag. For many applications, including prolonged work, the helmet is preferable. For working on aluminium or close to other welders, a helmet which encloses the head is necessary to protect the back of the neck, otherwise light reflected from another welding arc, or from the sheet metal, may burn the back of the neck or the scalp of the operator. All helmets and handshields are fitted with a filter glass and a protective cover glass. The correct grade of glass should be used as recommended in

B.S. 679, summarized in Table 13.1, but the original standard should be consulted for details.

Table 13.1

Arc welding currents (A)	Grade of glass	Shade no.
Up to 100	EW	8 or 9
100–300	EW	10 or 11
300–500	EW	12, 13, or 14
Over 500	EW	15 or 16

Screens

All electric welding operations should be screened to prevent the rays of the arc from affecting other persons working in the vicinity. Screens can be either fixed or portable, but where portable they should be of sturdy construction and yet sufficiently easy of movement, so that their use is not discouraged.

The surfaces of welding booths and screens should have a matt finish to reduce reflected light. Precautions against reflected light are important, particularly on heavy current welding, and high intensity arcs such as semi-automatic CO_2 welding. Helmets and goggles should be worn to protect the eyes when removing slag.

Ventilation

Wherever welding is performed adequate ventilation is necessary, especially when welding in confined space. Although welding fumes are not normally detrimental to a welder's health, heavy concentration over a prolonged period can cause discomfort. Particular care should be taken when welding galvanized materials as inhaling the fumes given off can cause temporary sickness. Such fumes should be extracted by exhaust fans suitably positioned to clear the fumes from the vicinity of welding. Certain gases such as argon, nitrogen, carbon dioxide, etc., which are used in the gas-shielded process, involve certain hazards because (a) there is no practicable way of detecting the presence of argon or nitrogen, (b) argon and carbon dioxide are heavier than air and so will accumulate in the bottom of a vessel which may appear safe because the top is open, (c) the gases are odourless and give no warning to workpeople, (d) men employed in confined spaces such as large tanks may be out of sight of their fellow workers, and there may be considerable delay before anyone appreciates that they have been affected by the gases.

If inert gas is being used or is liable to be accidentally discharged in any area, adequate ventilation should be provided to ensure that the gas cannot accumulate in that area. For example, if inert gas is being used in an open-ended cylindrical vessel, adequate ventilation could be achieved by lifting the vessel off the floor so that the heavy gases can escape through the bottom opening. As a further precaution, where argon or carbon dioxide is being used as the shielding gas in particularly confined spaces, breathing apparatus of the airline type should normally be worn.

Radiographic inspection

Examination of welds by X-rays is now common practice, and personnel undertaking this work are usually fully aware of the implications and the precautions necessary. However, harmful effects are delayed, and at the time when the harm is being done no one can either see or feel the radiations. The first and foremost precaution is to keep everybody not engaged in the radiography work away from the danger zone. The precautions to be taken by those engaged in the actual inspection are not the concern of the welder or supervisor, other than to adhere strictly to the instructions of the specialized technician in charge of the radiographic inspection work.

Summary

All the precautions outlined above are matters of commonsense, and for convenience here is a brief summary of them.

DO:

Make sure of adequate ventilation of equipment, operator, and material to be welded.

Make sure that the insulation of cables and ancillary equipment is in good condition.

Make sure that all electrical joints are sound.

Make sure that the welding equipment and the work to be welded is adequately earthed.

Wear adequate protective clothing.

Use handshield, helmet, or correct goggles.

Select the correct grade of filter glass for the job.

Turn off gas cylinder valves when not in use.

Open cylinder valve slowly.

Respect warning notices posted during radiographic inspection work.

DON'T:

Take risks.

Weld near flammable material.

Use an engine-driven generator in a closed confined building, unless the exhaust gases are led outside.

Attempt to repair a tank or other vessel that has held combustible material, until proper precautions have been taken to ensure safety.

Weld material cleaned with trichloroethylene until it is absolutely dry.

Weld galvanized or other coated metals without taking the correct precautions.

Use a compressed-gas cylinder as a work support.

Get too close to X-ray or γ-ray equipment when it is in use.

Leave welding plant electrically energized when not in use.

Let familiarity breed contempt.

Bibliography

1. Challen, P. (ed.) *Health and safety in welding*. Institute of Welding, 1965.
2. Ministry of Labour H.M. Factory Inspectorate and Safety, Health and Welfare Dept., *Electric arc welding*. H.M.S.O., London, 1969.

14. Running nips

S. F. Smith, O.B.E.

Formerly Chief Officer,
Safety, Health, and Welfare,
Dunlop Ltd

In manufacturing processes today we have gone some way towards complete automation—a phrase conjuring up visions of totally enclosed and completely safe operations. In the interim we see more and more continuous operations introduced—long lines of machinery manipulating the product to its final shape. Unfortunately, most of these long lines are not totally enclosed and it is in this type of machinery, using multitudes of rollers and conveyors, that thousands of running nip hazards are created.

Definition

The running nip occurs where a material runs on to, or over, a roller or similar device. A conveyor running over a roller creates a nip; material being wound on to a roller creates a nip; there are in-running nips between rolls on process machinery such as calenders and two-roll mills; nips on transmission machinery such as those created by flat belts or V-belts running over pulleys; and nips created between chains and sprockets or between gears.

We will deal here with those aspects of the problem which have required special study in the rubber industry: conveyors, materials winding onto rollers, two-roll mills, and calenders.

There is some disbelief that a piece of material running over a roller can be dangerous—an attitude of mind not unlike that often found towards the seemingly innocuous revolving shaft. This may be an over-simplification, but a study of the number of unguarded nips which exist and of accidents which have occurred shows more than ample evidence of an appalling ignorance of the hazard among management and men alike.

A former Chief Inspector of Factories, Mr T. W. McCullough, C.B., O.B.E., wrote: 'This type of accident is particularly distressing, for in most cases the result is mutilation; in many instances, when a man is caught in a nip, he is carried off his feet and may suffer additional injuries through contact with the floor or other fixed objects.'

Types of accident

A study of 'nip' accidents involving lost time in the rubber industry in 1966 showed that the nips occurred in many different locations: between belt and powered rollers, belt and idle rollers, belt and framework or endplate, nips between bowls, between material and bowls, between sheets of material and even in one case between the bowl and the 'safety bar'! The particular causes were varied and included:

(a) inadvertent starting,
(b) straightening of material, including tucking-in torn edges,
(c) clothing caught,
(d) slipping and falling into an unguarded nip,
(e) removing foreign bodies or jammed stock,
(f) cleaning rollers whilst in motion.

Conveyor belt nips

The common conveyor system takes a large toll and yet is probably the easiest running nip hazard to guard. It is possible to standardize the design to some extent; this is most easily demonstrated by reproducing a drawing used as a reference for plant engineers (Fig. 14.1).

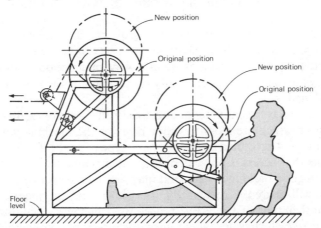

Fig. 14.1 This diagrammatic arrangement shows how a man could be trapped between the roll and the floor where the reeling takes place too close to the floor. It is a simple matter to raise the mounting of the roll.

With a little ingenuity all conveyors can be guarded, but they require frequent maintenance and adjustment. It is necessary in designing a guard to keep in mind the points of access required; otherwise it may be found that the guards are so damaged or distorted through constant removal and refitting that eventually they are left off altogether.

Nips between materials and rollers

The more complex problem of material being wound on to a roller is intrinsically a running nip. This, unfortunately, is only the beginning of the matter; the material being wound up creates a nip of ever-changing position in a radius which is increasing all the time. The periphery of the batch can in these circumstances continually create another nip between itself and a fixed structure.

Prior to the batching up it is inevitable that the material will have passed over or under a number of rollers which exist to maintain a certain tension in the material. Even if such rollers are not power driven, a dangerous nip can be created between the material and the roller at the point at which they meet. The 'nip' as an injury-creating situation exists at some point prior to the two surfaces actually coming into contact, e.g., the thickness of a finger.

A further complication increases the hazard when materials are being used as liners for other sheet material being processed. For instance, in the rubber industry linings are often used to give backing to rubber or rubberized fabrics being processed. The hazard is thus increased.

The most obvious method of minimizing the risk of a running nip accident is to arrange that the material runs *over* the top of the roller wherever possible, thus confining the hazard to the sides of the nip. With this arrangement no one can fall into the nip.

Where a series of rollers occur in rapid sequence they are perhaps easiest to deal with by *fixed* guards which enclose the side sections of the machine, because it is difficult to guard each one in such a way that nobody can gain access, whether they are the operators or simply people passing the machine who could slip and fall against it.

Individual nips between material and fixed roller, where regular access may be required or which are close to other parts of the machine which must be accessible, are best protected by the *interlock* type of guard. It is important with all interlocks that access is prevented until the machine has come to rest. This necessitates a good braking system, and may require the installation of mechanisms which will delay the removal of the guard until the machine has come to rest. Another type of protective device used in these circumstances is the *photoelectric cell* which is dependent upon the interruption of a light beam; but in such cases the machinery should be slow moving and must stop instantaneously (Fig. 14.2).

In the cases of both interlocked guards and the photoelectric cell, it is important to ensure that any failure of a mechanical or electrical nature will cause the machine to stop. In all cases the guard must be such that the operator cannot circumvent it and thus defeat its purpose.

The greatest problems occur when material is being handled at the feed or take-off ends, and it is particularly severe when a new reel is being started because the operator is virtually creating a nip with his own hands. The best method of dealing with this part of the problem is to use a *festoon* of material

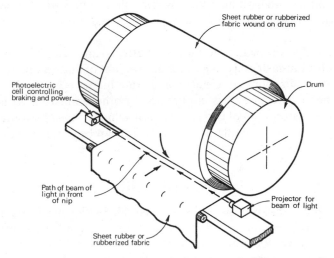

Fig. 14.2 This diagram shows the position of the light projector and the cell receiver, with the beam passing along the front of the nip.

which creates a reserve at both feed and take-off and, therefore, permits continuity of running whilst the reels at either end are stationary. This means that they can be changed in complete safety. In plant of this type the whole train of machinery should be kept continually threaded with a band of material.

In single batch runs, materials may have to be threaded through a number of rollers, and devices should be provided for inching or very slow running. A useful starting-up device consists of a pair of suitable short tapes attached to the corners. These, when fitted on to each end of the liner, enable the machine attendant to commence the batching operation without presenting his hand to the actual nip. Having effected one complete turn the tapes may then be disposed of by being thrown into the second turn. It is also recommended that the fullest use should be made of leaders and followers for connecting interrupted batches. This practice is very well known and can serve a most useful purpose by permitting the installation of fixed guards.

Another device used is to encage the whole of the end section on a sensitive floor which is interlocked so that the machinery stops as soon as anyone steps on to the boards in front of the batching nip. (Although not directly related to the problem of 'nip' accidents it is worth noting that the rolls must be securely mounted to prevent their jumping out. A falling roll is itself a serious hazard.)

Strict training in job methods and follow-up by supervision are vital aspects of combating the running nip problem but, since the human element is not always entirely reliable, they should not be regarded as satisfactory substitutes for safe conditions. Operators have frequently been drawn into nips whilst straightening running material or even just touching it for no apparent reason. There are several straightening devices which will help to obviate these dangers. It is also wise to insist that liner materials be removed if the edges are frayed or torn.

Apart from the nip of infinite radius created by material batching upon a roller, the rubber industry has long experience of nips of finite radius at machines consisting of two or more large steel rollers running together, and it is known that this type of machinery has application in other industries. It is possible therefore that a description of two particular machines—the horizontal two-roll mill and the calender—will offer guidance which can be useful in a much wider range of applications.

Two-roll mills

The two-roll mill consists of two rolls situated side by side in a horizontal plane through which rubber is squeezed into sheet form. Normally these rolls vary in diameter from about 400 mm upwards to about 700 mm. The nip between the two rolls has been a source of many accidents in the past—all of them serious and some of them fatal. In the early 1950's, R. W. Lunn, C.B.E., of the Leyland and Birmingham Rubber Company, calculated and demonstrated that it was possible to erect a sensitive safety barrier so placed that, if a man was entangled in material on the mill, his body would be pulled towards the mill and he would involuntarily actuate the barrier, which itself would operate a brake to stop the machine within such a distance that the man could not possibly be pulled into the nip (Fig. 14.3). In other words, although an 'accident' could occur, there could be no injury. This in fact is now the accepted standard method of guarding the two-roll mill.

To arrive at a solution offering complete safety a number of problems had to be solved, the main one being to create a state in which a man could still work comfortably at the machine, the normal job being to cut off sheeted rubber from the face of the roll.

There must be a limited distance which the roll may travel after actuating the brake if the operator is not to be drawn into the nip if he becomes entangled in the material being worked. The first requirement therefore is rapid braking, following the important criterion that the shorter the stopping distance, the greater the safe working accommodation available on the mill. Allowing for a finger gap of 20 mm prior to the actual nip, one can determine a point beyond which the operator must not be able to reach before actuating the brake. The location of the sensitive safety bar is fixed, therefore, with regard to the operator's reach (Fig. 14.3). It has been found necessary in existing installations to excavate a pit in front of the mill to achieve this condition. On the other hand, new mills are raised to secure the same conditions—that of taking the danger zone further away from the man.

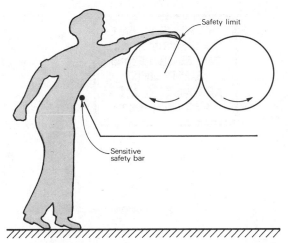

Fig. 14.3   Diagrammatic side view of a two-roll mill showing the position of the sensitive safety bar relative to the safety limit.

Having taken this practical care, one must then ensure that the operator cannot gain access either over the bar or under it, and this means removing footholds and enclosing the area under and at the sides of the safety bar. Care must be taken to avoid creating a nip between the safety bar and the fixed guards adjacent to it.

It will be seen that the whole principle is based upon the relative dimensions of the man to his machine. Having established the location of the safety bar on the machine, no operator may work at it if he can reach beyond the safety limit, and this test must be established and recorded for each man concerned.

Work on back rolls

Some mills are required to be worked on the back as well as the front roll. In such cases sensitive safety bars must be fitted at both locations. It will be obvious that where, due to unequal gearing or different diameters, the rolls run at different surface speeds, the position of the safety limit on the back roll will not be the same as on the front.

If work on the back rolls is not required then this area must be enclosed with interlocking guards to deny access when the mill is working.

It is necessary to conduct regular tests of the braking distance and of the pressure required to operate the sensitive safety bar (a horizontal pressure of about 16 kg has been found to be a satisfactory loading), and written records should be kept of these tests. It should also be standard routine for the bar to be tested at the beginning of every shift to prove that the mechanism is in working order.

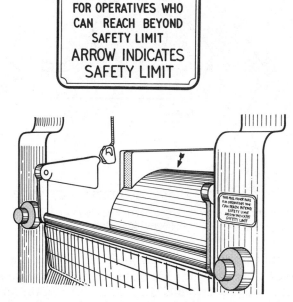

Fig. 14.4   Sketch of a two-roll mill showing sensitive safety bar and safety limit.

Laboratory mills

Laboratory mills, the rolls of which have a relatively small diameter, cannot be guarded in the same fashion. The most practicable type of guard prevents access from above the nip when feeding rubber into it. Where the guard lies adjacent to the face of the roll, there are 'knuckle' controls which will actuate the brake if the hands are pulled against them. Again good braking is essential.

Calenders

The calender is a much more complex problem and no common solution is possible because of the varying number of rolls and the varying ways in which they are arranged. The problem is even further complicated if the machine is capable of operating in reverse. Many calenders nowadays can operate at high speeds and can be set to produce very fine sheets of uniform thickness. They require many additional devices such as gauges for controlling thickness etc., and all these factors tend to complicate the fitting and operation of guards.

Feed nips

As always the fixed guard is of course the most reliable (Fig. 14.5), but this is only convenient when feed methods do not change from process to process and

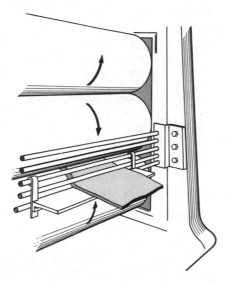

Fig. 14.5 *Fixed guards:* a simple system of fixed parallel bars; the bars must be strong and suitably mounted to prevent spreading. Sheet rubber compound is fed between the bars; a small horizontal table support is shown.

the rubber is delivered in sheet form. Unfortunately, feeding stock to a calender is not always so simple; consequently, the guards have to accommodate bulk but at the same time they must keep the operator at a safe distance. One method is to feed over a fixed distance board with a sensitive safety bar between the board and the operator.

The tunnel guard is another method of feeding; it does require that the tunnel is of adequate length to protect men with exceptional reach. It is usually necessary to provide a push stick (Fig. 14.6).

The manger guard is a similar distance guard and is used at heights over which the operator cannot normally reach. The guard is in the shape of a manger presented to the face of the rolls and rubber is fed into it. The rubber is assisted into the nip using a push stick between the grill bars of the guard. The push stick must be fitted with a hilt which limits the extent to which it can reach towards the nip.

Many manger guards are simple fixed fences but refinements have been introduced to make the guard pivot about its axis with trips to limit the amount of movement towards the nip in order to prevent access to the danger zone from the top of the manger. Because the shelf at the bottom of the manger must be clear of the roll surface to permit vertical adjustment, only the minimum clearance should be permitted so that fingers are not trapped between the edge of this platform and the roll surface. Upward movement of the platform also applies the emergency brake.

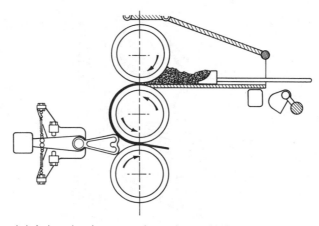

Fig. 14.6   *Top 'nip':* there is a long tunnel guard over the feed table with a trip bar to stop the machine, in front of, and slightly below the edge of the table. *Bottom 'nip':* A strong looped guard extends across the calender; the loops are sufficiently close to the roll and so spaced as to prevent access to the nip. The loop has limited movement up and down and may incorporate a device to operate the brakes.

The simplest and safest method of feeding is that done automatically either by direct gravity conveyor or by a pendulum conveyor moving backwards and forwards across the face of the roll. This introduces extra machines and thus extra hazards, not least of which is the danger of a nip between the pendulum moving to and fro and fixed parts of the calender.

Other calender nips

On nips away from the feeding point, two types of guard can be used. A fixed guard is desirable but it is not always possible to present it close to the face of the roll because of the varying thicknesses of material being processed and the danger of trapping between the guard and the roll. In these circumstances the guard has to be supplemented with knuckle trip bars at the two points nearest to the faces of the rolls (Fig. 14.7).

The 'loop' guard consists of a series of grille bars presenting a side view shaped like a loop. If the fingers approach too closely to the nip they can be moved to a position between the loop bars and withdrawn. The loop has limited movement up and down and it is set to trip the brake if pressure is exerted between the roll and the loop bars.

An important feature of operator protection using all these guards is the braking system. All modern machines can and should be fitted with highly efficient brakes which can stop the machine rapidly without damage even when it is operating at high speed.

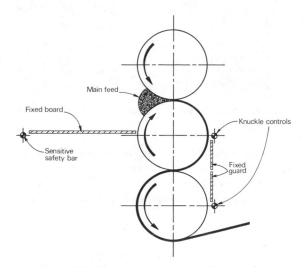

Fig. 14.7   Fixed guards and knuckle controls.

The threading of calenders can be a hazardous operation, and should only be undertaken by experienced men. However, the hazards can be eliminated or reduced by the use of threading leaders where fabric is involved and by the use of inching devices. Indeed, every calender should be capable of being 'inched' or run extremely slowly during the starting-up operations. Under these circumstances the braking is virtually instantaneous.

A device used to facilitate joining of materials in order to permit continuous running of the calender is the festoon, which has already been mentioned in connection with running nips.

Ancillary equipment causes 50 per cent of the accidents occurring at calenders. This equipment consists of rollers or has rollers built into it because of the nature of the work. We must guard the nips between these rollers and the work in hand just as much as the calender—these are running nips.

Many calenders form part of a large line of operations which present a

potentially dangerous situation at starting or at a change in routine. It is imperative to consider carefully the siting of controls. The starting control should be
under the direct guidance of a supervisor who should follow a predetermined
system of checking that all personnel are in positions of safety before com-
mencing the operation. Communications or signals between members of the
working gang should be established and understood by everyone joining the
team.

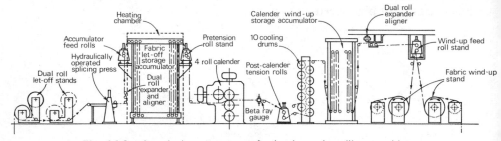

Fig. 14.8   A typical arrangement of calender and auxiliary machinery.

One of the lessons learned from accidents which have occurred at running
nips is that one cannot rely upon the potential victim to operate a stopping
device to save himself. When he realizes he is trapped he invariably tries to free
himself before being pulled further into the nip. Therefore, safety devices should
operate involuntarily.

Safety by position

Another lesson learned from hard experience is that one cannot rely on the
assumption that a nip is safe by position; nothing is safe by position if a person
has to go near it, even if only occasionally and even if access is difficult. One can
only assume a state of 'safety by position' if it is completely inaccessible.

Conclusion

Many of the circumstances described have a universal application and can be
found in most factories. They constitute hazards not only to process workers
but to maintenance workers who might become ensnared in circumstances with
which they are not familiar. Running nip hazards need to be sought out and
they are not always obvious. When they have been located there are two
principles to be observed which serve as a useful guide to the effectiveness of
guards.
    (a) The guard or system of guards should be such that it is impossible for
        an operative, during the legitimate and proper working of the machine

under normal speed and power, to be able to reach with his hand or any part of his body or clothing the dangerous running nip.

(b) When the guards preventing access to the running nip are opened or moved away, so that such access is no longer completely denied, the machine must not be capable of moving under normal power or any other power that would cause other than relatively minor injury by trapping in the nip.

Bibliography

National Joint Industrial Council for the Rubber Manufacturing Industry. *Safe working on horizontal two-roll mills,* 1952.

National Joint Industrial Council for the Rubber Manufacturing Industry. *Guide to 1952 mill report,* 1960.

National Joint Industrial Council for the Rubber Manufacturing Industry. *Running nip accidents,* 1959.

National Joint Industrial Council for the Rubber Manufacturing Industry. *Safe working on calenders,* 1967.

15. Compressed air

A. G. Paterson

*Formerly of Air Mover Research Syndicate,
Glasgow*

Most workshops and factories are equipped with a comprehensive system of compressed air connections for general use. It is not always appreciated that compressed air, at the normal factory pressure of 5–6 bar, can cause grievous injury to workers, whether they are actually operating the compressed air equipment, or merely standing within a range of up to 12 m from it. General cleaning down of machines and work surfaces, or cooling of parts, are among the common usages of compressed air in factories.

When 'blowing down', the danger lies in particles of metal and swarf which, propelled at high velocity, can get into the operator's eyes—when cleaning out a blind hole, for example. Or, and this is common, the particles can be blown into the eyes of a person standing nearby, whose reflexes are not conditioned to the danger.

There is danger in allowing compressed air to enter the blood stream through a cut or abrasion on the skin. This has been known to happen, with fatal results, when a workman was using compressed air equipment to clean his clothes after work: a highly dangerous practice which must be strongly discouraged.

Apprentices and young factory workers not made aware of the dangers, can injure themselves severely, and injure other people, if they are allowed to indulge in 'horseplay' while using compressed air equipment. They must be well and fully trained in the correct use of all such equipment.

There is a high and sometimes dangerous noise level from compressed air jets. This noise may occur during a cooling or drying-off operation; generated over a long period it can cause damage to hearing.

Harm can also be done to machinery. When cuttings are being blown from a surface, for example, they may become lodged in gearing or under slides and, if they remain unnoticed, can cause damage.

There are available special units of compressed air equipment which were specifically designed for general factory use, and which have a safety advantage over the conventional compressed air jet. They transform the high-pressure, relatively high-volume normal air jet into a higher volume, high-velocity, low-pressure air stream. Consequently, all the normal factory functions can be carried out without the danger inherent in high-pressure systems, and with an economy in the use of compressed air.

A normal air gun with a 3 mm dia. jet uses approximately 0.68 m^3/min of air at 5.5 bar. These special units fitted to the air gun can use as little as 0.23–0.28 m^3/min to do the same job. They do not propel cuttings at high speed away from the surface and so the safety of work people in the immediate vicinity is assured. Wherever compressed air is used open-ended for cooling, removal of moisture, and cleaning, this new design of unit can be used with economy. It employs the Coanda effect, and is manufactured in various sizes, from 3–254 mm orifice. Because of the throat contours, large volumes of ambient air are sucked into the units and discharged at high-velocity low-pressure from the outlet. For an input of 0.28 m^3/min free air at 5.5 bar, an entrainment factor of 18 to 1 can be achieved. The discharge of the unit would be 5.38 m^3/min at, approximately, for the smaller sizes, 0.05 bar.

The units can be manufactured in any machinable material for specific purposes, including equipment specially designed for fume extraction. They are small and light in weight, and so are suitable for removing fumes in enclosed spaces, e.g., welders working in vessels, closed columns, etc. They have a low noise rate and so are not a continuous distraction to the operator. Men working in hot conditions, such as in stripping the lining from a pouring ladle, can be supplied with large volumes of cold air to enable them to work safely for a much longer period.

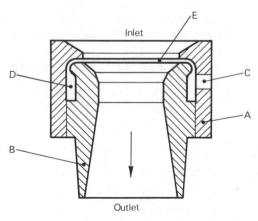

Fig. 15.1 Annular airmover.

The installation of these units throughout a factory ensures a supply of high-volume low-pressure air at any point, and provides a vacuum line wherever required. All this helps to reduce the hazards associated with the use of compressed air. As an example, the equipment described below is designed to convert a normal factory high-pressure compressed air supply at low-volume, to high-volume, low-pressure uses. The components, consisting of an outer chest and inner core, may be non-ferrous, or of stainless steel/mild steel, depending on the use and individual requirement.

Air from the factory enters at (C), charges (chest D), and is ejected through controlled gap (E). Because of the profiles of both component parts, the air follows the contours of the inner core, accelerates towards the outlet, and creates an intense partial vacuum over the entire inlet. This vacuum is filled by ambient atmospheric air which accelerates with the primary compressed air, passes through the unit, and is discharged through the outlet.

Since neither the primary nor the secondary air is in contact with the profile of the inner core, but is riding on a partial vacuum, friction losses are minute, and the pressure drop usually associated with conventional venturi systems does not apply.

Dependent on input pressure and the gap setting, volumetric advantage of up to 20:1 can be achieved. For example, with an input pressure of 6·9 bar, and a volume of 0·28 m^3/min of primary air, an output of 5·66 m^3/min is obtained. To ensure maximum efficiency when dealing with high volume applications, it is essential to minimize pressure losses in supply lines. These must be of minimum length, maximum bore, and free from regulating restrictions. There are no electric motors or moving parts, so this equipment is suitable for use in hazardous areas where dust, gas laden, or heated atmosphere constitute real problems.

The application potential for this type of equipment is high. The absence of external mechanical parts means that it can be used in a wide range of industries, e.g., for dust extraction, toxic gas drainage, swarf removal, light materials handling required in plastics moulding, blasting fume evacuation, tank defouling, and cooling and carbon dust removal from electrical switchgear.

In this second type of equipment, a strong flow of air is emitted, not a jet as from a conventional nozzle. The units are specifically designed to eliminate the

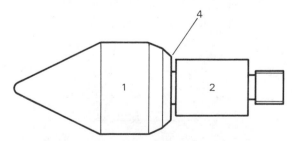

Fig. 15.2   Air miser safety blowing nozzle.

kind of accidents which are caused either by careless handling of normal air jet equipment despite stringent plant regulations, or simply because there has always been some risk inherent in the use of such equipment.

Compared with the average blow gun nozzle which, with a 3 mm jet, uses 0·74 m³/min at 6·9 bar, or 0·43 m³/min at 3·45 bar running continuously, this special equipment, under laboratory test conditions, uses 0·23 m³/min at 6·9 bar, or 0·12 m³/min at 3·45 bar running continuously. This economy is explained by the small size of the annular orifice through which the compressed air is vented to atmosphere, and its efficiency by the fact that, on release to atmosphere, the air entrains to itself a large volume of ambient air, giving a volumetric advantage of some 20:1 at the nozzle point.

It can cost up to 5p to compress 28 m³ of free air to 6·9 bar. The use of a 3 mm jet running continuously would consume the air in 39 min. Using a safety nozzle to equal effect, the same supply of air would last for 2 h 5 min.

It is now possible to eliminate from air lines the regulators which were formerly required to step down pressures. When the working air flow is too strong to use (see (c) below), then reduced velocity can be achieved by fitting a simple tap into the system while still using line pressure.

Suggested uses for this equipment are:

(a) blowing down machinery of various kinds,
(b) dust removal from components prior to plating or painting,
(c) blowing out of delicate instruments, or equipment under assembly or overhaul, where high-pressure air might do damage,
(d) blowing down electric motors and switchgear,
(e) cooling of parts being machined,
(f) removal of fumes.

It should always be remembered that compressed air is dangerous, compressed air is expensive, and compressed air is not a toy. Operators must be trained to recognize the dangers so that they may work in personal safety and not endanger by-standers or machinery. Taking precautions to reduce the accident risk and costs (the direct cost of the compressed air and the indirect cost of industrial injuries) is a common sense, managerial responsibility.

Of current interest is the Flixborough disaster. This resulted not from the detected fault in a process vessel, but from the manner of providing a by-pass connection when the vessel was removed for repair. The failure to design and support the temporary pipe adequately resulted in its breakdown under normal service conditions, the resulting leakage of flammable gas giving rise to explosion and fire, which killed 28 men, destroyed the works, and caused widespread damage to villages several miles away. The total cost of the direct damage was in the region of £36 000 000.

16. Steam boilers and pressure vessels

R. G. Warwick

Chief Engineer,
National Vulcan Engineering Insurance Group Ltd

The most serious case of steam boiler explosion of recent years is that which occurred in an office building in the U.S.A. in 1962, whereby 23 people were killed and over 90 injured. The boiler was of the Scotch multitubular type rated at 5 500 lb/h (2 500 kg/h) and designed for 15 lb/in² (1·03 bar), the overall length being 16 ft (4·9 m) and the diameter 6 ft (1·8 m). The boiler was oil-fired and amongst other controls there were fitted two overpressure cut-out switches to the burners. The boiler was unattended at the time of the accident although steam was being raised with the stop valve closed in preparation for connection to the main steam range. The closed stop valve isolated one of the pressure cut-outs and the second failed to operate. An unexpected rapid increase in steam pressure took place and, consequent on the safety valves also failing to function, disruption occurred due to overpressure. The subsequent investigation revealed:

(a) the boiler was not subject to periodical independent inspection;
(b) the controls to the burners were defective;
(c) it had not been the practice to test the safety valves whilst the plant was working.

Of current interest is the Flixborough disaster. This resulted not from the detected fault in a process vessel but from the manner of providing a by-pass connection when the vessel was removed for repair. The failure to adequately design and support the temporary pipe resulted in its breakdown under normal service conditions, the resulting leakage of flammable gas giving rise to explosion and fire which killed 28 men and destroyed the works and caused widespread damage to villages several miles away. The total cost of the direct damage was in the order of £36 000 000.

At the other end of the scale considerable damage was done by a 3 ft (0·9 m) diameter 5 ft (1·5 m) high welded-steel hot water cylinder which was projected upwards with great violence. This cylinder was directly connected to the jacket of a chemical vessel through which cooling water was circulated by means of a pump at a pressure of 60 lb/in² (4·14 bar). Investigation made after the accident proved that the vessel should never have been used for pressure purposes. The welded seams viewed externally appeared to be of excellent quality but in fact no edge preparation of the plates had been carried out prior to welding which was undertaken from one side of the plate only. In addition, the junction of the end plates to the shell was effected by corner welding without any knuckle radius, giving rise to serious stress concentration in this area. Failure occurred through the bottom end plate being blown out of the vessel owing to the rupture of this corner welding. It is interesting to note that this extensive damage was done by a vessel working at a reasonably low pressure, 60 lb/in² (4·14 bar), and containing liquid rather than gas under pressure.

As will be seen from the examples quoted above a large amount of potential energy is contained in boilers and pressure vessels which, if released suddenly, will cause serious damage to the surrounding equipment and injuries often fatal to personnel. It is therefore essential that suitable precautions are taken to ensure that failures do not occur.

Accident prevention in this field depends on attention to detail throughout a pressure vessel's life and it would be useful to consider each stage in more detail.

Design

The initial and probably the most important stage in the design of any pressure vessel is to determine the conditions to which it is to be subject. This should take into account system fault and test conditions as well as normal working. The main conditions likely to be encountered are:

(a) pressure,
(b) temperature,
(c) cyclic conditions of temperature and pressure which may cause fatigue failures,
(d) the effects of the contents of the vessel,
(e) the effect of induced loads from pipework, etc.,
(f) the effects of weather, snow, wind, earthquake, etc.

Although with simple vessels it may be satisfactory to design using only some of the above criteria it is often necessary to consider all the above aspects.

The British Standards series of design codes as listed in the *British Standards Yearbook*[1] has been developed over the years to provide design information on numerous categories of boilers and pressure plant and deal with materials of construction ranging from carbon steels, through the alloy and stainless steels, nickel, aluminium, and copper and their alloys, etc., to non-metallic materials such as glass-reinforced plastic (G.R.P.).

At the time of writing this article in mid-1975, an extensive revision of British Standards covering unfired pressure vessels is nearing completion. The resulting *Master Pressure Vessel Code* will in its first edition cover pressure vessels manufactured from carbon, ferritic alloy and austenitic steels, and will supersede B.S. 1500 Part 1 and B.S. 1515 Parts 1 and 2 etc. It is anticipated this will be published in 1976. Later editions will incorporate appropriate supplements covering pressure vessels in other materials, e.g., aluminium and pressure vessels such as air receivers, nuclear reactor vessels, etc., for which individual special standards currently exist. Although reference is made to British Standards, other internationally accepted codes for the construction of boilers and pressure vessels are available such as the American A.S.M.E. Power Boiler and Pressure Vessel Codes and the German T.R.D. and A.D. Merkblatter Codes.

Although these codes are comprehensive for the majority of vessels it is sometimes necessary to revert to a fundamental approach to arrive at suitable scantlings for the vessel and many suitable publications are available giving guidance on this. We suggest references such as *Process Equipment Design*, *Pressure Vessel and Piping Design*, *Pressure Vessel Design and Analysis*, and *Theory and Design of Modern Pressure Vessels*, etc., full details of which are given in the References at the end of the chapter.

Temperature

The use of materials at high temperatures is limited by the reduction in their mechanical properties and their susceptibility to creep, but it should be appreciated that certain materials especially in the higher alloy steel ranges can be subject to metallurgical effects such as loss of ductility in certain temperature ranges. With ordinary mild steels amongst other materials low temperatures are just as undesirable as high temperatures, as the materials are liable to be subject to brittle fracture, and special care should be taken if the working temperature of the material is to be less than 0°C. British Standards such as B.S. 1515: Part I: 1965 give guidance on this problem.

Resistance of the material to corrosion

Corrosion can take effect on pressure vessels in the following way and these principles must govern the choice of material.

(a) Component parts suffer loss of material by corrosion to such an extent that their load-carrying capacity is reduced to an unacceptable degree. If the loss of material is localized, this loss of material can also result in penetration and consequent leakage.

(b) Even without loss of weight, the material may become brittle or crack due to such effects as stress corrosion cracking.

(c) The corrosion products can cause inadmissible impurities to the product or impair heat transfer properties.

(d) The appearance of components can deteriorate to an unacceptable extent or corrosion products can prevent the operation of mechanical parts.

The different safety philosophy required to combat chemical attack of types (a) and (b) is now considered but attack of type (d) is also important if the operation of control valves or systems is affected. When corrosion cannot be eliminated by the choice of suitable materials or protection, acceptable safety can be achieved by designing with a sufficient excess of material to ensure that the necessary load carrying capacity is available throughout the expected life of the component. It should, however, be stressed that the rate of loss of material and the capacity of the component to carry the required load must be checked by inspection at suitable intervals throughout its life. This general concept is well understood but localized wastage can occur due to the build-up of corrosive conditions and this must also be considered.

The embrittling or cracking of components due to any cause cannot be tolerated where they affect safety, and the design philosophy must be to eliminate these possibilities as far as possible by the choice of suitable materials, welding methods and design details.

Construction

The safety of a well-designed pressure vessel in service can be adversely affected by the standards of construction used, but in considering the standard of workmanship and inspection necessary, care should be taken to think of the use to which the vessel is to be subject and the code to which it is to be designed. This can vary from a large water tube boiler or a heat exchanger used in a chemical process where the cost of the materials involved and the difficulties in construction make it necessary for a very detailed consideration to be carried out of the design with low factors of safety being used on the material stresses and a high degree of control of manufacture by independent inspection in accordance with a code such as B.S. 1515: Part 1:1965. At the other end of the spectrum a simple air receiver may be built to B.S. 487: Part 1: 1960, where much higher factors of safety are involved, and for some vessels built to this code very little independent inspection is necessary. A lower standard of weld preparation, etc., can be used but this code is limited to low-strength mild steels, which are in the normal way more easily fabricated.

The jointing of different parts of a boiler or pressure vessel, even in the days of riveting, caused problems and the modern welded vessel has not made this jointing any easier. Although welding appears at first sight to be a means of joining two different parts homogeneously, this is far from the case. It must be remembered that welding is a casting process whereas pressure vessel components are normally constructed from wrought or forged materials, i.e., plates or forgings, and it is essential to appreciate the metallurgical implications of this. It follows that welding is a complex process and the procedures used can have a very considerable effect on the final suitability of the welded joint. This is particularly so when alloy steels are used or different materials are joined. It is therefore essential that the procedure used is specified in detail and steps are taken to ensure that the procedure is followed rigorously throughout the construction of the vessel. These considerations apply equally to the welding of nozzles, branches or other attachments to the shell of a pressure vessel, as to the pressure vessel's main seams. A study in which the author has been involved indicated that the vast majority of cracks which have developed in boilers and pressure vessels built to a high standard of construction have been associated with discontinuities of one kind or another—branches, tube ends, attachments, etc.— rather than with the main shells of the vessels. It is gratifying to see that design codes are now paying more attention to the suitability and quality of welding for these components than they did in the past.

To ensure that a welded joint has comparable properties to the original materials and is metallurgically satisfactory, it is necessary for the actual physical properties of the weld to be investigated by destructive as well as non-destructive tests. The former cannot, of course, be carried out on the actual seams and has to be done using test plates. The approach to the problem used by modern codes is to approve the manufacturer's procedures for the type of welding and the materials used in the vessel manufacture and to approve individual welders for the correct class of work by the mechanical testing of separate test plates prepared by them before the actual production welding is commenced. This general procedure applies both in the case of hand and machine welding. It is essential that the inspecting organization concerned ensures that the weld procedure used for the actual production is the same as that proved by the procedure test in *every detail*, including, for example, the exact geometrical set-up, welding currents, etc.

Figure 16.1 shows a typical weld-procedure certificate and indicates the detail required. These weld procedures should be carried out for all welding— not only the main circumferential and longitudinal seams.

Depending on the requirements of the pressure vessel code used, the thickness of materials, the difficulty of welding the materials selected, the use to which the

Associated Offices Technical Committee

British Engine Boiler & Electrical Ins. Co. Ltd.
Commercial Union Assurance Co. Ltd.
National Vulcan Engineering Insurance Group Ltd.
Scottish Boiler & General Insurance Co. Ltd.

PLATE WELDS IN PRESSURE VESSELS

MANUFACTURER'S WELDING PROCEDURE

Manufacturer's name X.Y.Z.Fabricators	Test record of shop*/xxx* weld procedure *delete as necessary	Procedure no. 16.1

Welder's identity A.N.Other No.xx

Weld preparation   Double butt
(sketch)

(State method and fit-up)

Run sequence and completed weld dimensions

(sketch)

2-d side

Parent material(s)   BS 1501 – 622
                     Grade 31

Dimensions of test piece 1½" thk

Welding position   Downhand

Position of test piece   Flat

Pre-heating and interpass temperature, method and control
200°C Submerged arc
250°C Manual metal arc

Post-weld heat treatment temperature, method and control
680° – 720°C for 4 hours min.
Electric furnace – contact thermocouples

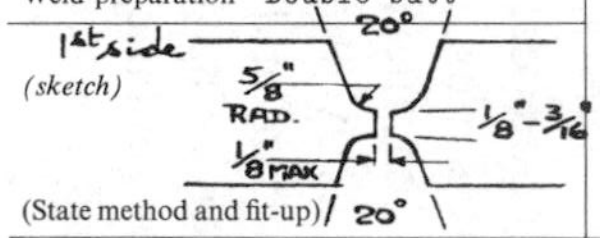

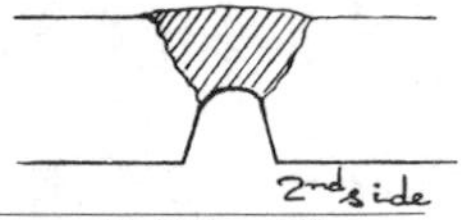

Filler material
Make   ) BS 2901-A32
Type   ) 2¼ Cr 1 Mo
Composition)
Size   5/32"dia. 1st side
        6 SWG 2nd side
Any special baking or drying

Welding consumables
Shielding gas/flux
Type of flux   OP41
Composition of gas   –
Flow rate   –
Electrodes BS 2493 – E 614 HJ
           6 SWG 2nd side

Travel speed (mechanized welding)
15" – 20" per min.

Back gouging
Mechanical chipping

Welding process(es) Submerged arc 17 beads; Manual metal arc 3 beads

Information for particular process Sub-arc 30-32 V, DC positive 400-575 amps.
xxxxxxxxxx MMA 75V o/c DC positive 140/260 amps.

Test results for procedure no. 16.1

Type of test (Delete any or not required)	State: Satisfactory, unsatisfactory or not approved except where numerical results are obtained

Non-destructive tests

Visual	Satisfactory
Magnetic particle	–
Penetrant	–
Radiography/Ultrasonics	Satisfactory

Destructive tests

Transverse tensile	Tensile strength	Yield stress	Location of fracture	Test temperature
	33.8 tsi	–	HAZ	

All-weld tensile	Tensile strength	Yield stress	0.2% proof stress	Elongation	Test temperature
	39.5	–	19.2 tsi	*24.0%	20°C

Guided bend tests

Root	Side	Xxx	Satisfactory
Xxx	Side	Face	Satisfactory
Root	Xxx	Xxx	Satisfactory
Root	Xxx	Xxx	Satisfactory

Fillet weld fracture	–
Macro-examination	Satisfactory
Hardness survey	Satisfactory
Additional tests Impact 2 off Centre weld	Satisfactory

Remarks   Tests in accordance with BS 1515 : Pt.1 and BS 4870 : Pt.1

A.N.Official
for Manufacturer

Date   31 JUN 1976

A.Norman
NATIONAL VULCAN
ENGINEERING INSURANCE
GROUP LIMITED
Engineer Surveyor
for Member Company

(H.O.)

To be completed in duplicate by the Manufacturer and sent to the Head Office of the Member Company for countersignature, the original to be returned to the Manufacturer.

Fig. 16.1

vessel will be put and the factors of safety used in the design code, non-destructive testing will be called for on the materials of construction and on the welding carried out. There is general guidance on this in most modern codes, but the detailed specifications of the non-destructive testing required are a matter for experts. The requirements can vary from radiography of the junctions only of the longitudinal and circular seams on a simple vessel to a case such as a large boiler drum constructed from alloy steels, where the high yield-strength properties of the material are used to keep the weight of metal as low as possible. In this case many penetrations will be fitted and the non-destructive testing required will probably include a total ultrasonic inspection of all plates used in the construction, radiography and/or ultrasonic inspection of all main seams, and crack detection of all fillet welds and penetration welds by a combination of magnetic dye penetrant and ultrasonic inspection as may be appropriate in the particular case.

Although radiography has been used for the non-destructive testing of pressure vessels for many years and produces a permanent record (radiograph), its effectiveness for finding defects is very dependent on the techniques used and the radiograph by itself should never be taken as evidence of quality unless full details of the procedures are also known. Ultrasonics are now widely used and in the construction of nuclear reactor vessels are much more effective in certain cases than radiography. However, in the case of ultrasonics there is only limited scope for a permanent record of the results of an inspection. Very often the assessment of the ultrasonic images produced can only be done by the ultrasonic operator viewing images of the echo produced on the ultrasonic equipment whilst the probe is in actual contact with the area of the vessel in question; it is therefore important that the ultrasonic operator must be an expert in interpretation of the results, since without further tests no verification of these can be obtained. In the United Kingdom various schemes of ultrasonic operator certification run by inspecting organizations such as A.O.T.C. and large users, such as C.E.G.B., have now generally been superseded by the C.S.W.I.P. Scheme, which certifies ultrasonic operators as competent only after a rigorous theoretical and practical examination.

In considering the inspection of boilers and pressure vessels by N.D.T. methods, it is first essential to decide what defects are being looked for and to choose the method to suit. As the most serious defect likely to be encountered in the pressure vessel is usually a crack, it is most important to remember that radiography *only* shows a fine crack if it is parallel to the direction of radiation, i.e., approximately perpendicular to the X-ray film, ultrasonics does *not* show a crack in the direction of the ultrasonic beam and dye penetrants only show a crack which breaks the surface. The foregoing statement regarding radiography and ultrasonics is a general one for guidance; it will be appreciated that in practice there are degrees of latitude and in specific cases a N.D.T. specialist should be consulted.

Installations and fittings

The design and construction of boilers and pressure vessels is largely carried out by experts in this field but the installation of these is often left to personnel who have little appreciation of the problems involved or the dangers that the lack of provision of fittings can present. The requirements for boilers and pressure vessels vary but for ease of reference Table 16.1 summarizes the minimum requirements of fittings for various classes of pressure vessels and boilers. Further general guidance is given in the approprate standards, B.S. 759:1975 *Valves gauges and other safety fittings*, for boilers and their connected piping, and B.S. 1123:1961 for compressed air installations, but the following additional notes will serve to enlarge on the outline given in the table.

Safety valves

The Factories Act requirements for safety valves are summarized in Table 16.1 but in addition the Act requires that the safety valves on a steam boiler shall be separate from any stop valve, and if the valve is of a lever and weight type the weight must be secured on the lever in the correct position. The important considerations which should govern the choice of safety valves are summarized as follows.

(a) The full discharge capacity of the safety valves fitted to any boiler should, at maximum evaporative capacity, be such that the maximum pressure reached is not in general more than 10 per cent in excess of the highest safety-valve set pressure.

(b) The safety valve should always be fitted with hand-operated easing gear so that the free movement of the valve disc can be checked without the necessity of raising the pressure to blow-off. This is particularly important when the safety valve is down stream of a reducing valve and it would be necessary to alter the reducing valve setting to blow off the valve normally.

(c) Safety valves should be locked, and where appropriate ferruled, so that unauthorized personnel cannot tamper with them.

(d) No packed glands should be fitted to a safety valve spindle.

(e) The vent pipe fitted to a safety valve should be satisfactorily drained to avoid condensate collecting in the pipe. The runs of the pipe should also be checked to ensure that no blockage of the pipe is possible due to an accumulation of deposits, formation of ice, etc. Finally the vent pipe should have sufficient area so that it will not present an unacceptable back pressure at the safety valve.

(f) Where, due to the nature of the product involved, seizing of the safety valve is likely to occur, it is essential that steps be taken to ensure that

Table 16.1   Summary of requirements for safety fittings of boilers and pressure vessels

	Safety or relief valve	Pressure gauge	Reducing valve	Water or liquid level gauge	Low-water alarm	Fusible plug	Flame failure device	Level-operated feed pump control
STEAM BOILERS Shell, automatically controlled	F.A.R. *	F.A.R.	—	F.A.R.	F.A.R. One or both essential for safety. Fusible Plug not usually recommended		Yes	Yes
Shell, hand controlled	F.A.R.	F.A.R.	—	F.A.R.	F.A.R. One or both		No	No
Water tube automatically controlled	F.A.R.	F.A.R.	—	F.A.R.	F.A.R.	No	Yes	Yes
Coil, automatically controlled	F.A.R.	F.A.R.	—	—	Automatic fuel cut off required if feed water is not delivered at sufficient capacity to coil	No	Yes	—
Economizers	F.A.R.	Necessary in some cases	—	—	—	—	—	—
Superheater	F.A.R.	Necessary in some cases	—	—	—	—	—	—
High-pressure hot water boilers	Yes	Yes	—	Yes	Yes	In some cases	Yes	Possibly but not essential
Low-pressure water boilers, cast iron or steel	Yes	Yes	—	—	—	—	Yes if automatic	—
Electrically heated water cylinders	Yes	Yes	—	—	—	—	—	—
Steam receivers	F.A.R.	F.A.R.	F.A.R. if supply at high pressure	—	—	—	—	—
Air receivers A: directly connected to compressor	F.A.R.	F.A.R.	—	—	—	Yes	—	—
B: supplied by air line	F.A.R.	F.A.R.	F.A.R. if supply at high pressure	—	—	Yes	—	—
Chemical reaction vessels	Yes	Yes	—	—	—	—	—	—
Storage tanks	Yes unless fitted vents	Yes unless fitted vents	—	Yes	—	—	—	—
Storage tanks where contents are discharged by gas pressure	Yes	Yes	F.A.R. if supply at high pressure	Yes	—	—	—	—

* F.A.R.—Factories Act requirement.

Pressure operated fuel supply control	Low-water fuel supply cut off	Independent extra low-water fuel supply cut off with alarm and manual reset	Fuel supply cut off on failure to relight	Vacuum valve	Vent to atmosphere	Inspectors test gauge connection	Outlet stop valve	Drain and blow-down
Yes	Yes	Yes	Yes	—	—	F.A.R.	F.A.R.	Yes
No	No	No	No	—	—	F.A.R.	F.A.R.	Yes
Yes	Yes	Yes	Yes	—	—	F.A.R.	F.A.R.	Yes
Yes	—	Yes, worked on steam temperature	Yes	—	—	F.A.R.	F.A.R.	Yes
—	—	—	—	—	—	—	—	Yes
—	—	—	—	—	—	—	F.A.R.	Yes
Yes in self pressurized systems	Yes	Yes	Yes	—	—	—	Yes	Yes
No. Temperature operated	—	—	Yes	—	Yes	—	Yes	Yes
—	—	—	—	—	Yes	—	Yes	Yes
—	—	—	—	Yes if not strong enough for vacuum	—	—	F.A.R.	Yes
—	—	—	—	Yes if not designed for vacuum	—	—	Yes	F.A.R.
—	—	—	—	Yes if not strong enough for vacuum	—	•—	Yes	F.A.R.
—	—	—	—	Yes if not strong enough for vacuum	—	—	Yes	Yes
—	—	—	—	Yes if not strong enough for vacuum	Yes unless pressurized	—	Yes	Yes
—	—	—	—	Yes if not strong enough for vacuum	—	—	Yes	Yes

the safety valve will work correctly when called upon to do so. Fig. 16.2 shows a suggested arrangement with a bursting disk protecting the safety valve but it is essential that a pressure never builds up in the interspace between the safety valve and bursting disk and this is ensured by either fitting a vent or a pressure gauge to the interspace. If a pressure gauge is fitted steps must be taken to ensure that it is read regularly and investigations made if it reads above zero.

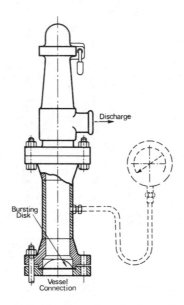

Fig. 16.2 Arrangements for the protection of safety valves by a bursting disk.

Pressure gauges

In general, the important safety requirements for pressure gauges are as follows.

(a) The range of the gauge used should be suitable for the working pressure. It is recommended that the graduation used should be not less than one and a half times but not greater than twice the operating pressure.

(b) A stop pin should be provided below the scale zero position to prevent multiple revolution of the pointer.

(c) The gauge should be clearly marked and in a position so that it can be read easily. There are also certain statutory requirements concerning the distinctive indication of maximum permissible working pressure.

Water and liquid level gauges

In addition to the requirement that at least one water gauge be fitted to a steam boiler, the Factories Act requires that its type be approved by the chief inspector and, where the gauge is of the glass tubular type and the working pressure normally exceeds 2·75 bars that it be protected by an efficient guard so as not to obstruct the reading of the gauge. At least one and preferably two independent water gauges should be fitted direct to the boiler without intervening cocks or valves other than those required for testing purposes, and it is preferred that no other fittings be taken off the connections for a water gauge; it is particularly undesirable for water gauges to be attached to the automatic control float chambers.

Fusible plugs

A fusible plug consists of a brass, gunmetal, or steel holder, in which closure of the orifice is effected by a low melting point white metal alloy seal. These devices are fitted to both boilers and air receivers and perform different functions.

(a) Boilers

When fitted in the furnace of a shell boiler the melting point of the alloy is chosen so that it will not melt when the boiler is operating normally but if the water level falls below the level of the fusible plug the alloy will melt with the result that water and steam will escape giving warning to the boiler operator of the low water condition. However, it must be stressed that the fusible plug will only give warning of a low water condition in the boiler and is unlikely to extinguish the fire even in a hand-fired solid-fuelled boiler and certainly not in a gas- or oil-fired boiler. As a low water safeguard it is only effective if the boiler is continually manned and if the boiler operator is fully trained to recognize the distinctive hissing of the fusible plug blowing.

(b) Air receivers

When air is compressed its temperature rises and is liable to become mixed with oil vapour and an explosive mixture can be produced. However, if a fusible plug is fitted which melts at a temperature below the self ignition temperature of the compressed air/oil mixture, warning will be given if an explosive condition is likely to arise.

It is possible that a charged air receiver may be disconnected from its source of compressed air which is fitted with the necessary safety valves, etc. and if it is subject to a fire in the locality it may explode due to overpressure combined with the weakening of the materials of construction by heat. This can be prevented by the fitting of a fusible plug to the air receiver which will melt when it is subject to the heat of a fire.

Reducing valves

The purpose of this fitting is to control the pressure downstream of itself to a set value lower than that of the steam or gas upstream of the valve. It is then possible to use equipment which is suitably constructed only for the lower of these two pressures. It is essential for safety that a relief valve and pressure gauge are fitted downstream of the reducing valve. The relief valve should be sized so that it is able to relieve, at the lower pressure, the full discharge of the reducing valve in the fully open position. This will involve the use of a relief valve of larger diameter than the reducing valve. In designing the pipe runs for reduced pressure lines it is essential to ensure that the steam or gas flow through the relief valve is not restricted by an opening of lesser diameter than the relief valve size.

Water level controls and alarms

As the majority of serious accidents to industrial pressure plants are due to the failure to maintain the correct water level in boilers, too much stress cannot be laid on the need to ensure that this condition does not arise in either hand- or automatically-controlled boilers. Because of this, the Associated Offices Technical Committee in 1958 made a series of recommendations concerning the minimum requirements for the controls to be fitted to automatically-controlled boilers where they are not constantly supervised. There is now also a pamphlet by H.M. Department of Employment issued in the Technical Data Note series (T.D.N. No. 25).

It is strongly recommended where controls embodying external float chambers are employed that a sequencing type valve be employed on the water connection; this co-ordinates the blowing down and testing for clear waterway function and ensures the control cannot be left in a non-operational condition. Many boilers operate in boiler houses which are manned on a full-time basis but are fitted with automatic feed control equipment. In this case it is considered essential for safety purposes that an audible low water alarm is fitted. However, it should be stressed that with an automatically-fed solid-fuel stoker the air supply as well as the fuel supply must be cut off when any of the automatic low water controls operate.

Fuel supply and flame controls

These controls are essential to prevent flue gas explosions and are considered in more detail in the A.O.T.C. and T.D.N. Recommendations referred to above.

Vacuum valves

Many vessels, although quite suitable for an internal pressure of 50 lb/in^2 (3·45 bar) or more, are not suitable for vacuum which can be produced by condensation of vapours, draining of liquids, or the connection to a pump discharge amongst other things. In these cases it is essential to fit vacuum valves. It is important that these valves are placed in such a position that they cannot be covered with liquid. On rotating vessels, such as drying cylinders, where it is impossible to ensure a single vacuum valve would always be above the liquid level, it is necessary to fit a number so that at least one valve is above the liquid surface at all times.

Vents to atmosphere

In many low- or non-pressurized systems it is usual for an atmospheric vent to be fitted. Where this method of protection is fitted the following points should be noted.

(a) The vent should be connected directly to the boiler or vessel concerned.
(b) The vent pipe should be protected from frost or blockage throughout its run.
(c) No valves should be fitted in the vent line.
(d) If a combined vent and feed make up is fitted no non-return valve or non-return type stop valve should be fitted.
(e) The vent should be of sufficient area considering the use to which the item is put.

In the interests of safety a safety valve should always be fitted in addition to the vent.

Drain or blowdown

When constructing drain systems to handle hot liquids, and especially when considering the blowdown from boilers, care must be taken to ensure that adequate allowance is made for expansion of the pipes. Where a common blowdown is fitted to a range of boilers, the blow-off valve or tap in each boiler should be so constructed that it can only be opened by a key which cannot be removed until the valve or tap is closed and there should only be one key. This is to ensure that only one boiler can be connected to the drain system at a time. Although the case of boiler blowdowns has been cited it is recommended that it is followed for other common drain systems.

Maintenance

There is a school of thought that the only preventive maintenance necessary to a boiler or pressure vessel is the 14- or 26-monthly thorough internal and external examination which is carried out by the Insurance Company's Engineer

Surveyor, but this is certainly not so and the preventive maintenance of pressure vessels should consist of:

(a) regular functional checks on all safety controls;
(b) regular visual examination looking for early signs of trouble such as slight leaks, discoloration of vessel or lagging, slight distortion, unusual sounds;
(c) thorough inspections to locate any actual growing or latent defects which could affect the plant's safety.

Regular functional checks on safety controls

The frequency at which these should be carried out will vary with the type of plant and the use to which it is put. However, a suggested schedule of tests with explanatory notes based on an automatically-controlled shell boiler would be as follows, and could serve as a guide for other boilers and pressure plant.

After carrying out each series of tests, the qualified mechanic or competent attendant making them should ensure that the water level is restored and all the valves are in the correct operating position. He should not leave the boiler until he is satisfied that it is operating normally, with all safety controls operational; this means that he should remain at least another twenty minutes.

Daily

This should be carried out by a qualified mechanic or competent attendant.

(a) Blow down and check the operation of all water gauges—Fig. 16.3 illustrates the correct method of doing this.
(b) Check the operation of the water level control by blowing down the feed control float chamber with the feed pump stopped and note that it starts and then stops when the drain valve is closed again.
(c) Check the normal and independent firing controls and alarms by blowing down the requisite float chambers with the burner in operation and note that the cut-out operates.

Note: the control chambers should be blown down in a similar way to the water gauges.

Weekly

This should be carried out by a qualified mechanic or competent attendant.

(a) Check the operation of all water level controls, firing controls, and alarms by lowering and raising the water level in the boiler.
(b) Check the pressure-operated firing controls by raising and lowering the boiler pressure.
(c) Check the safety valves by lifting with the hand easing gear.

Maintenance and testing of water gauges

It is dangerous to operate a steam boiler unless the water gauge or gauges are in good order and the true level of the water within the boiler can be observed at all times. Personnel in charge of boiler plant should be familiar with the operation and testing of the water gauges.

Testing water gauges directly attached to boiler

The objects of testing are to ensure that there is a clear steam way at the top and also a clear water way at the bottom.

With ordinary water gauges the procedure is as follows—

Open the Drain Cock (D) allowing a clear blow through of steam and water. Close the Steam Cock (S). There should be a clear blow through of water, thus testing the water passage. Open the Steam Cock (S) and Close the Water Cock (W). There should be a clear blow through of steam, thus testing the steam passage. Close the Drain Cock (D), thus allowing the water gauge to be first filled with steam. Open the Water Cock (W) and allow the water to return to the proper level in the glass. Check that all cocks are set correctly.

Sequence of operation

(1) Open Drain Cock (D)
(2) Close Steam Cock (S)
(3) Open Steam Cock (S)
(4) Close Water Cock (W)
(5) Close Drain Cock (D)
(6) Open Water Cock (W)
(7) Check that all cocks are set correctly.

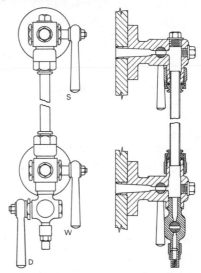

WATER GAUGES SHOULD BE TESTED AT LEAST TWICE EACH EIGHT-HOURS SHIFT

The authorised attendant should systematically test the water gauges several times each day and he should be provided with suitable protection for the face and hands, as a safeguard against scalding in the event of breakage of glasses.

Water gauge glass guards

The water gauge glass guard should be kept clean. When the guard is being cleaned in place, or removed for cleaning, the water gauge should be temporarily shut off. *Make sure there is a satisfactory water level* before shutting off the gauge and take care not to touch or knock the water gauge glass. After cleaning, and *when the guard has been replaced*, the water gauge should be tested and the cocks set in the correct positions.

Maintenance

The water gauge should be thoroughly overhauled at each annual survey. Lack of maintenance can result in hardening of packings and seizure of cocks. If a cock handle becomes bent or distorted special care is necessary to ensure that the cock is set full open. A damaged fitting should be renewed or repaired forthwith, water gauge glasses often become discoloured due to water conditions, also they become thin and worn due to erosion. Glasses, therefore, should be renewed at regular intervals.

A stock of spare glasses and cone packings should always be on hand.

Remember that—

If the steam passages are choked a false high water level may be given in the gauge glass. After the gauge has been tested a false high water level may still be indicated.

If the water passages are choked a false high or low water level may show in the glass. After the glass has been tested it will remain empty for a time unless the actual water level is dangerously high.

If the water gauge is connected to a hollow column, when testing the water gauge steam connections, any cock on the connected water pipe should be closed with the water gauge water cock and when testing the water gauge water connections, any cock on the connected steam pipe should be closed with the water gauge steam cock.

Water gauges should be well illuminated

Fig. 16.3   Instructions for the testing of boiler water gauges.

Quarterly

This should be carried out by a service engineer under maintenance contract.

(a) Service all controls.
(b) Examine and clean out any float or control chambers.
(c) Lift the safety valve under steam pressure.

A log should be kept of all the tests made and any repairs carried out.

Regular visual examination

Looking for early signs of trouble such as slight leaks, discoloration of the vessel or its lagging, unusual noises, etc. is important. With a boiler, the inspection frequency should be at least once a day, but with equipment such as steam or air pipework, once a week may be quite adequate.

Thorough inspections

Generally, the maximum interval between these examinations is governed by the requirements of the Factories Act or other legislation. Such inspection consists of the removal of all access doors into the inside of the vessel and a detailed visual examination of all the internal and external surfaces of the vessel to locate any corrosion, cracking, or any latent defect. This examination may be supplemented by the removal of lagging or brickwork, accurate thickness measurements by drill tests or preferably ultrasonic means, non-destructive flaw detection, sampling for chemical or metallurgical examinations.

The Factories Act and kindred legislation lay down that these inspections must be carried out by a 'competent person' and the actual inspection is best left in the hands of a specialist such as an insurance company engineer surveyor.

Access for maintenance and manholes

Particular attention is drawn to the dangers due to fumes and lack of ventilation when working in pressure vessels and boilers. A separate section of this manual refers to these dangers.

The removal of a manhole door or cover whilst the vessel is still under pressure has killed at least 2 people in the past few years. A pressure gauge reading zero is not a satisfactory way to check this, and any pressure vessel should be open to the atmosphere before any cover bolts are removed. This is also particularly important in the case of autoclaves and other vessels where it is necessary to use rapid opening doors, and it is recommended that interlocks be fitted to prevent the door-opening mechanism being operated until the vent is open to the atmosphere and all pressure released.

The blowing out of the packing from manhole doors has also caused a number of deaths in recent years due to scalding, which could have been prevented if the excessive clearance between the manhole cover spigot and its frame had been rectified.

Legislation

From 1 April 1975 the legislation in the United Kingdom (other than Northern Ireland where the Factories Act (Northern Ireland) 1965 is still in force) governing boilers and pressure vessels has been the Health and Safety at Work etc., Act 1974. As far as boilers and pressure vessels are concerned, this Act has had the following effects:

(a) It lays the general duty on employers to provide and maintain safe pressure plant and the necessary instructions in its use to prevent danger to their employees (Section 2) (similar in effect but with wider implications than the duties of the occupier in the Factories Act 1961).

(b) Lays a general duty on employers and other persons concerned with premises to conduct their activities so as to prevent risk to health and safety to any person other than employees (Sections 3 and 4).

(c) Lays down a duty on the designer, manufacturer, importer or supplier of equipment (including boilers and pressure vessels) to provide safe plant (Section 6).

(d) Extends the area covered by the Health and Safety at Work Act as compared with the Factories Act 1961 to include all areas where anyone is employed (including self-employed persons) except where the employment concerned is as a domestic servant in a private household.

(e) The Act gives power to the Minister to make regulations over a wide area in the Health and Safety Field.

As far as the detailed legislation of boilers and pressure vessels is concerned, the existing legislation continues in force by virtue of the Health and Safety at Work, etc., Act 1974 with minor alterations by Regulations. The major detailed legislation governing the use of steam boilers and pressure vessels is, therefore, the Boiler Explosions Acts 1882 and 1890 and the Factories Act 1961 modified by Regulations made under the new Act in 1974.

Other legislation covering boilers and pressure vessels is found in the Mines and Quarries Act 1954–71 (now modified by (Repeals and Modifications) Regulations 1974) under which were made Coal and other Mines (Steam Boilers) Order 1956, the Quarries Order 1956 and the Miscellaneous Mines Order 1956. As the application of these is limited to certain specialized industries they will not be considered in detail.

In addition to the government legislation on boilers and pressure vessels, many local authorities have bye-laws governing boilers in places of entertainment. These, of course, vary from authority to authority and the local regulations should be consulted.

Factories Act 1961

This is now modified by the Factories Act 1961 etc. (Repeals and Modifications) Regulations 1974. The more important provisions of this legislation dealing with boilers and pressure vessels are summarized as a guide to the provisions but it must be stressed that the full text should always be consulted when a definite ruling is required.

Section 30 details the practice to be followed prior to and during entry to any confined space where there is a risk of dangerous fumes being present, or of there being a lack of oxygen for respiration. It also prohibits working in hot boiler flues or furnaces.

Section 31 details the precautions to be taken with flammable gases and vapours.

Section 32 specifies the fittings to be attached to steam boilers, superheaters and economizers. It also specifies that every part of every steam boiler shall be of good construction, sound material, and adequate strength and free from patent defect.

Section 33 specifies that a steam boiler shall be properly maintained and lays down the requirements for regular inspection by a competent person. Under this section, Examination of Steam Boiler Regulations 1964 have been made, which govern the periodicity and the extent of examinations to be made.

Generally, a thorough examination will be required every fourteen months but this period may be extended for some welded water-tube and waste-heat boilers.

Section 34 restricts the entry into any steam boiler which is one of a range unless all inlets through which steam and water may enter are disconnected or the valves or taps are closed and securely locked. This section also requires that the blowdown system shall be safeguarded against opening more than one blow-off valve (see section on 'Drain or blow-off') at a time.

Section 35 requires that receivers not suitable for the maximum permissible working pressure of the connected boiler or pipe connecting the receiver with any other source of supply shall be fitted with reducing valve, safety valves, etc. Every steam receiver shall be of good construction, sound material, adequate strength and free from patent defect. It shall also be properly maintained and thoroughly examined by a competent person at least once in every period of 26 months.

Section 36 requires that every air receiver shall be fitted with a safety valve, pressure gauge, drain, a suitable manhole or handhole to enable the interior surface to be cleaned and, if connected to a compressor supplying air at a pressure above that for which the receiver is suitable, a reducing valve. Every air receiver shall be of sound construction and properly maintained and shall be thoroughly cleaned and examined at least once in every period of 26 months by a competent person except that in the case of a solid drawn receiver this period may be extended to 4 years.

Section 38 defines 'steam boiler' and other terms used in the Act.

Under the Factories Act 1961 there is no detailed legislation on the construction or maintenance requirements for boilers and pressure vessels, and the only requirement is a general one that all items must be of good construction, sound material, of adequate strength and free from patent defect, and be properly maintained, but it must be remembered that these are absolute in context and place a great responsibility on the user. The general responsibility on the supplier, etc., under Section 6 of the Health and Safety at Work etc. Act 1974 has been referred to earlier.

The implications of this general requirement are well illustrated from the following extract from a judgement given in 1949 following a fatal lift accident.

> '. . . the words "shall be . . . properly maintained" in s. 22(1) of the Factories Act 1937 read in conjunction with the definition of "maintained" in s. 152(1) (meaning maintained in an efficient state, in efficient working order, and in good repair) were imperative and imposed on the occupiers of a factory an absolute and continuing obligation, and there is nothing in the context or the general intention of the Act which could lead to the inference that there should be any qualification of that obligation; to succeed, the respondent need only prove that the mechanism had failed to work efficiently and the failure had caused the accident; and that burden she had discharged.'

It should be noted that in some places the new Health and Safety at Work etc. Act 1974 does use the words 'so far as is reasonably practical' in relation to some of the general duties it imposes on the employer.

Boiler Explosions Acts 1882 and 1890

These are now modified by the Boiler Explosions Act 1882 and 1890 (Repeals and Modifications) Regulations 1974.

The definition of a boiler under the Boiler Explosions Act is as follows:

> 'any closed vessel used for generating steam or for heating water or for heating other liquids or into which steam is admitted for heating, steaming, boiling or other similar purposes.'

It will be noted from this that the Act in general applies to all closed vessels subject to the containment of fluids and the application of heat. Under these provisions the owner of such a vessel which has suffered explosion must report this to the Health and Safety Executive, following which an enquiry, under Section 14 of the Health and Safety at Work etc. Act 1974 can be held. Prior to 1975 these enquiries were conducted by the Department of Trade and over 3 400 official reports have been published since 1882, containing an extremely useful amount of information and providing a number of illustrations of lessons to be learnt from explosions. In the Bibliography a number of the more important and interesting recent reports have been particularly referred to. In addition to reports on individual enquiries, an annual report was prepared and published by Her Majesty's Stationery Office which gives general statistics and also brief details of explosions which have been reported under the Acts but have not been the subject of published reports.

References

1. *British Standards Yearbook*. British Standards Institution, London.
2. Brownell, L. E. and Young, E. H. *Process equipment design*. John Wiley & Sons Inc., New York, 1959.
3. *Pressure vessel and piping design* (collected papers 1927–1959). American Society of Mechanical Engineers, New York.
4. *Proceedings of the symposium on pressure vessel research towards better design 1961*. The Institution of Mechanical Engineers, London.
5. *Requirements for automatically controlled steam and hot water boilers*. 5th edn. Associated Offices Technical Committee, 1972.
6. *Safe Operation of Automatically Controlled Steam and Hot Water Boilers*. Technical Data Note 25. Department of Employment. H.M. Factory Inspectorate.
7. *Prevention of Explosions of Water Heating Systems in Launderettes*. Technical Data Note 34. Department of Employment. H.M. Factory Inspectorate.
8. Bickell, M. B. and Ruiz, C. *Pressure Vessel Design and Analysis*, Macmillan, 1967.
9. Harvey, J. F. *Theory and Design of Modern Pressure Vessels*, Van Nostrand Reinhold, 1974.
10. *The Report of the Court of Inquiry into the Explosion in June 1974 at the Nypro U.K. Ltd chemical plant at Flixborough*. H.M.S.O., London, 1975.

Bibliography

Chuse, R. Unfired pressure vessels. *The A.S.M.E. Code Simplified*, 4th edn, McGraw-Hill Publishing Co. Ltd, 1960.
A.S.M.E. Boiler and pressure vessel code. American Society of Mechanical Engineers, New York.
Samuels, H. *Factory law*, Charles Knight, 1961, supplement, 1965.
Redgrave's Factories Acts. 21st edn. Butterworth, London, 1966.
Boiler Explosion Act, 1882. H.M.S.O., London, 1958.
Boiler Explosion Act, 1890. H.M.S.O., London, 1960.
Factories Act 1961. H.M.S.O., London, 1961.

Health and Safety at Work, etc. Act, 1974. H.M.S.O., London, 1974.

Various Reports of Inquiries held under the Boiler Explosion Acts 1882 and 1890. In particular reports of recent preliminary inquiries numbered: 3437—1965; 3439—1965; 3441—1965; 3447—1966; 3448—1966; 3450—1967; 3451—1967; 3452—1967; 3456—1969; 3464—1972; 3466—1973. H.M.S.O., London.

Accidents and how to prevent them. H.M.S.O., London.

Proceedings of a convention on steam plant availability, Dublin. Institution of Mechanical Engineers, London, 1965.

British Engine technical report. British Engine Boiler and Electrical Insurance Co., Ltd, Manchester.

Vigilance. National Vulcan Engineering Insurance Group Ltd, Manchester.

Brown, J. C., Nickels, P. J., and Warwick, R. G. *Periodic Inspection of Pressure Vessels*, (A.O.T.C.) Institution of Mechanical Engineers Conference, London, May 1972.

17. Manual handling

J. Halliday

Lifting and Handling Instructor,
Central Electricity Generating Board (N.W. Division)

Training in manual handling

What is the problem?

In one year 200 000 reportable injuries were caused by wrong physical behaviour, or by some circumstances leading or contributing to such behaviour. Translate this figure into man-hours and wages cost, cost in administration, national insurance, etc., and there is a considerable national problem.

These were reportable accidents, occurring at work; there are many more similar injuries incurred outside working hours. Aches, pains, discomforts and postural deformities which interfere with production, and inconvenience the individual, never appear in accident statistics.

If we accept the fact that everyone over the age of 30 suffers from some physical disorder incurred by wrong behaviour, the extent of the problem is clearly indicated.

What has been done?

This is not a new problem, but it is an increasing one. This is because the full scope and implications of the injuries have not been realized, and the preventative steps taken have been extremely limited. Of the total injuries, 26·8 per cent are directly attributable to the handling of objects. A large portion are back injuries and hernias, caused whilst lifting and carrying. It is natural to assume that these types of injuries may be caused by handling excess weights, yet the only regulation that exists in connection with this is the Factory Act rule which says 'a person shall not be employed to lift, carry, or move any load so heavy as to cause an injury'.

Who determines whether a load is too heavy or not? The employer, the employee? Until we have a specific directive in this respect, it is impossible to impose any adequate limitations. Even when this is done, the reduction of injuries will be slight. This is because the main cause of these injuries is the adoption of incorrect body positions and faulty handling techniques.

For a long time we have had instruction on the kinetic principles of handling. Unfortunately, instruction has been confined to specific kinds of heavy occupations. Heavy work done in incorrect fashion may lead to immediate injury, but most hernias and disk lesions are incurred by cumulative strain over long periods and happen to people in every walk of life, even sedentary workers. This means that limiting instruction to the moving of heavy objects will only result in a small gain.

What considerations are required?

Our considerations must be aimed at preventing accidents in all kinds of work, not only in industry, but also outside working hours. This means that the worker must be educated so that he is working in correct fashion at all times. There is little point in teaching a man to handle a 45 gal drum in a safe way, if he later hurts his back changing a wheel on his car. Training must be comprehensive and repetitive, especially in the case of older workers who have become deeply indoctrinated into bad practices. Because injuries are incurred cumulatively (slight stress imposed often enough will result in injury), and because injuries happen at home as well as at work, everyone can benefit from this instruction. We must be concerned about weight and also size and shape of objects, at least until people are fully educated on this subject. We must also investigate circumstances and conditions which make it difficult for people to employ themselves in correct fashion. Injuries may be caused by factors which are outside the worker's control. The worker is often forced into difficult and dangerous postural attitudes by wrongly positioned apparatus. The same applies when there is restricted access to machines for repair and maintenance work. Lack of handles, or some proper means of carrying cases, boxes, etc., enforces dangerous holds and leads to tense, overstrenuous work. Wrongly designed benches and chairs, which force poor body positions, lead to aches, pains, postural deformities, and probably serious injuries. Long periods spent in such wrongly seated positions is one of the main contributory causes of disk troubles.

There must be time for safety; rush jobs often lead to a lessening of concern. A lack of knowledge and a completely wrong attitude to manual work not only leads to faulty techniques but also to a seemingly easy, 'lazy' method of working (which actually involves harder, more dangerous practices) and to acceptance of certain standard methods which have remained unchanged for years and which are absolutely contrary to mechanical and physical laws. This individual thinking, assisted by wrong assumptions, leads to a lot of dangerous work being done physically that could be done more safely, easily, cheaply, and efficiently, by employing some mechanical aid.

The main points for consideration then, are:

(a) to introduce some specific regulations governing permitted weights that may be handled, including such points as lengths of time in which such work may be done, distances such weights may be carried, etc.;

(b) to investigate circumstances and conditions which prohibit the correct application of the principles; this involves deeper considerations, but the modification of existing conditions may be possible—a recognition of adverse circumstances is essential, so that the worker can adapt himself with less detrimental results until they can be rectified;

(c) a more scientific approach to all manual work is required; physical work is a skill and must be applied as such—we must eliminate the 'it's always been done that way' outlook and look for methods that incorporate the principles of correct working;

(d) time must not be a barrier—it just is not true that safe methods of working lead to loss of production;

(e) we must consider the greater use of mechanical aids; this does not necessarily mean expensive equipment—skilful working demands a reduction in extreme physical effort, and the use of yokes, slings, chocks, rollers, etc. with maximum effect can induce this;

(f) the reduction of injuries is more likely to result from proper training of the individual in correct working methods and in the application of correct techniques; an examination of the causes of hernias and back injuries, and a look at the principles, will prove that if the latter are applied diligently, at all times, such injuries will seldom occur.

The injuries

Injuries caused by handling are many and varied. Abrasions, cuts, etc., result from the failure to use protective appliances, falls result from non-observance of good housekeeping rules, and direct instructions on these aspects are required.

There is no doubt that a contributory cause of falls, dropping objects, etc., is the wrong use of muscles over long periods. A loss of elasticity and reaction speed in the muscle, created by faulty application, makes a person more prone to this type of accident.

The observance of correct principles will lead to a reduction of all injuries, not the least contribution being the lessening of fatigue. Training must consist of a full discussion on injuries and the causes; remember, it is not only necessary to instruct, but also to convince. This is one aspect of injury prevention that cannot be enforced. There are no rules or regulations that can be laid down; each individual must realize that the avoidance of injuries like hernias and disk lesions lies mainly in his adoption of the correct methods, and he must be convinced of the necessity of their adoption. Because there must be voluntary co-operation and because it is a lack of knowledge that is responsible, the more information that can be presented the better the results must be.

What is a hernia?

A hernia is a protrusion of one of the internal organs through a gap in the walls of the cavity in which it is contained. The commonest point for hernias is at the lower part of the abdominal wall immediately above the inner part of the inguinal ligament, where there is no muscular protection and support.

How is it caused?

Any compression of the abdominal contents towards these naturally weak areas may result in a loop of the intestines being forced into one of the ducts, and a hernia is created. When the body is bent forward from the waist, there is a lessening of the abdominal cavity and this compression is enforced. Common advice, and practice, is for the legs to be kept straight and the feet together when lifting. This *must* result in the bent over position, and this practice has been responsible for more hernias than any other. Lifting a heavy object in this fashion may result in immediate rupture, but regular bending this way may, by a cumulative further weakening of these regions over periods of time, lead to a similar condition. Positioning the body and applying it in a way that does not create this compression will permit hernias in this region to be completely avoided.

Disk injuries

Not long ago, all back troubles were defined as 'lumbago' and preventative and remedial measures were based on this assumption. It is now found that 90 per cent of back troubles are attributable to disk lesions. In order to see how these lesions occur, let us look at the construction of the spine, its purpose and use.

It is a flexible structure composed of 33 bones or vertebrae, 9 of which, in the base, are fused, the remaining 24 permit movement. Between each of these movable bones are pads of fibrous tissue known as intervertebral disks (Fig. 17.1). The main function of these disks is to act as cushions, or shock absorbers.

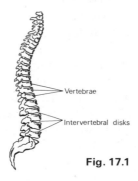

Vertebrae

Intervertebral disks

Fig. 17.1

They are composed of a hard outer cover of cartilaginous tissue and have a soft, gelatinous centre. Man's assuming of the upright position means that the spine is intended to hold a man upright and balance his head on top. In this upright position, any compressional force placed upon the spine whilst lifting, carrying, etc., is distributed throughout the whole length of the column, each disk being able to act in proper fashion, no special stress at any one point.

In Sweden, it was found that tremendous stress may be applied with no detrimental results, providing this position is maintained. However, when the spine is bent, most of the stress is thrown upon the part where bending occurs, usually at the lumbar region where movement is most free, particularly on the fourth and fifth lumbar vertebrae. Merely bending the spine to touch the toes imposes a compressional force of 225 kg at this point. Every 23 kg of additional resistance increases this stress by 136 kg. Due to the bending, all this stress is on one side of the disks, pinching them between the vertebrae. This pinching effect may scar and wear the hard outer cover of these disks so that at some time they become weak and, under pressure, burst. The soft centre can then extrude and this is known as a prolapsed intervertebral disk lesion.

Heavy work done in this fashion may result in rapid, perhaps immediate injury; again, it is mostly the performance of regular, repetitive work of this nature that leads to such injuries. This bending of the spine has become habitual due to exercises like touching the toes, etc. It is possible that the performance of this exercise, especially in the jerky fashion that is usually employed, is a main contribution to disk troubles. To remove it from the curriculum in schools and training routines would result in nothing but benefit and there would be a chance to eliminate the bad habits that invoke great distress.

Muscular injuries

Using muscles statically to hold fixed positions is the most tiring form of work, brings on fibrositis, leads to more serious injuries and promotes postural deformities. It is safe to say that every man in the U.K. over 30 years of age suffers some disorder produced by this type of work. Again, bad habits and wrong attitudes towards work are responsible.

In using the back muscle to hold the body in a bent over position to sweep up, weed the garden, tie shoe laces, etc., the muscles are only holding the body in a fixed position, they are not employed in the actual work, yet they are the first to tire; they are the ones that produce the aches and pains. This way of working also leads to stress on the intervertebral disks and presumed muscular pain may actually be coming from this region.

Carrying with arms bent, using weak arm muscles statically, and poor posture involving an imbalance of muscles, lead to early fatigue, discomfort and resultant deformity. There is sure to be some loss of elasticity in the muscle, the reflexes are impaired and the man is left prone to torn muscles, falls, and every other accident and injury.

Much of this injurious type of work can be avoided by the provision of correctly designed apparatus. Chairs and benches of proper construction, long-handled brushes and spades, correctly placed valves, switches, etc., will permit the adoption of proper body positions and the use of correct working techniques. The elimination of static use of muscles at all times is a certain way to avoid muscular injury and eliminate much of the fatigue.

In the following discussion of the six principles, only a few of the resultant advantages can be listed. Even if they are dealt with fully, this does not exhaust the possibilities of factors that can make work easier and safer, but if training results in the adoption of them it will have justified itself.

The principles

Correct grip

An incorrect grip may lead to an object being dropped; this, in turn, may lead to an accident. There is, at least, a complete waste of effort; nothing having been accomplished for the work expended. A less obvious result is that a finger-tip grip forces the arms to bend. This throws the bulk of the work onto relatively weak arm muscles, which leads to early fatigue, tension, and injury. A full palm grip not only makes for security, it enables the arms to be kept straight and permits the load to be distributed over the whole of the body.

Arms close to the body

When lifting or carrying, this keeping of the arms close to the body is of vital importance. The whole body can now be employed to carry the load and the framework can support tremendous weights without stress. The use of harness and similar aids can be advantageous, yet their employment in manual work is limited. Workers in the east use these methods to carry exceptionally heavy loads for long distances, with negligible fatigue and minimum resultant stress. In the absence of mechanical aids, carrying on the back or shoulders is the best procedure. If the objects must be supported by the arms in front of the body, the arms must be kept straight and as close to the body as possible.

Chin in

Disk troubles are not confined to the lumbar region. The component parts of the spine are very small in the cervical region and disk injuries can easily be incurred here. Slightly elongating the neck and tucking in the chin locks the vertebrae and prevents injury. Direction of movement is controlled by the position of the head in relation to the body. With head and body in proper line any action can be done in proper fashion. Breathing during effort is also facilitated by this position.

Flat back

This is the important part of the whole instruction. The adoption and maintenance of a flat back during effort eliminates compression of the abdominal
contents, there is no uneven stress on the disks and the back muscles are not
being employed in severe and dangerous fashion. In conjunction with the next
principle, this position provides a definite mechanical advantage that reduces
the work done to an absolute minimum, and even heavy jobs can be done in
safe and easy fashion.

Foot positions

When performing any action the body needs to be balanced, the position must
be comfortable, the ability to move must be facilitated. The strong muscles of
the legs must be employed to the best advantage because they form the strongest
part of the body, mechanically intended to do the work of lifting, pulling,
pushing, and carrying. Even to move one's own body is very difficult if done
incorrectly and can result in fatigue and injury; in fact it is the regular moving
of one's own body in incorrect fashion that cumulatively leads to most troubles.
Placing the feet a hip-breadth apart gives a large base, and balance is assisted in
a lateral direction. Putting one foot forward and to the side of an object to be
moved gives an even bigger base and balance in all directions. The performance
of movements (and the ability to prevent unwanted movement) is facilitated.
Most important, the knees can now be bent to lower the body vertically, allowing
the back to be kept flat and in fact, permitting all other principles to be employed
in proper fashion.

A look at Fig. 17.2(a) and (b) will show how a mechanical advantage results

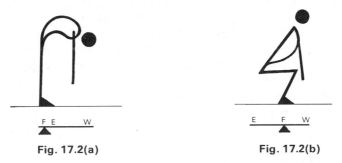

Fig. 17.2(a)   Fig. 17.2(b)

from the adoption of these two principles. In Fig. 17.2(a) where the back is bent,
the worker is off balance forward, additional resistance increases the possibility
of falling. Muscles must be used statically to preserve the balance, and a state
of physical and mental stress is imposed. A third-order lever is being used. This
gives a distance advantage but an effort disadvantage. To lift an object in this
position, one must raise the body back to the upright position. Presume a 23 kg
weight must be moved and the upper body weight is 46 kg, giving a 69 kg load.

If the ratio of the lever is F.W.-6 to F.E.-1, the back muscles must lay out 408 kg of effort merely to raise a 23 kg weight. In Fig. 17.2(b) the strong leg muscles are now being employed and a more advantageous lever is being used. With weight and fulcrum now close together, and the effort behind them, there is a tremendous mechanical advantage.

Use the body weight

Using muscles, especially in disjointed action to start movement, stop, and re-start, is a very uneconomical, fatiguing and dangerous procedure. All movement should be done ballistically. This means that the strongest muscles are employed to overcome the inertia of the object, then they relax and permit the

Fig. 17.3 The illustrated position shows how the object can be lifted without bending the spine or creating abdominal compression. The foot position gives balance and permits immediate movement. Arms are close to the body, the head is in the correct position and a full palm grip has been taken.

momentum of the object to complete the work. Employing movement, to allow the body weight to be used to the best advantage in such action, results in heavy loads being moved with minimum amount of effort and least possibility of cumulative strain. This is efficient work.

These principles are employed in varying degrees in all sporting activities and allusions to this can be used advantageously in training. Weightlifters employ all the principles except the one foot forward. Gymnasts must consider head and body positions, boxers adopt the recommended foot positions for required balance and movement. Wrestlers and exponents of judo use all the factors of the sixth principle to a high degree. This enables 44 kg girls to throw 100 kg men.

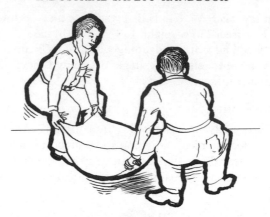

Fig. 17.4   Where more than one person is employed, it is vital that there should be unison in the action. One person should give the orders. As can be seen, the adopted position permits a ballistic movement, allowing the work to be done economically and safely with the legs bearing the bulk of the physical effort.

The fact that these things are taught, at enormous expense of time, energy, and money, so that people 'working' for pleasure can get the best results from effort expended, with minimum possibility of injury, and that there are no such considerations given to work that is essential so that we can survive, creates a ridiculous situation.

We are morally bound to provide this training, which is such a vital factor in the attempt to reduce injuries.

18. Mechanical handling

J. V. Fairclough

*Formerly of British Engine Boiler
and Electrical Insurance Co. Ltd*

Preface

When this chapter was written in 1969, reference was made in a number of instances to the Factories Act 1961, the Construction (Lifting Operations) Regulations 1961 and the Docks Regulations 1934. In updating the chapter in June 1975, it is pointed out that whilst these references are again quoted, Part 1 of the Health and Safety at Work etc. Act 1974 came into operation in January 1975 and Part 2 came into operation in April 1975.

Reference to *Protecting People at Work, An Introduction to the Health and Safety at Work, etc., Act 1974*, by the Department of Employment, indicates under Statutory requirements that:

> 'The existing health and safety legislation listed in Schedule 1 to the Act will be replaced in due course by an improved and updated system of regulations, and codes of practice approved under the Act. All existing statutory provisions will, however, continue in force until new regulations are made.'

The Introduction also indicates under Codes of practice that:

> 'Where appropriate, regulations will be supplemented by approved codes of practice, which will have a special legal status. They will not be statutory requirements, but they may be used in criminal proceedings, as evidence that a statutory requirement has been contravened.
>
> Existing codes of practice and other forms of practical guidance will not automatically be approved. Before doing so the Commission is required to consult appropriate bodies.'

These points should therefore be borne in mind when referring to the Regulations quoted in this chapter.

Statutory requirements

The Factories Act 1961, Section 27(1) and (2) requires that cranes and other lifting machines shall be properly maintained and shall be thoroughly examined by a competent person at least once in every 14 months. The Construction (Lifting Operations) Regulations 1961, in addition to a thorough examination by a competent person every 14 months, requires in Section 10(c) 'as far as the construction permits, be inspected at least once in every week by the driver, if competent for the purpose, or other competent person' and in Section 28(1) it is stipulated that 'no crane, crab, or winch shall be used unless it has been tested and thoroughly examined by a competent person within the previous 4 years'.

It will, therefore, be appreciated that on cranes and other lifting machines coming within the purview of the Construction (Lifting Operations) Regulations 1961 the crane driver is, in many cases, expected to be able to carry out working inspections.

Thorough examinations are normally carried out by suitable qualified outside specialists, e.g., insurance company surveyors, and are clearly more searching than an inspection, and demand far more skill, knowledge, and application than those mentioned above.

The inspection by the driver referred to in Section 10(c) is taken to mean a visual check of an item without any dismantling, measuring, or using of special techniques. Tower crane, mobile crane, and excavator drivers' training courses usually include in their syllabus instructions on the correct and safe use of lifting machinery and the maintenance and weekly inspections by the drivers of these machines; the need to check the following matters is stressed:

(a) tyre pressures on cranes mounted on a pneumatic tyred chassis;
(b) adjustment of controllers, clutches, driving chains, and brakes, etc., to maintain smooth and accurate control of all crane motions;
(c) condition of ropes, chains, and their anchorages;
(d) testing the automatic safe load indicator circuit by means of the test button provided;
(e) the carrying out of checks detailed in the maker's operating instructions;
(f) reports to the plant department of any matters which in the opinion of the driver require special attention.

Site engineer's responsibility for hired cranes

Contrary to general belief, it is the hirer's responsibility to ensure that a hired crane complies with statutory requirements before being taken into use on any site: supervisors should study the *General conditions for the hiring of plant* on the Hire Agreement/Contract and in particular Section 6 'Delivery in Good Order and Maintenance (Inspection Reports)', also Section 8 'Handling of Plant'.

As a result of the House of Lords decision 'Arthur White (Contractors) Ltd. *v*. Tarmac Civil Engineering Ltd', there is no doubt that this clause protects the owners of the equipment hired and it is the hirer, when he acquires responsibility for the safety of others, who must make sure that he is adequately covered by his own insurance.

Rules and procedures

If lifting machinery and tackle are to be used and maintained with safety, certain fundamentals must be observed. These should be incorporated in codes and procedure for safe working drawn up to suit the requirements of the various works and the relevant regulations applicable to sites where the plant is operating. The co-operation of plant manager and site agents is therefore of first importance.

General requirements

Thorough examination by the competent person

The thorough examination of lifting machines begins at the crane wheels or crawler tracks and finishes at the load hook to accord with Section 27 of the Factories Act. Chains, ropes, and lifting tackle suspended from the crane hook are regarded as coming within Section 26 of the Act. Although magnets, grabs, sucker gear, weighing machines, etc. do not fall within the definition of cranes and other lifting machines or lifting tackle, they should be thoroughly examined.

The reports on these examinations are split into three groups—defects, observations, and matters outside the scope of the inspection service. In accordance with the requirements of the relevant regulations, where the examination shows that a lifting machine cannot continue to be used with safety unless certain repairs are carried out immediately, or within a specified time, the person making the report of the examination must, within 28 days of the completion of the examination, send a copy of the report to the inspector for the district. In addition, the competent person should make an immediate report to a responsible official on site.

So far as the testing of cranes is concerned the Associated Offices Technical Committee have published a booklet entitled *Guide to testing of cranes and other lifting machines*, the primary object of which is to provide guidance on the subject of retesting. The information can also be applied to initial testing; in such cases care must be taken in respect of safety margins relative to both strength and stability.

Safe operation of lifting machinery and tackle

Electric overhead travelling cranes

No person should be permitted to drive a crane unless he or she is over the age of 18 years, has passed a medical examination and has been approved at the completion of his training by the examiner appointed by his employer. For the safe operation of electric overhead travelling cranes, each driver should be issued with a manual drawn up by his employer. Drivers should acknowledge receipt of the manual and confirm that the contents have been fully explained to them. Guidance on the medical standards for crane drivers, together with advice on their selection and training, will be found in the Recommendations on the Training of Crane Drivers (Cab-Operated Overhead Travelling Cranes) published by the Foundry Industry Training Committee.

During their training, drivers should be taught to appreciate the fact that electric overhead travelling cranes should be used only for making direct lifts and should not be used for dragging loads from bays on either side of the crane track, nor for dragging wagons; this can cause the load ropes to leave the hoist barrel grooves, with serious results. If the overwinding switch is operated by the loadblock coming in contact with an operating lever hanging below the crab it can interfere with the operation of the overwinding switch with subsequent damage to the ropes, loadblock, bearings and gears.

Failure of the load ropes can lead to the load and loadblock falling, with serious or even fatal results to men working immediately below the crane.

If the cross leads are situated below the crab and have not been effectively guarded and if the main leads are situated below the end carriage and they also have not been effectively guarded, serious damage to the load ropes, through coming into contact with the live conductors, can occur. Drivers should note particularly that if *Danger board(s)* is/are in position on the main isolator switch and the switch is in the 'off' position, they are forbidden to operate the crane until the board(s) have been removed by the owner(s).

Procedures to ensure safe working should be stated in printed rules which must be clearly understood by crane drivers, maintenance engineers, and any other person who has to go in or on a crane, or on to a crane track, for any purpose. The Personnel or other appropriate department should obtain a signed acknowledgement slip in respect of each set of rules issued and the rules should be strictly enforced.

On a crane provided with a driver's cage/cabin and drum type reversing controllers, the driver should check the movement of all controllers with the main switch/isolator in the 'off' position at the commencement of each shift. Care should be taken to ensure that all controllers are returned to the 'off' position before closing the main switch/isolator. If the controllers are provided with 'off' position interlocks, then all controllers must be in the 'off' position before the circuit breaker can be closed.

Reversing the controllers for checking any of the crane movements should be forbidden because of possible damage to the motors. Before leaving the crane unattended on outdoor tracks, the driver should apply the storm brakes provided.

Cross traverse and long travel limits, and overwinding switches, should be tested at the beginning of each shift. Overwinding switches, and overlowering switches, should also be tested after any adjustments to the switches or rope renewals. When carrying out these tests on overwinding switches it is not uncommon to find them 'shorted' out or adjusted to give increased height of lift. In case they have been rendered inoperative it is advisable to hoist slowly, stopping a foot or so from the normal extremity of hoisting, then 'inching' up until the switch operates. Overwinding switches are intended to operate under emergency conditions only, not as a normal means of stopping the hoist motion, and drivers must not on any account 'short' out or alter the setting in any way in order to obtain a greater height of lift.

Electric overhead travelling cranes fitted with sucker gear, used mainly for handling sheet glass, are normally provided with a brake release so arranged that by moving a hand lever in the cage, the pressure exerted by the brake arms on the hoist brake drum is released and the load is lowered by gravity. In the event of a failure of the power supply, loads suspended from the crane hook can be lowered to safety before the vacuum falls sufficiently to allow the load to slide off the suckers. This point should be fully explained to drivers during their training; they should also be instructed to report to the maintenance engineer any failure of the return spring in the hose reeling drum to wind the flexible hose back on the drum as the load is being hoisted. This allows the hose to festoon and become trapped between the load rope and the loadblock pulleys, causing loss of vacuum at the suckers and the fall of the load; this could result in a fatality. Daily testing of sucker gear is recommended.

On cranes employing a two-speed hoist motion using a sliding clutch to select fast or slow speeds it should be stressed to drivers that although it is usual to fit a barrel brake which exerts just sufficient pressure on the hoist barrel to sustain the loadblock when the clutch is in the mid-position, it may not, however, sustain any additional weight, for example slings, and care should be taken to change speeds only under light hook conditions.

Signals

There are two codes of hand signals recommended for use in crane operations. The one published by the Sheffield Area Safety Group (Industrial and Commercial) applies to electric overhead travelling cranes. The other is published by the Federation of Civil Engineering Contractors and applies to cranes having the additional motions associated with scotch derrick cranes, tower cranes and mobile cranes, etc. which incorporate derricking and slewing motions. Drivers must only take signals from the person responsible for the lift and must make no

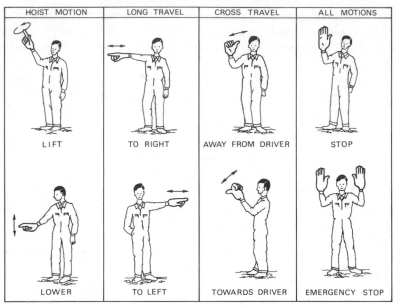

Fig. 18.1 Hand signals from the Slinger's Handbook of the Sheffield Area Safety Group (Industrial and Commercial).

KNOW YOUR CRANE SIGNALS

Signaller should stand in a secure position where he **CAN SEE** the **LOAD** and **BE SEEN CLEARLY** by the driver

Face the driver if possible. Each signal should be distinct and clear.

Fig. 18.2 *Know your crane signals.* The Federation of Civil Engineering Contractors.

movement until such a signal is given. Only signals in accordance with the relevant code should be used.

Overloading

Overloads are forbidden except for the purpose of a test and the driver should demand a weight check on any suspected load. The slinger is responsible for ensuring that the load is properly slung before giving instructions to the driver.

Drivers must not lift any load unless satisfied that it is properly slung and they should be authorized to refuse to handle a load which in their opinion is not safely slung.

Riding on the load

Under no circumstances must any person be allowed to ride on the load or on the empty hook.

Fire

Drivers should also ensure that the fire extinguisher—which should be of a type approved by the fire authorities for working in a confined area—is full and in its usual position, and in case of fire must, if possible, dispose of any load as safely as is practicable before switching off the current and using the extinguisher. If they are for any reason unable to leave the cage by the usual means they should use the emergency escape apparatus which should be provided for this purpose.

Magnets

Particular care is necessary in the operation of cranes fitted with electromagnets. A power cut or a blown fuse may cause a load to fall and it may be thrown a considerable distance. Loads carried by magnets must, therefore, be well clear of places where persons work or pass, and of recognized passages. No person, unless authorized, should be allowed to stand near places where magnets are used.

The following points are of particular importance.

(a) Except when in contact with the load, power to the magnet must always be cut off.
(b) Any pole used to guide an electromagnet must be non-conducting.
(c) No one should be allowed inside a wagon or lorry when it is being loaded or unloaded by means of a magnet.
(d) Care must be taken to see that the magnet cable does not become caught in any projection.
(e) The shunting of wagons by magnets must be avoided.
(f) Swinging of magnets to reach distant places, or the using of magnets as a tup to line up a number of blooms, should not be permitted.

(g) The multilegged chain sling used for suspending the magnet should be periodically inspected and subjected to heat treatment where appropriate.

Grabs

Grabs take many different forms but can be broken down into two basic parts: the jaws or shells which are designed to suit the different materials to be handled, and the top mechanism which is supplied to suit the crane operating the grab.

The simplest grab has two jaws and a single rope-sheave closing mechanism. These grabs may have one holding rope and one closing rope (double-rope grabs) or two holding and two closing ropes (four-rope grabs).

Any grab working from a single line crane must contain mechanism for discharging the material collected and these grabs require careful maintenance. Grease gun lubricators are usually incorporated and regular attention to greasing is essential.

In addition to examinations required under any statutory provisions, ropes and chains require regular inspection and should be replaced when wear exceeds safe limits.

Scotch derrick cranes

Investigations of crane accidents over a number of years have shown that many fatal and serious accidents could have been prevented by the observance of reasonable precautions.

As crane drivers are in some instances expected to carry out weekly inspections, it is suggested that, when preparing the syllabus for derrick crane drivers' training courses, the section dealing with design principles should refer to the fact that while modern scotch derrick cranes are of all steel construction, and that cranes with timber members have been prohibited since December 1939 on building operations, cranes with timber members are still operating in sawmills, foundry yards, and engineering works yards as the Factories Act 1961 does not prohibit their use.

It should be stressed in training that timber members may rot internally and show no visible external signs of this and the shanks of bolts embedded in timber can also waste without any visible signs. Any signs of rot, or bolts attaching gland irons to the back stays pulling through impoverished timber, should be reported to the plant department immediately.

The stability of these cranes usually depends on ballast in the form of concrete blocks sunk in the ground, or surface ballast placed across the end of the two sleepers. As cranes have been known to topple over due to insufficient ballast it is advisable to draw attention to the requirements of the Construction (Lifting Operations) Regulations 1961, Regulation 19 *Stability of lifting appliances*, which requires that after each erection of a crane or alteration affecting the anchorage or ballast, tests should be applied to establish its adequacy.

These tests should be the imposition of a 25 per cent overload above the maximum load to be lifted by the crane as erected at the position where there is the maximum pull on each anchorage, or of the imposition of a lesser load arranged to provide an equivalent test of the anchorage or ballasting arrangements. Caution should be exercised, as it is possible under certain conditions to considerably overstress various components.

In the latter connection, where the stability of a crane is secured by means of removable weights a diagram or notice indicating the position and amount of such weights must be affixed on the crane where it can readily be seen. A report should be made forthwith in the prescribed form and containing the prescribed particulars of every such test and the results signed by the person carrying out the test.

The plant department of most firms arranges for its insurance company to carry out the necessary tests before cranes are taken into use, but it may be overlooked, particularly if cranes are working on isolated sites. If the driver is made aware during his training of the correct procedure he will be able to draw the attention of a responsible official to this requirement before the crane is taken into use.

One of the commonest and most serious type of accident occurring in the use of scotch derrick cranes is the fall of the jib. Such a fall may be due to failure of the derricking rope or part of the derricking mechanism. This may be due to any of the following causes:

(a) the derricking clutch not being properly engaged with the clutch wheel at the beginning of the derricking operation and afterwards slipping out of gear;

(b) lifting the pawl before the clutch is properly engaged with the clutch wheel;

(c) disengagement of the derricking clutch whilst derricking out under power when the jib is at a steep angle and does not react quickly as the derricking rope is paid out;

(d) the load catching under a fixed structure during hoisting and then being suddenly released, throwing the pawl out of engagement with the ratchet wheel;

(e) sudden arrest of the load during lowering, causing a surge which throws the derrick pawl out of engagement;

(f) inefficient interlock between the derricking clutch and pawl.

Falls due to any of these causes usually occur on cranes with no independent brake for the derricking drum, and can be prevented by the provision of an efficient interlocking mechanism between the derrick clutch and the derrick drum pawl or by a self-locking mechanism on the derricking drum. The interlocking mechanism must prevent the pawl being lifted until the clutch is fully engaged, and must also prevent disengagement of the clutch unless the pawl is fully engaged. Interlocking gear is necessary on both power- and hand-operated cranes.

An interlock is not necessary if the hoisting drum and derrick drum can be driven independently with each motion controllable by an independent brake. Various mechanical interlocking arrangements are fully described in Safety Pamphlet No. 15, *The use of derrick cranes* published by H.M.S.O. In addition this pamphlet contains a considerable amount of useful information about these machines.

The gland irons connecting the back stays to the sleepers and mast top pin require careful examination for possible fractures and the bolts connecting the gland irons to the back stays should be withdrawn for thorough examination at least once every 2 years.

Facilities should be afforded for landing the jib to enable the rope to be slacked off for examination and lubrication and also to check carefully the extent of possible corrosion due to the ropes being exposed to the weather.

Splintering of steel wire ropes

Guidance on this subject can be found in The Docks Regulations 1934 (Part III) which states in

20(c) No wire rope shall be used in hoisting or lowering if in any length of eight diameters the total number of visible broken wires exceeds ten per cent of the total number of wires, or the rope shows signs of excessive wear corrosion or other defect which, in the opinion of the person who inspects it, renders it unfit for use.

and The Construction (Lifting Operations) Regulations 1961 (Part IV) which states in

34(3) No wire rope shall be used in raising or lowering or as a means of suspension if in any length of ten diameters the total number of visible broken wires exceeds five per cent of the total number of wires in the rope.

Special care must be taken when examining load ropes on overhead cranes to lower the loadblock so that the section of the rope which is normally inaccessible because it lies round the equalizer pulley—usually situated on the underside of the crab framework—may be examined. Possible damage may also occur on ropes on overhead and mobile cranes on construction sites due to coming in contact with live conductors. Bear in mind also the age of ropes and the question of possible fatigue.

Rope Anchorages

A correct rope anchorage is effected by no less than three bulldog grips each arranged with the ' U ' bolt on the tail or short end of the rope with the pitch of the grips spaced at six times the rope diameter.

When wire ropes are anchored by means of a pear socket anchorage, if the wrong type of wedge is used the rope is liable to pull out of the socket when the machine is under load. As this has led to fatal accidents it is stressed that:

(a) wedges which will pass through the socket should not be used;
(b) there should be between $\frac{1}{16}$ in. and $\frac{1}{8}$ in. clearance between the flat faces of the wedge and the inside of the socket at both sides and matching wedges and sockets should be suitably marked to ensure that they will will be used as a pair.

When the wedge has been fitted and driven home it should project slightly from the socket at both ends. It is still acceptable if the larger end of the wedge projects from the socket when under full-load conditions. A further check should be made after a short period of working.

To prevent the rope pulling through the socket the tail end of the rope should be clamped to the standing part by a correctly fitted bulldog grip or, alternatively, the tail end should be formed into a loop clamped by a bulldog grip.

Mobile cranes

These machines have one feature in common; they all incorporate their own prime mover. The most commonly used systems of power transmission are diesel–mechanical, diesel–electric, diesel–hydraulic, and steam driven. They can be divided into four classifications, self-propelled, lorry-mounted, crawler-type, and rail-type.

Manufacturers of these machines issue comprehensive instruction manuals, and it will be appreciated that each system has its own particular peculiarities of operation and maintenance. These manuals should be studied by the crane driver to ensure that he can operate the crane with the maximum efficiency and safety, is aware of the crane's characteristics and limitations as regards stability and strength, and has the basic knowledge to enable him to carry out the weekly inspections referred to under Statutory requirements. Adequate training, on the lines indicated at the end of this chapter, is also most essential.

The modern mobile crane is built to a safe standard of stability providing the crane is used in accordance with good practice and under good conditions which do not adversely affect stability and are within the manual's service conditions.

The load ratings for mobile cranes apply to firm and level ground only; soft or waterlogged ground is generally unsuitable, unless special steps are taken to provide a satisfactory foundation. It may be necessary to compact the ground, and provide timber sleepers, steel mesh mats, or a concrete raft.

The operation of mobile cranes on demolition sites where the ground may not have been adequately compacted due to the existence of voids such as cellars in demolished buildings, embankments of excavations, and work on the seashore or river beds are matters which should be carefully investigated.

The operation of 'tandem lifting' can be hazardous, particularly with mobile cranes; the detailed procedure should be carefully planned by a competent person beforehand, and in this connection attention is drawn to C.P. 3010: November 1972 U.D.C. 621 873:6418. Code of Practice for the safe use of cranes (mobile cranes, tower cranes, and derrick cranes) which indicates, under tandem lifting, that the safe working load of each crane for the required jib length and operating radius should be at least 25 per cent in excess of the calculated share of the load to be handled by each crane during the tandem lift.

When carrying out investigations of accidents arising from the toppling of cranes involved in these special lifts, it was found that in some cases, when expensive machinery was being lifted and had sustained extensive damage as a result of the accident, that insurance cover against damage to lifted goods, own surrounding property or third party liability and loss of use of the cranes whilst undergoing repairs following the mishap had either not been taken out, or if it had, the sums insured were inadequate. It is therefore strongly recommended that the insurance position be investigated when planning special lifts.

The regulations relating to mobile cranes travelling on highways should be observed, and, in the case of hired cranes, photostat copies of the weekly inspection report, the 14 monthly report of the thorough examination by the competent person and the certificate of test in respect of the crane, should accompany the crane for perusal by the hirer who is responsible for ensuring that the crane complies with the relevant statutory requirements.

Tower cranes

Since tower cranes were introduced into the U.K. there has been a rapid increase in their use. Some of the users have failed to consider the consequences of high-velocity winds, and have taken no measures beforehand to safeguard their cranes by providing wind-speed alarms and issuing instructions regarding precautions which should be taken. Different makes of cranes have been involved in accidents: some with horizontal jibs, some with luffing jibs. The height of the masts have varied from 65–180 ft (20–55 m). The wind speeds recorded at the time of the accidents were 28–90 miles/h (45–144 kmh^{-1}).

Contributory causes on some of the sites were unevenness of the crane track, slewing of the jib and its load into the wind, and failure to anchor, brake, or scotch the travelling motion so as to prevent uncontrolled movement of the crane along its track. In some cases partial failure of the mast structure occurred before the final collapse; in others the effect of the wind on loads with large surface areas caused collapse of the jib, followed by overturning of the crane. Several accidents have occurred where collapse or damage happened during erection, dismantling or alteration of the crane, for example while the height of the mast was being increased.

All the accidents emphasize the importance of the competence and qualifications of those employed to install or control the cranes, and of the supply of essential information to these workers. The value and importance of the statutory requirements designed to check strength and stability is also clearly shown. Too little attention is paid both to installation and to operation of these complicated and valuable machines (see Reports of H.M. Chief Inspector of Factories 1962 and 1966).

Technical data notes

The Department of Employment, H.M. Factory Inspectorate, have published a number of Technical Data Notes which are obtainable free on application to H.M. Factory Inspectorate.

No. 26. *Erection and dismantling of tower cranes*
The aim of this particular note is to set out in general terms recommendations for the procedures to be followed and safety precautions to be observed when erecting or dismantling all types of tower cranes under the following headings:

> Manufacturers' instructions
> Supervision
> General precautions
> Foundations, rail tracks, and hard standings
> Installation of crane base or chassis
> Erection of mast
> Assembly and erection of counter job, counterweight and main jib
> Extending or retracting the mast
> Electrical installation
> Testing and examination
>> (Under this heading reference is made to the effect that a Technical Data Note on the testing and examination of tower cranes is in course of preparation)
> Checks on completion of erection
> Dismantling
> General
> Statutory provisions

No. 27. *Access to tower cranes*
In the introduction to this note it is pointed out that the need for proper means of access to enable the driver of a tower crane to reach the cab safely is generally recognized, but safe means of access to other parts of the crane is also necessary for the purpose of erection, dismantling, inspection, maintenance and repair.

It is stressed that modifications made to cranes for the purposes detailed in the note may affect their strength and stability, and it is therefore important that the makers' advice is sought before any such modifications are made.

It draws attention to the need to provide safe means of access to and egress from any place where persons may be required to work, and of a safe place of work, which is required under the Factories Act 1961 and regulations made thereunder. The statutory provisions are listed and the aim of the note is to indicate means by which these requirements may be satisfied.

It refers to the fact that many tower cranes have jibs which are too shallow for an interior walkway; in such cases it suggests that consideration be given to the use of lightweight mobile cages which can be constructed to run on the jib saddle rails, or slung from the saddle itself, in which case it recommends a remote-control system so that movement is controlled from the cage itself, and that the remote-control system must incorporate interlocks so that if the cage control is operative the saddle traverse control in the cab is locked 'off'. It also deals with the provision of a manual operated cage for access to the counter jib.

For cases in which it is not reasonably practicable to provide a protected walkway on the jib or fit a traversing cage, the note deals with the use of a safety harness with two safety lines so that the wearer can be attached at all times either to the structure or to a taut wire fixed along the length of the jib. It mentions that specially slotted brackets are available which will allow the safety-line connection to slide through them, so that the harness safety line can remain anchored to the taut horizontal wire whilst the workman is on the jib.

Danger of underground and overhead 'live' power lines

In view of the number of accidents on building sites due to excavators coming into contact with 'live' electrical cables and crane jibs coming into contact with 'live' overhead power lines, attention must be paid to the requirements of the Construction (General Provisions) Regulations 1961 (Electricity 44(1) and (2)).

Before excavating work begins the Area Electricity Board's Engineer should be asked for the location and depth of any 'live' underground cables within the working area. So far as 'live' overhead power lines are concerned the safest procedure is to ask the Electricity Board to cut off the power. When this is not practicable, and cranes, excavators, or drag lines have to pass under 'live' overhead power lines, the provision of gauges of 'goal post' type bearing notices marked 'Danger' together with the voltage of the overhead lines and the head-room, is an effective means of preventing jibs coming into contact with the lines. Remember that on lines carrying high voltages, 'flash over' from the lines may take place without actual contact.

There are devices available which utilize either a stout mechanical reinforced fibreglass guard mounted on high-voltage epoxy resin insulators designed to be easily and quickly fitted to the jib of most mobile cranes and similar plant to prevent any contact with overhead power lines or, alternatively, devices designed to sound a warning when the jib gets within a dangerous proximity of overhead power lines.

In the North West region of the Post Office, damage claims amounting to £100 000 per annum are made against Civil Engineering Contractors in respect of damage to underground telephone cables and at least as much again has to be spent in repairing damage when they have been unable to identify the responsible party.

Any contractor about to dig, tunnel, or blast anywhere should enquire by using Freefone 111, before they begin, if there are any underground cables in the vicinity of the place where they intend to work. The Freefone 111 call will be connected to the External Plant Maintenance Control who can usually respond very promptly through their Plant Protection Officers who are specially trained for the job. They are equipped to locate and mark their cable routes and will be responsible for site liason work.

Training courses

Mobile cranes are becoming more complex and the modern crane driver must have a much greater degree of skill than in the past. The safe operation of the machine lies with him. Drivers are often expected to be able to erect the jib to a height suitable to site conditions, reeve the hoist and derrick ropes and check that the height of the 'A' frame (supporting the derrick rope anchorage) is correct. They may also be expected to fit the correct balance weight, correct load radius indicator and set the automatic safe load indicator; it will be appreciated that if drivers are expected to do this, adequate training is essential if accidents are to be avoided.

Large firms operate their own training courses and a number of safety groups are now operating training courses which are essentially practical and extend over a period of 2–3 days. Speakers are drawn from H.M. Factory Inspectorate, crane manufacturers, rope manufacturers, chain manufacturers and testers and insurance companies, to deal with the wide range of subjects covered by the course syllabus.

The syllabus of a course for crane drivers by the London Construction Safety Group is given below as an example.

1. Crane accidents   *Day one*

Accidents caused by overloading
Accidents caused by failure to use safety devices
Accidents caused by abnormal conditions
Accidents caused by faulty slinging, signals, placing, etc.
Accidents caused by faulty erection, alteration, and dismantling procedures.

2. Design principles

Stability
Safety factors

Effects of height, jib length, wind pressure
Types of crane, their advantages and limitations
Rail mounted cranes—special problems
Safe load indicators and cut outs
Questions and discussion

3. Erection, testing, and dismantling

Only trained people to undertake
Adherence to manufacturer's recommended procedures
Base, ground conditions, track level, and correct gauge
Obtain manufacturer's approval for any structural modification
Wind conditions during erection and dismantling
Dismantling procedures of equal importance to erection procedures
Use of correct tools, safety equipment and access
Jibs, strut, cantilever, telescopic and fly
Indicators including Weighload, Wylie, Vickers Nash
Correct use of outriggers
Test loads and procedure to be adopted in testing
Competent person to supervise
Rigging excavator as crane

4. Steel wire and fibre ropes

Construction, inspection, selection, and use
Hoist ropes
Luffing ropes
Slings and their correct use
Storage and maintenance of ropes
Correct method of fixing bulldog grips
Fibre ropes, S.W.L., inspection, and use
Making endless sling from fibre ropes

5. Chains, hooks, and lifting gear

Construction, inspection, and storage
Materials used and their recognition
Correct use of hooks, shackles and eyebolts
S.W.L. and effect of angles of legs for multileg slings
Improper use—shock loads, knots, etc. in slings
Incorrect positioning of eyebolts
Special lifting gear, brickcages, block fork skips
Swivel eyebolts, etc.

6. Maintenance of cranes *Day two*

Maintenance of lifting appliances (general)
Access to cranes
Periodic inspections and examinations
General lubrication
Care of tyres and tracks
Brakes and clutches
Hooks, sheaves and return blocks
Questions and discussion

7. The safe operation of cranes

Crane controls
Use of load indicators and other safety devices
Overhead obstructions
Travelling and positioning
Raising and lowering loads with safety
Overlapping jibs
Tandem lifting
Travelling over sites
Questions and discussion

8. Regulations

The Construction (Lifting Operations) Regulations 1961
Legal duties of contractors
Legal duties of crane operators
Reporting of accidents and dangerous occurrences
Penalties
Responsibilities of driver on highways
Questions and discussion

9. Methods of slinging

Signalling and other methods of communication
Loads—dealing with shape and weight
Stability
Selection of suitable lifting gear
Connecting the load to the hook
Controlling the load
Questions and discussion

10. Practical work *Day three*

Students are separated into parties
Students examine serviceable slings and tackle
Assessing lifting capabilities under varying circumstances

Faulty lifting gear examined and faults discussed

Assessment of weight and vertical centre of gravity of various loads

Correct method of slinging of loads

Practical work on crane with strut type jib

Folding lattice jib including folding, unfolding, addition and removal of intermediate section and the associated tasks, reeving of ropes, selection of radius plate, setting of audible/visual safe load indicator

Crane operations and signalling

Questions and discussion

References and Bibliography

Protecting People at Work. An Introduction to the Health and Safety at Work, etc., Act, 1974, H.M.S.O., London.

The Factories Act 1961, H.M.S.O., London.

The Construction (Lifting Operations) Regulations 1961, H.M.S.O., London.

The Docks Regulations 1934, H.M.S.O., London.

Erection and Dismantling of Tower Cranes. Technical Data Note 26. H.M.S.O., London.

Access to Tower Cranes. Technical Data Note 27. H.M.S.O., London.

The Use of Derrick Cranes. H.M. Factory Inspectorate Safety Pamphlet No. 15. H.M.S.O., London.

Syllabus for Mobile Crane Drivers/Safety Training Course. London Construction Safety Group.

Crane Safety in the Iron and Steel Industry. Technical Report. British Steel Corporation, London.

Recommendations on the Training of Crane Drivers (Cab-Operated Overhead Travelling Cranes). Foundry Industry Training Committee, London.

Hand Signals from the Slinger's Handbook. Sheffield Area Safety Group. (Industrial and Commercial.)

Know your Crane Signals. The Federation of Civil Engineering Contractors, London.

Damage to Post Office Underground Cables. North West Telecommunications Board, Manchester.

Code of Practice for the Safe Use of Cranes (Mobile cranes, tower cranes and derrick cranes). C.P. 3010: November 1972. U.D.C. 621.873:6418. British Standards Institution, London.

19. Mechanical safety and electrical control gear

R. D. Haigh
Chief Engineer,
UML Ltd

Introduction

With the advent of the Health and Safety at Work Act of 1974, the basic obligations of employers to ensure health, safety, and welfare for all their employees at work are clearly specified; those affecting mechanical and electrical safety are to be found in sections 2 and 6, where among other requirements the employer is obliged to provide:

(a) Healthy and safe systems at work.
(b) Healthy and safe working environment.
(c) Safe plant, machinery, equipment and appliances which are maintained in good order.

Furthermore, definite obligations are placed upon manufacturers, sellers, importers and installers.

So far as safe systems of work are concerned, equipment safety can be ensured in the manner discussed in this chapter. However, it should not be forgotten that there are those systems of work which can only be effectively safeguarded by the adoption of permit to work systems which are referred to in Chapter 37 of this book.

Mechanical safety

A general acceptance of the importance of mechanical safety can only be achieved if everyone in the organization is given training in the principles involved. On the engineering side safety consciousness must be engendered in the design department, emphasized in the installation or erection department, and continually demonstrated in the maintenance department. The maintenance department also has a responsibility to feed back safety experience to the design department.

The effectiveness of safety measures in the production department is increased if the line production manager is responsible for safety there. He can then lay down procedures for his operators to carry out routine checks on safety devices and immediately report any failures or abnormalities in their operation.

Safety in the machine shop

The proper layout of machines is an obvious starting point for ensuring safety in the machine shop. Adequate space between machines and between the machines and walls must be provided, taking due account of the type of work to be handled. Although this statement seems to be elementary it is nevertheless not long ago that a case was reported in one of the safety journals in which an apprentice was trapped between a wall and the traversing table of a milling machine while being instructed in its use by a skilled tradesman.

On a number of machines the traversing handles rotate under power whilst the machine is operating, and this can be a hazard even if regulation overalls are worn. Disked wheels instead of handles are useful in avoiding this hazard.

Controls on machines are sometimes placed in unsafe positions. A typical example is a case in which a lathe operator used a bar as a lever to remove the face plate at the conclusion of a job. The control lever for stopping the lathe was near the headstock and the man's body accidentally pressed it, causing the lathe to rotate and trapping his hand between the bar and the machine framework. The control lever should have been designed so that it had to be pulled towards the operator rather than pushed away from him to start the machine. In any case it should have been repositioned in a safer place.

A similar point arises in connection with electrical start buttons on machines. To avoid any confusion, one *start* push-button only should be provided on any machine and should be shrouded so that if any person or object fell against it the machine would not start inadvertently. Conversely, because of the need to make the *stop* push-buttons easily accessible for quickly stopping a machine in an emergency, such buttons should have large exposed mushroom heads. The colours of push-buttons should be in accordance with British Standards Institution recommendations. They should, in certain circumstances, be of the non-resetting type. Regular cleaning of control buttons is essential so that not only do the colours remain clearly visible but also to ensure that the movements are not impeded by layers of grease or dirt.

Guarding of the work on milling machines in a jobbing workshop can be a problem, but various effective adjustable types of guard are now available.

These are illustrated in the Health and Safety at Work Publication No. 42, *Guarding of Cutters of Horizontal Milling Machines.*

Drilling machines can also present problems. As opposed to the telescopic guard which fits over the chuck and twist drill, a device comprising a vertical trip bar located close to the drill and suspended from a snap-action changeover switch has advantages as far as visibility and general accessibility are concerned. This device is usually used in connection with reverse plugging braking of the drill motor, and any displacement of the trip bar by more than a few degrees will actuate the safety circuit. The switch is not self-resetting so that the machine does not restart when the trip bar is restored to its normal position.

Safety in the factory department

Although care is necessary in guarding machines in the machine shop the situation there is helped by the fact that the operators are generally skilled craftsmen and therefore aware of the hazards of the equipment which they are using. In the factory department the process operator may be put more at risk by hurrying to meet production schedules. It is therefore necessary to take even more care in these situations to provide adequate guarding which will not easily be defeated by maloperation.

A great variety of devices is employed but most of the important principles are illustrated in the two following examples of protection of an automatic soap-wrapping machine and a paper-cutting guillotine.

The wrapping machine illustrated in Fig. 19.1 wraps soap tablets which are fed in on the left-hand side of the picture. The machine is fitted on four sides with an open-top welded mesh guard. The 'high wall' feature makes access to moving parts extremely difficult and these can be considered to be safe by position. With the exception of the front hinged guard doors, all guards are fixed and can only be removed with the use of a spanner. Access to the machine for the purpose of effecting repairs or removing a blockage is attained only *via* the two front hinged doors. These doors can only be opened after first raising the guard rail, initial movement of which breaks a magnetic switch (seen in Fig. 19.1 at the top left-hand corner of the fixed guard) and the machine is immediately stopped. The magnetic switch is preferred to other types because it is less easy to defeat its normal operation.

The welded mesh guard depicted in this illustration has the advantage that it affords the operator a reasonably good view of the interior of the machine without producing any reflections. In some cases, however, where total enclosure is necessary, e.g., to protect against dust, a transparent plastic guard may be appropriate in spite of the fact that it does produce some reflection. Care should be taken to use a material which is not easily flammable. The two extension pieces fitted to the side guard protect the belt terminal pulleys and prevent the operator from being 'nipped' between the belt and pulley.

Complementary to conventional mechanical guarding, some modern

Fig. 19.1 General view of automatic wrapping machine with fixed and hinged guards.

machines use a sophisticated system of light beams impinging on photoelectric cells for guarding areas where a high degree of visibility or accessibility is required. On breaking the beam the machine stops. With the guillotine, illustrated in Fig. 19.2, the stopping time is just over $\frac{1}{50}$ s represents a vertical movement of the knife of 3·5 mm. On the same machine there is a requirement for 'hand cranking' for knife adjustment purposes. The machine is equipped with a large flywheel which requires some $3\frac{1}{2}$ min to reach standstill after switching off, and it is essential that the hand-cranking operation be performed with the flywheel at standstill. A process timer is inserted in the machine control system to ensure that the hand-cranking circuit cannot be energized until after the flywheel has ceased to rotate.

For information on the proper use of limit switches and interlocks in association with mechanical guards reference should be made to two H.M.S.O. publications: *Electrical accidents and their causes, 1958* and *Electrical limit switches and their applications* a Department of Employment and Productivity Safety Health and Welfare New Series No. 24 booklet.

Important as it is to provide a high standard of guarding on machines it must also be emphasized that all operators should be adequately trained in the operation of machines on which they work. Systematic training schemes in which the departmental supervisor participates are of great value to make sure that this training is kept up to date.

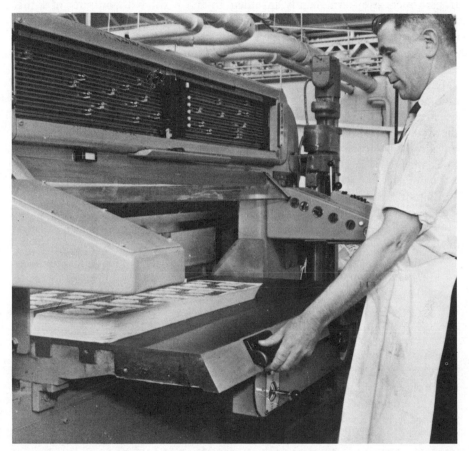

Fig. 19.2   General view of guillotine guarded by photo electric cells so that there is no loss of visibility or accessibility to the machine.

Electrical tools

Portable electrical equipment

Items of portable hand-held electrical equipment such as drills, sanders, and inspection lights are of great help in increasing factory productivity but care must be taken to ensure the use of the correct current rating of cable and h.r.c. fuses, and that adequate inspection and maintenance is provided. The hazards of this type of equipment are greater than those obtaining with fixed equipment because the supply is changed from a permanent installation to a temporary one by the use of flexible cables and plugs and sockets.

To minimize these hazards it is common practice to use some form of reduced voltage supply and two typical systems are worthy of mention. In the first, transformers are permanently installed in the plant with outputs of either 110 V or 50 V a.c. at 50 Hz, the secondary being centre tapped and solidly connected to earth thus reducing the shock risk to 55 or 25 V respectively. The second employs special portable transformers of similar outputs connected on the primary side to appropriately positioned mains voltage socket outlets.

Each system has its advantages. The first, that no flexible connection carries a voltage of more than 55 V to earth, and that no heavy transformers have to be carried about together with the tools. In the second, the total capital cost of the system is less (particularly where there are socket-outlets at mains voltage) but it must be borne in mind that some flexible cables are live at mains voltage to earth.

To prevent any risk of plugging a low-voltage tool into a mains voltage supply a different type of socket-outlet should be used. Suitable industrial metal clad interlocking plugs and sockets are available to B.S. (British Standard) 196.

When using hand-held tools the operator may be subjected to one or more of the following four hazards:

    (a) working in contact with earthed metal,
    (b) working in an inaccessible position,
    (c) working in hot or damp conditions,
    (d) working at a height above floor level.

In any system using Class I tools as defined in B.S. 2769 (tools operating at low voltage and which require to be earthed), socket outlets situated in areas where three or more of these hazards are present should be regarded as high risk and fitted with earth leakage protection, so that in the event of an earth fault occurring the supply is automatically disconnected. In places where two of the hazards exist, the situation can be considered on its own merits to determine whether or not an earth leakage protection should be provided. If only one hazard exists the use of low voltage equipment coupled with a regular inspection procedure of the type outlined in the following paragraphs is considered adequate without the use of earth leakage protection.

The introduction of all-insulated and double-insulated tools (B.S. 2769, Class II) is likely to do away with the need for earth leakage protection, but many installations of the type described will continue in use for some time. Whatever types of steps are taken to achieve safety, regular inspection of all portable equipment, including tools, lamps, or transformers, is essential and the following simple arrangements are recommended.

All equipment should be kept in a central store where there are also facilities for testing. Each item should have a registered number and record card and be issued to a user department only against a duly authorized requisition. Requisitions should be valid for 2 weeks, after which time the equipment must be returned for test unless it has been returned earlier on completion of use.

To ensure that all equipment not returned in this time is followed up, a simple 14-partition filing cabinet may be used. All record cards of equipment issued on a given day will then be filed together with the requisition in one section of the cabinet, marked with a date 14 days in advance of the issue date. Hence on any day the cards in the container marked with that date will show the equipment still out on loan and due for test. The storekeeper will send out a request for the return of the equipment and if this should produce no result in 3 days it will be followed with an 'urgent' form printed in red and prohibiting further use of the equipment. Should this also produce no result the storekeeper will send a man to collect the equipment from the user department. This procedure has proved to be necessary for a large company and experience has shown that only 10 per cent of initial reminders and 1 per cent of 'follow-up' notes need to be sent. It could of course be modified to suit the smaller user.

When equipment is received back in the store after use it should be placed with the record card in a 'for inspection and test' section. The test procedure should include visual inspection of the equipment, cable, and plug, for external damage. For Class I equipment these tests should also include a heavy current resistance test between the earthed metal of the plug and the metal body of the equipment, and an insulation test between the live pins of the plug (shorted together) and its earthed metal. In carrying out the last test it must be remembered that the switch on the portable equipment should be in the 'on' position. For other classes of equipment the appropriate tests are made.

The results should be entered on the record card together with a note of any repairs carried out, and the card initialled by the electrician carrying out the work.

Mobile electrical equipment

There is often equipment in factories which is not permanently wired but which by virtue of its size or position has to be inspected on site rather than in the test room. Voltages of this type of equipment should be kept as low as practicable but because of the power required may involve the use of a two- or three-phase supply. Where mains voltage equipment is essential the cable feeding it should be fitted with duplicate earth paths, one of which is a metallic screen sheathed overall. Earth leakage protection should be provided on all this type of equipment.

The frequency of inspection of the equipment will vary with its type, the situation in which it is used, and the frequency of use. The factory electrical engineer should decide the inspection intervals in the light of these factors, but they should never exceed 3 months.

Each item should have a record card and the cards be filed in a system which provides reminders when inspections are due. The tests carried out will generally be similar to those for portable equipment, but in this case the continuity of each of the duplicate earth paths in the cables must also be tested.

Electrical control gear

The manufacturers of electrical control gear subject all new designs to more stringent conditions than are met in actual service, before putting them into production. Subsequently they test all production equipment before dispatch to the purchasers. If these precautions are supplemented by regular testing after installation the user can be assured of safe and reliable operation. He should also ensure that operators are given training in the correct method of using the equipment.

A simple example is in the use of 'sequence' starting, in which the pressing of one start push-button may initiate the operation of not one but several conveyors. To ensure safety during cleaning and general maintenance, isolating switches should be locked in the 'off' position and their keys retained by a responsible person until work is finished. This precaution should also be supplemented by hanging a notice on the switch worded *Caution, do not start; man working on equipment.*

Where the isolating switches are associated with motors they should be fitted with a fourth pole, isolating the control circuit. This will ensure that if a remote start push-button is inadvertently pressed whilst the switch is open, the motor will not restart when it is closed.

In sequence schemes there may be live contacts in the isolating switch associated with the motor even though it is in the 'off' position. A prominent notice to this effect should be placed on the isolating switch cover as a reminder to the maintenance electrician when he removes it for inspection purposes.

The designers of fuse gear and switchgear also pay considerable attention to the provision of safety interlocks to prevent unauthorized tampering. For example, on combinations of fuses and switches, interlocks are provided so that firstly the cover cannot be opened when the switch is in the 'on' position, and secondly the unit cannot be operated with the cover open. Fuse-carriers and fuse-bases are also designed so that live metal cannot be touched whilst withdrawing or replacing the fuse-carrier.

The placing of electrical equipment is important, and positions beneath water or steam pipes must be avoided. Where it is impossible to avoid such positions adequate screens should be provided to protect the electrical gear because any protective enclosure is vulnerable when doors are opened for maintenance purposes.

To supplement the continuing improvements in writing of national standards and the subsequent care taken by the manufacturing industry, the user must play his part in ensuring safety throughout the life of the equipment. Correct training and strict supervision are therefore essential for all electricians responsible for maintenance work on electrical control gear.

Acknowledgements

The author wishes to thank the Directors of Lever Brothers Ltd for their permission to publish this information, much of which appertains to procedures in their factory at Port Sunlight. He also thanks a number of colleagues in his own and other Unilever companies for many helpful suggestions which have been included in the text.

Bibliography

1. *Portable electric motor operated tools.* B.S. 2769:1964. British Standards Institution, London.
2. *Motor Starters for Voltages up to and Including 1 000 V a.c. and 1 200 V d.c.* B.S. 4941: Part 1: 1973 Direct-on-line (full voltage) a.c. starters.
3. *Plugs, Sockets, Outlets, Cable Couplers and Appliance Couplers with Earthing Contacts.* B.S. 196:1961 British Standards Institution, London.
4. *I.E.S. Code for Interior Lighting.* The Illuminating Engineering Society, 1973.
5. *A Ready Reference to Plugs and Sockets, Low-voltage Transormers and Earth Monitoring Units.* Engineering Equipment Users Association, 1958.
6. *Drilling Machines.* Health and Safety at Work Publication No. 20.
7. *Electrical Limit Switches and their Applications.* Health and Safety at Work Publication No. 24.
8. *Safety in the Use of Guillotines and Shears.* Health and Safety at Work Publication No. 33.
9. *Safety in the Use of Woodworking Machines.* Health and Safety at Work Publication No. 41. Woodworking Machine Regs. 1974.

20. How to minimize fork truck accidents and reduce operating costs

H. Partridge

Managing Director,
Fork Truck Training Ltd

In the perfect situation, industrial accidents would never occur, but as nothing is ever perfect the Industrial Safety Officer is forever striving to reduce accidents to an absolute minimum. In this article I propose to suggest steps that must be taken to minimize accidents with but one item of industrial equipment—the fork lift truck.

Properly driven by a trained operator, the fork lift truck is a willing and most useful tool, but, in the hands of a man who is not properly trained, it can injure, maim and kill. In 1973 there were 21 people killed in accidents involving fork trucks. What is being done to reduce this toll?

Firstly the truck itself. All major manufacturers, world wide, build into their machines a stability factor, both lateral and longitudinal, which is to an internationally agreed standard. This is a minimum requirement and in most cases is exceeded. Attention has been given to protecting the operator by overhead guards, load guards, limb guards etc. The machine is adequately braked in many cases by two systems: electric and mechanical. In short, the machine has been developed over the past 30 years to be as safe as the manufacturer can make it.

But the machine is only as safe as the man or woman operating it, and the question to ask is: what steps must be taken to make the operation of the fork truck as safe as possible?

Assuming that the environment and the ancillary equipment is to an acceptable standard, for example: no damaged pallets, loads properly bonded and stacked, driver and pedestrian hazards identified and dealt with, this article is confined to the person who drives the truck—the operator. It is suggested that in order to achieve the goal of minimizing accidents and damage the following steps are necessary:

1. The potential operator should be properly *selected* in the first place.
2. He should be properly *trained*.
3. He should undergo a practical *end test* which validates his skills training.
4. He should undergo a theoretical test which validates his *knowledge* of safe operating practices.
5. He should be properly *supervised*.
6. He should be medically fit and possess good hearing and eyesight, and should wear his glasses at all times if they are normally worn.

Selection

The days when just anyone was allowed to operate a fork truck are, thankfully, rapidly drawing to a close. However, some of the methods which are superseding this situation are just as hazardous, let us examine and comment on some of these.

    (a) The subjective opinion of a supervisor or manager that 'such and such a man would make a good fork truck operator'.

    (b) The oral interview.

The people, and there are many, who use this approach to selection forget one elementary fact. Fork truck operating is a *manual skill* and as with all manual skills an opinion or an oral interview can do very little to determine a person's aptitude for that skill. It is rather like trying to pick the England football team on an interview basis!

    (c) The practical test.

Some companies insist on a practical test of the candidate in order to judge his aptitude. There are nine critical skills necessary to operate a fork truck safely. A critical skill can be defined as a skill without which a person cannot possibly carry out the total manual skill.

The critical skills necessary to operate a fork truck safely and competently are:

1. Memory for a sequence of operation.
2. Recognition and correction of mistakes.
3. Accuracy.
4. Sensitivity to controls.
5. Smooth co-ordination of hand, eye and foot.
6. Knowledge of safety factors.

7. The ability to relate the amount of steering lock to the radius of the arc described by the actual and potential outermost points of the truck.
8. The ability to judge height, tilt and position of forks.
9. The ability to approach a load squarely and centrally.

I have personally never known any practical test (with one exception which I will deal with shortly) which is designed to test an operator's aptitude and which contains more than four of the nine critical skills.

In 1971 this problem of fork truck operator selection was studied by The Industrial Training Research Unit of University College of London, with assistance from my own company and the Road Transport Industrial Training Board. From this work emerged, in my opinion, a foolproof method of accurately determining a person's potential to make a safe and competent fork truck operator. The technique is called *trainability assessment,* and the unit has since extended the technique to select such diverse skills as sewing machinists, electronic assemblers, carpenters and even dentists (after all, dentistry is largely a manual skill too!). Any reader wanting a full description of trainability assessment can obtain it from the unit. For the purpose of this article it is sufficient to say that trainability assessment is a *practical interview.* It allows the candidate to express himself practically instead of verbally. The candidate is given a short period of instruction in the truck controls, steering and hydraulic controls and how to transport a pallet. At the end of this period of instruction the candidate is given a short practical end test which must be carried out without help. It covers all that the candidate has been taught and the errors noted as they occur. From the number of errors the tester can place a man in one of five potential categories:

(A) Exceptional
(B) Above average
(C) Average
(D) Below average
(E) Unsuitable.

We have now tested many hundreds of trainees by T.A. and not once has the end test result been contrary to our prediction.

Even for applicants who have no experience of operating fork trucks, the period of instruction only takes 35 minutes and the practical test 10 minutes. At the end of that time the potential ability of the applicant can be accurately gauged. It is *essential* that trainability assessment is carried out by a qualified fork truck operator instructor who has been trained in the technique; this training takes one day.

Training

With a proven method of selection the applicant must be effectively trained. Two requirements are absolutely vital: an effective training programme *must* teach all the nine critical skills and the trainee must be given all the knowledge necessary; in short each skill is analysed and everything the trainee needs to know and everything he needs to do in order to perform that skill safely and competently.

Luckily for British Industry this quite considerable task has been done for them by the Joint Committee of the Industrial Training Boards in their book *The Selection and Training of Fork Truck Operators*. The book outlines step by step how to train a fork truck operator and how to end test him, as well as giving information on trainability assessment. It is an excellent training manual, in fact the only one in my opinion for general distribution. I only quarrel with one aspect of its recommendations. The book suggests a training course should last 5 days with a ratio of trainees to instructor at 2:1. My own company's experience from training many thousands of fork truck operators is that, with *fully experienced instructors*, a trainee portraying sufficient aptitude can be brought to a degree of skill to pass easily the quite difficult end-test in $2\frac{1}{2}$ days at the trainee–instructor ratio of 3:1. However, for the average part time instructor found throughout the U.K., my suggestion would be to stick to the R.T.I.T.B.'s recommendations.

The instructor

The instructor is *the* key man in any training programme. Far too many employers labour under the delusion that anyone, from the supervisor to another fork truck operator (particularly the latter) can train the potential operator. He just cannot do this. Another operator can *show* the trainee how to operate but he cannot teach him simply because he has no training in *methods of instruction*. Worse than this, if he in turn was shown how to operate by yet another operator, then the odds are heavily in favour of him showing the new trainee all the usual tricks of the fork truck cowboy. A properly trained instructor will *teach* the trainee. He will constantly involve him and make him think his way through the problems as they arise. You may have an instructor and would like to know how good he is. A very good rule is to listen to him giving a period of instruction; if when the trainee fails to, for example, turn the truck square to the pallet, the good instructor always stops the trainee and *asks* him to say what went wrong, whereas the incompetent instructor *tells* the man what he did wrong. The memory retention factor of good instruction is many times that of poor instruction.

Testing

In order to validate the training it is essential, as we are dealing with a manual skill, that the trainee takes a practical end-test at the conclusion of training.

It is advisable that he takes a theoretical test on safe handling practice to test what he knows. As stated previously, there are nine critical skills in fork truck operation; it therefore follows that the practical end-test should test each of these at least once. It is also essential that it is a test of *safe* operating as well as *competent* operating. The practical end-test, recommended by the committee of the Combined Industrial Training Boards and available from The Road Transport Industrial Training Board, tests each of the critical skills twice in each test. It is simple to set up and mark but it provides a real test of the trainee's skill and knowledge. The pass mark is 60 but the average operator should obtain a mark of 72; a mark of 80 plus indicates a trainee should make an above average operator, and those that obtain 92 plus invariably make exceptional operators if properly supervised.

The written test should always be of the multi-choice type with at least three and preferably four alternative answers to each question. Again the R.T.I.T.B.'s booklet contains an example of this type of test.

Certification

When, and only when, a trainee passes the practical end-test should he be issued with a certificate or company licence.

This should indicate the following:

(a) The employee's surname and full initials

(b) The *type* of truck on which he was tested, e.g., reach, counterbalance, etc.

(c) The categories of truck which he is allowed to operate

(d) The date he was end-tested

(e) The instructor's name

(f) The company stamp.

It is advisable to number all licences or certificates keeping a register of employees' names and certificate numbers. It is also advisable to re-test on a triennial basis or when the operator reaches the age of 50.

It is the writer's experience that space should be left for endorsements for unsafe practices; I have yet to meet a fork truck operator who does not want to keep a clean licence! My company issues its test certificate in a pocket-sized plastic folder with the certificate on one side and the safety rule book (Fig. 20.1) on the other. If an operator is required to drive over the public highway, it is essential that he has a driving licence for a car.

Supervision

If we select, train and end-test properly, all we need now to achieve our goal of reducing fork truck accidents to an absolute minimum is *well-trained supervisors*. By supervisor I mean the man or woman to whom the operator directly responds, be he chargehand, foreman or warehouse manager.

This is the area most neglected in the U.K. and it is here where proper training

Safety is one of the most important aspects of your training programme. For this reason we have collected together in this pocket sized reminder the most important safety rules you have been taught. Never forget that a safety minded operator protects both himself and others and eliminates the risk of damage to the truck and its load.

Learn these rules thoroughly.

● NEVER—lift loads which exceed the truck's maximum capacity shown on the nameplate.

● NEVER—tilt the mast forward when carrying a maximum capacity load, except when over a stack.

● NEVER—travel forwards with a bulky load obscuring your vision—travel in reverse.

● NEVER—travel on soft ground.

● NEVER—carry passengers.

● NEVER—block fire-fighting equipment by parking the truck or stacking the load in front of it.

● NEVER—attempt to carry out repairs, leave this to a qualified maintenance engineer.

● ALWAYS—observe floor loading limits—find out what is the unladen weight of your truck.

● ALWAYS—watch out for overhead obstructions

● ALWAYS—ensure the load is not wider than the width of the gangways en route.

● ALWAYS—when driving on inclines follow these rules:

   ● when a load is carried, the **load** must always face uphill

   ● when no load is carried, the **forks** must always face downhill

   ● adjust tilt to suit gradient and raise just enough to clear the road.

● ALWAYS—travel at a speed consistent with road and load conditions.

● ALWAYS—sound the horn and slow down for corners.

● ALWAYS—avoid sudden stops.

● ALWAYS—travel with the forks lowered—maintaining ground clearance.

● ALWAYS—ensure that bridge plates are secure and strong enough to withstand the weight of the truck plus the load.

● ALWAYS—carry out a pre-shift check.

● ALWAYS—take note of the load capacity indicator when fitted.

● ALWAYS—lower loads as soon as they are clear of the stack.

   ● lower heavy loads slowly.

   ● leave your truck with the forks fully lowered.

● ALWAYS—remove key when you leave the truck.

FINALLY, REMEMBER ONLY AUTHORISED DRIVERS SHOULD OPERATE TRUCKS.

FORK TRUCK TRAINING LTD.
Hook, Hampshire
Telephone Hook 2700, 2936, 2750, 2752

Fig. 20.1

is most effective in cost and accident saving. The rule is quite simple but so often ignored:

The supervisor must know when an operator is performing in an unsafe manner and *he must know what the consequences will be if it continues*.

Why is this so? The operator, like all of us, is human; we have trained him both on the truck and in the lecture theatre. He has passed a difficult practical end-test and been tested on his knowledge of safe operating practices. Who can blame him if some supervisor who has never sat on a truck starts telling him what to do?

The action of the average fork truck operator when told that he must not operate in a certain manner is to ask the simple question, 'Why not?' If the supervisor cannot give an acceptable answer or says: 'Because I tell you', he is lost.

To take a typical example, one of the most common errors to which fork truck operators are prone is turning the truck with the load elevated. If a supervisor pulls the operator up and is asked: 'Why?' he should say something like: 'When the load is in the air the truck is far less stable than when the load is in the carrying position. Turning with the load elevated can lead to the truck overturning sideways, probably with you under it.'

It is my experience that operators always take notice of a *supervisor who knows what he is talking about*. So train your supervisors and cut accidents and costs.

Costs can be cut in five ways:

1. A reduction of accidents to truck operators and factory personnel
2. A reduction of damage to goods handled
3. A reduction of damage to the fabric of the workplace
4. A reduction of damage to the truck
5. A reduction in truck down-time.

Our goal is to reduce accidents and damage involving fork trucks to an absolute minimum, so to sum up:

Select by proven methods. *Train* by proven methods using a *qualified instructor*. *End-Test* to the R.T.I.T.B.'s standard. *Supervise* by supervisors who *know* the *consequences* of unsafe operating techniques.

Follow these simple but essential rules and your accident and damage rate will most surely fall.

21. The construction industry

J. A. Hayward, M.B.E.

Formerly Superintending Safety Officer,
John Laing Construction Ltd

During the past 20 years the number of people killed each year in the construction industry has averaged one man for each day worked. The casualties for the year 1973 were 230 killed and, in addition, 35 947 accidents reported to the Factory Department. How many others were not reported is a matter for conjecture. In his annual report, H.M. Chief Inspector states that the construction industry has the worst record of any with which he has to deal. This is despite new regulations which have been changed, revoked, and revised, until we have probably as fine a set of regulations for the prevention of accidents as any industry in the country.

We cannot deal here at any length with all these regulations in detail, but the main causes of accidents are shown, together with preventative methods, and mention made of the number and sets of regulations which might easily apply. The Health and Safety at Work Act 1974 and certain sections of the Factories Act 1961 (Section 127) apply to the construction industry. The other main codes of Regulations and Orders which apply are listed at the end of this chapter.

The Construction (General Provisions) Regulations 1961

This code of regulations came into operation in 1962. It applies to building operations and to works of engineering construction, and to lines and sidings used in connection with them, which are not part of a railway.

The responsibility for carrying out the regulations rests on every contractor and every employer of workmen engaged on building and civil engineering work. Regulation 3(1) of this code specifies who is responsible for carrying out particular regulations. Unless otherwise stated, the duty is an absolute one, if the work cannot be done in accordance with the regulations, then the law says it must not be done at all. The employees have a duty under Regulation 3(2) to co-operate in the carrying out of the regulations and to report any defect on the plant or equipment to their employer, foreman, or safety supervisor.

Every contractor and employer with more than 20 employees, whether or not they are all on one site, or are all at work at any one time, must appoint a suitable qualified safety supervisor (Regulation 5). He need not be employed full-time, but his safety duties must take priority over any other job which he has to perform (Regulation 6(1)).

The regulations deal with a wide range of subjects which relate to such general matters as:

1. Safety in excavations, shafts and tunnels, the supply of timber or other suitable material to be used in shoring them and the need for inspections and examinations by competent persons at the commencement of every shift, etc.
2. The type of supervision on such work.
3. Dangerous or unhealthy atmospheres and ventilation of excavations etc.
4. The need to prevent inhalation of dangerous fumes from grinding, cleaning, spraying, etc., at all places of work, the danger of fumes from internal combustion engines in enclosed or confined areas.
5. Transport of workers by water and the prevention of drowning.
6. The methods of laying rails and rail trucks, competence of drivers of machines on construction sites (they must have attained eighteen years of age); riding on machines not properly adapted is prohibited.
7. The section on demolition states that all supervisors must be competent in such work and further deals with fire and flood risks and dangers caused through over-loading structures during demolition work.
8. In the miscellaneous section, fencing of machinery is clearly laid down, the need for protection from falling material, the lighting of working places and protection from projecting nails and loose materials, the lifting of excessive weights.

The Construction (Lifting Operations) Regulations 1961

Cranes

All working parts, drums, ropes, anchorages, and fixing devices of cranes should be well constructed of sound material, and be of sufficient strength for the job. They should be erected under the supervision of a competent person experienced in that type of work, and should be well maintained.

The stability of a crane is of paramount importance at all times. Sloping ground, soft uneven surfaces, and weather conditions all affect the stability of a crane, and must be allowed for. Cranes must be securely anchored or ballasted. Ballast should be secure against accidental displacement.

All cranes must have a current test certificate, which requires renewal at times varying with the type of crane. All cranes must be retested by a competent person after structural alterations.

All jib cranes must have an automatic safe-load indicator which is correctly set to give the driver both visual and audible warning when the crane begins to operate above the safe working load (s.w.l.). The s.w.l. at varying radii must be clearly marked, and be visible to the driver.

All dangerous moving parts must be guarded and all control levers and handles should have locking devices to prevent accidental displacement and/or falling of the load.

The driver's cab must be provided with adequate protection against the weather and afford ready access to all parts requiring periodic maintenance.

When a crane runs on rails, the rails should be laid on firm, level ground on a properly constructed and tied track with stops or buffers provided at each end. There must be no risk of derailment. The crane must have effective brakes.

Lifting gear

All lifting gear must be of good construction, made from sound material and of adequate strength. Hooks should be fitted with safety catches to prevent displacement of the sling or load. All lifting gear must be marked with means of identification, and with the s.w.l. unless this can be found on the test certificate. They must be inspected, tested and annealed if necessary at specified intervals.

Crabs and winches must be marked with the s.w.l., as should pulley blocks and gin wheels if used for more than 1 ton load. All parts of the framework of a winch should be metal and it should have effective brakes capable of controlling the load.

The beam of a gin wheel must be of adequate strength. It must be secured at two points so as to support the load without undue movement, and the wheel must be adequately secured to it.

The drivers of cranes should be fully trained and experienced in the operation of the appliance they control, and must be over 18 years of age, except when they are under training.

Whenever there is the slightest possibility that the driver does not have a completely unrestricted view of all movements of the load and the crane, a fully trained banksman should be provided. The banksman must give distinctive and clearly understood signals, and be over 18 years of age, unless he is under training.

Hoists

It is a statutory requirement that all accessible parts of hoistways be fully enclosed, and as all parts are usually accessible at some time or another, the whole of the hoistway should be enclosed in a substantial gauge wire mesh. Gates must be provided at each access point at least 2 m high and kept closed except when loading or unloading. Operation of a hoist should only be possible from one position, and the operator should have a clear view of the full travel. If this is not the case an arrangement for signals to be made from each landing should be instituted.

Automatic devices must be fitted to prevent overrunning of the hoist platform above the highest point to which, for the time being, it is constructed to travel, and safety devices must be fitted to support the platform with full load in the event of the failure of the hoist gear or ropes. The brake must operate automatically when the hoist is not in operation. The s.w.l. must be marked on the cage, and notices, prohibiting the carrying of passengers, should also be fixed to the cage.

When it is permitted to carry passengers on hoists certain extra requirements are necessary. The cage must have gates so constructed that they afford complete protection to people within. Interlocks must be fitted so that landing gates can only be opened when the cage is at that landing, and so that the cage is prevented from moving if the hoistway gates are open. Similar interlocks should be fitted to the cage gates. An automatic stopping device must be fitted so that the cage stops before reaching the lowest point of travel. In addition to the s.w.l. the maximum permitted number of passengers must be clearly shown in the cage, and this is not to be exceeded.

If loaded trucks or wheelbarrows are carried, their wheels must be secured to prevent any movement, and their load must be so stacked that no part of it can fall off. Wherever possible the truck or barrow should be so placed on the platform that the person off-loading it has no need to enter the cage.

Any person using a passenger hoist has a statutory duty to close the landing gates immediately after he has used it, and his employer must see that this is obeyed.

Drivers of all lifting appliances must be eighteen years of age and trained and competent to operate that appliance. A lifting appliance is clearly defined and means a crab, winch, pulley block or gin wheel used for raising or lowering and a hoist, crane, sheer legs, excavator, dragline, piling frame, aerial cable way, aerial ropeway or overhead runway.

The Construction (Working Places) Regulations 1966

This code of regulations came into operation on the 1st August, 1966, and applies to building operations and works of engineering construction, and to lines and sidings used in connection with them, which are not part of a railway.

The regulations are concerned mainly with scaffolds (dealt with as a separate subject below), safe means of access, and roof work.

Regulation 3 allocates responsibility among various persons for complying with these regulations. Employers are responsible for their own men for any breach of the regulations. Contractors are also liable for breach of the regulations by their sub-contractors. Persons employed also have a duty under this regulation to comply with the regulations so far as their own acts are concerned, and to report defects to their employer, foreman, or safety supervisor.

The Construction (Health and Welfare) Regulations 1966

This code of regulations came into operation on the 1st May, 1966. It applies to building operations and to works of engineering construction, and to lines and sidings used in connection with them, which are not part of a railway.

The responsibility for carrying out the regulations rests on every contractor employing workmen on a site.

The regulations permit other contractors to share the facilities provided by the main contractor in respect of first aid, ambulances, shelters, and accommodation for clothing and taking meals, provided the main contractor gives a certificate containing the approved particulars.

The regulations also lay down when a trained first-aid attendant should be employed on the site; the scale of washing facilities, and sanitary conveniences to be provided, and the provision of protective clothing for use in inclement weather.

The main causes of accidents for a normal year are shown in Figs. 21.1 and 21.2 which show first, the causes of fatal accidents that were due to persons or materials falling, and second, the causes of the reportable accidents.

Whilst the percentage of falls of the reportable accidents does not spotlight to the same extent the percentage of falls of the fatal accidents, the pie chart of the reportable accidents clearly shows that the highest percentage is due to falls.

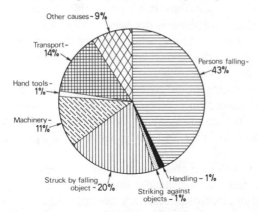

Fig. 21.1   Analysis of causes of fatal accidents in the construction industry.

The best way to deal with this problem of accidents is to follow the natural order of the progress of its construction, from the time a contract commences, through its various phases until it is completed, and having served its purpose its eventual demolition.

A contractor must, before starting a building operation or a work of engineering construction, notify the factory inspector of the district of the nature of the operation being undertaken, on the official Form No. 10. This notification is not necessary if the work is likely to be completed within 6 weeks.

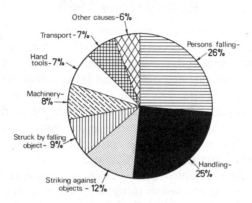

Fig. 21.2 Analysis of causes of reportable accidents in the construction industry.

Excavations

After the erection of the temporary huts and offices etc. on any building site, large or small, foundations must be prepared. This invariably means digging holes in the ground. Experience shows that when accidents occur in excavations they are usually extremely serious. In fact probably 1 in 10 turn out to be fatal.

Therefore, when excavating it is necessary to understand that the Construction (General Provisions) Regulations 1961 apply, and both employers and employees have certain duties and obligations to fulfil. They are detailed in Regulation 3 of this code.

It is essential to ensure the safety of workpeople who may be working in or around the excavations, and protection for employees of other firms working on the same site. Protection must also be provided to ensure that the general public are not liable to suffer and that nearby buildings will not be adversely affected.

There are many misconceptions in regard to safety of excavations, and it should be known that there is hardly any type of ground whatsoever which, given the right set of circumstances, may not collapse.

It is necessary to prevent this, either by battering the edge of an excavation so that the ground slopes to its natural angle of repose, or to supply and use suitable materials to stop the sides from sliding in. The nature and extent of these materials vary according to the nature of the ground and prevailing weather conditions. Some types of ground are likely to collapse in wet weather, others are more likely to collapse when they dry out. One or two obvious examples of this are with sandy and chalky types of ground, which tend to collapse when they become wet, whereas the reverse process occurs in clay soils.

The nature and the lie of the ground will also affect the conditions considerably. This applies particularly when digging on a slope, as pressure from the higher part will obviously be exerted in such a manner as possibly to cause collapses.

Experience has proved quite conclusively, that even apart from injury or death which may result, trench collapses are costly. A study undertaken some years ago also showed that one cannot forecast with any degree of accuracy when a trench is likely to cave in. This study showed that in certain circumstances trenches have caved in within seconds or minutes, and that on other occasions they have not caved in for weeks or months after the original cut has been made.

Besides making sure that the sides of excavations do not fall on to people who may be working in them, there is a need to ensure that people, and materials, do not fall in from the top. Materials must be kept well back—1 m from the edge of a trench, and this includes the soil which has been dug out. There have been many serious accidents when large stones or boulders have fallen because this elementary precaution was not taken.

Proper barriers should be provided to prevent people falling into the holes. Obviously the best type of barrier is something which is solid enough to prevent a person from accidentally stumbling and falling. Ropes and spigots are considered adequate provided they are far enough away from the edge to warn people effectively that there is an excavation; also, that the spigots are not driven into the ground in such a way as to cause damage or weaken the side of the excavations. Any buildings which are close by should be properly shored if there is any possibility of them being adversely affected. Such shoring needs to be done before the excavation starts and it must be remembered that the sole purpose of the shoring is to prevent movement of the building. Considerable damage may be done if shoring is too tight. By the same token a movement can occur if it is not tight enough.

Scaffolds

The vast majority of accidents in the construction industry are as a result of falls, of persons and of materials. Falls of persons from heights can be prevented if scaffolds and working places are properly constructed and protected by the provision of guard-rails and toe-boards.

The base

Just as the foundations for a building are most important, so are the foundations for a scaffold. Before erecting any scaffolding it is necessary to prepare the ground which is to be the base. The ground must be levelled off and compacted so that there will be no movement once the scaffold is erected.

Sole-plates acting as bearers should then be placed in position to receive the standards. These sole-plates should be of bulk timber, such as old railway sleepers. On them the baseplates should be placed and fixed and a nail driven into the hole provided for this purpose. It is necessary to take care in setting out the position of standards, and obviously this has to be remembered when placing the sole-plates.

The span between standards should never be greater than 2·1 m; they may be placed closer according to the load the scaffold will have to carry. All standards should be fixed in a vertical position—ledgers should be clamped to the standards using right angle couplers (load-bearing). All ledgers must be horizontal, and either putlogs or transoms fixed to them, depending on the type of scaffold being erected.

Decking

For decking it is normal to use $1\frac{1}{2}$ in. × 9 in. (38 mm × 229 mm) scaffold boards. The maximum span permitted by law for this board is 5 ft (1·524 m); it is safer, however, to have less than this span.

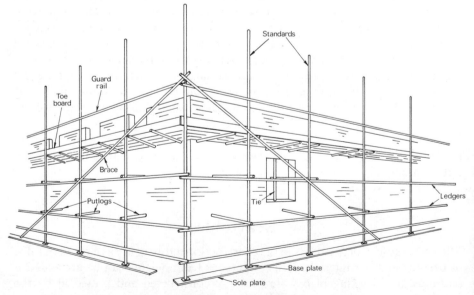

Fig. 21.3   Putlog Scaffold.

The usual width of a putlog scaffold, sometimes referred to as a bricklayer's scaffold, is five boards wide. As the boards are 9 in. (229 mm) wide this gives a width of 45 in. (1·14 m), but allowance must be made on the inside for the use of a level or plumb rule. This distance should not exceed 4 in. (102 mm), consequently when setting out, one must position the standard 49 in. (1·24 m) from the wall. At any time the space between the platform and face of a building should be as small as practicable. If workmen have to sit on the edge of a platform, the gap should not exceed 12 in. (305 mm).

For a working place the minimum width permissible for the passage of persons is three boards, for the passage of materials it should be four boards wide. For a gangway or run, the minimum width permissible is 17 in. (431 mm), i.e., two boards, and for the passage of materials it is 25 in. (635 mm), i.e., three boards.

Guard-rails and toe-boards

Any scaffold from which a person can fall a greater distance than 6 ft 6 in. (2 m) must be provided with guard-rails and toe-boards. The guard-rails must be of adequate strength and be between a height of 3 ft and 3 ft 9 in. (0·9 m and 1·14 m) above the platform. The distance between any toe-board and the guard-rail above it should not exceed 30 in. (762 mm).

Scaffold construction

All scaffolds must be built using material of adequate strength suitable for the job and be free from patent defect, and properly maintained. Timber used for scaffolds must be of a suitable quality and be in good condition with the bark completely stripped off. Timber used whether for scaffolds, trestles, ladders and folding step-ladders must not be painted or treated in any way that defects cannot easily be seen.

All scaffolds must be properly braced and adequately tied-in to ensure stability. In the case of independent tied scaffolds or putlog scaffolds, it is as well to ensure that the building is strong enough to carry the loading that this imposes.

All scaffolds must be adequately tied into the building throughout their length and height, to prevent movement either towards or away from the building. This tying-in should be done using double couplers. As far as possible all bracing should also be fixed using double couplers at alternate pairs of standards. If this is absolutely impossible, then a swivel coupler may be used for bracing, but no other type of fitting.

The general principles mentioned above should also apply to all other types of scaffold.

Ladders

All ladders must be long enough for the job; they should extend a minimum distance of 3 ft 6 in. (1·06 m) beyond the landing they serve. They should be properly lashed near to their upper resting place in order to prevent sideways movement, and if such fixings are impracticable they must be secured at or near their lower end.

If these precautions are impracticable, then somebody must foot the ladder whilst it is in use. It must have a firm and level footing and should not be stood on loose bricks or other loose packing. It should not exceed the maximum height of more than 30 ft (9 m) between landings and in any case should be supported to prevent any undue sagging. The correct angle for a ladder should be 1 ft (0·25 m) out for every 4 ft (1·0 m) of vertical. Care should be taken whenever fixing a ladder to ensure that there is a sufficient space between rungs and possibly ledgers or any other obstructions, to give an ample foothold on each rung which is to be used. Ladders should be fixed by their stiles, never by the rungs.

Cantilever scaffolds

It is sometimes necessary to build a scaffold over gateways or openings where it is impossible to build traditional scaffolding because means of access is needed for traffic. This situation is sometimes met by building a cantilever scaffold, on others by a hanging scaffold.

If a cantilever scaffold is used, then the work should be done by people who specialize in it.

Before the work starts a thorough examination must be made in order to

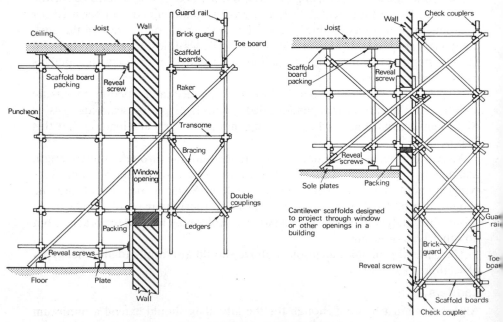

Fig. 21.4 Cantilever scaffold.

ensure that the building from which the cantilever is to be erected is strong enough and in good enough condition to bear the loadings.

Fig. 21.4 illustrates how the cantilever should be erected.

Woodworking Machinery Regulations

There are a number of changes and additions in the recent regulations compared with those brought out in 1922, 1927 and 1945. Details of the major changes and additions are as follows:

1. *Regulation 13(2)—Training*
   The extent of the training to be given to a woodworking machinist is far greater than in Regulation 9(a) of the 1922 Regulations.
2. *Regulation 13(3)—Training of persons under the age of eighteen*
   The training of persons under eighteen years of age on woodworking machines is approved only if they have been trained at a college of further education and to a City and Guilds syllabus in machine woodworking and also in carpentry and joinery. This is laid down by the certificate of approval given by H.M. Chief Inspector of Factories.
3. *Regulation 14(2)—Duties of persons employed*
   As in the construction regulations, every person who discovers any defect in any woodworking machine, or in any guard, device or appliance pro- vided for in the regulations, or who discovers that the floor or surface around the machine is not in good and level condition, etc., must report the matter to a supervisor or the site or works manager.
4. *Regulation 15—Sale or hire of machinery*
   The provisions of Section 17(2) of the Factories Act 1961 are extended to any woodworking machine, in that the sale or hire of any machine must comply with the regulations detailed in Regulation 15. If it does not, then the seller or hirer can be fined.
5. *Regulation 16(4)—Guarding of circular sawing machines*
   The top adjustable guard of the circular saw must have a guard on each side of the saw blade, not single sided as now. For machines manufactured before 24th November 1974 this will not apply until 24th November 1976. The whole of the regulation 16 is included in the requirements set out in paragraph 4 above. So manufacturers will have to fix this double-sided guard to their machines if they are manufactured after the 24th November 1974.
6. *Regulation 17—Sizes of circular saw blades*
   There has to be fixed to every circular sawing machine a notice specifying the diameter of the smallest saw blade which may be used in the machine, as circular saw blades must never be less than six tenths of the diameter of the largest blade for which the machine is designed.

7. *Regulation 18(2)—Limitations on the use of circular saws*

No circular saw shall be used for a ripping operation (other than those specified in the regulations) unless the teeth of the saw blade *project* throughout the operation through the upper surface of the material being cut.

8. *Regulation 20(1)—Removal of material cut by circular saws*

When a circular saw is being used, the person taking off must stand at the delivery end of the machine.

Regulation 20(2)—Removal of material cut by circular saw

If a person (other than the operator) is employed in removing material which has been cut, the machine table must be extended over its whole width (by provision of rollers or otherwise), so that the distance between the delivery end of the table and the up-running of the saw blade shall not be less than 1·2 m (4 ft).

9. *Regulation 42(2)—Maintenance and fixing*

Every woodworking machine, other than a portable one, must be securely fixed to a foundation, floor, or to a substantial part of the structure of the premises. Where this is impracticable, other arrangements shall be made to ensure its stability. When a skill saw is placed in a bench, then this regulation applies.

10. *Regulation 44—Noise*

When woodworking machines are used in a factory or in a joiners shop and any person employed therein is exposed continuously for 8 hours to a sound level of 90 dB(A) or an equivalent or greater exposure, then suitable ear protection must be provided. The operative word here is 'continuously'; nevertheless if the machine operators require ear protection for exposures less than 8 hours then they should be provided.

11. *General*

11.1 All measurements, etc., given in the regulations are in metric terms.

11.2 Temperature—The only difference between the 1922 regulations and the 1974 ones is that when working in the open air on woodworking machines, there must be, if it is reasonably practicable, an effective means of warming persons working there.

11.3 Copies of the regulations should be displayed where persons are operating woodworking machines. The abstract form F988 is revoked by the new regulations.

Bibliography

The Building (Safety, Health and Welfare) Regulations 1948. S.I. 1145 Regulations 1–4 and 99.

The Construction (General Provisions) Regulations 1961. S.I. 1961 No. 1580 as amended S.I. 1966 No. 94.

The Construction (Lifting Operations) Regulations 1961. S.I. 1961 No. 1581.
The Construction (Working Places) Regulations 1966. S.I. 1966 No. 94.
The Construction (Health and Welfare) Regulations 1966. S.I. 1966 No. 95.
Diving Operations Special Regulations 1960 S.I. 1960 No. 688.
Work in Compressed Air Special Regulations 1958 S.I. 1958. No. 61 as amended by S.I. 1960 No. 1307.
Electricity Regulations 1908 (S.R. and O. 1908) No. 1312. Amended by Electricity (Factories Act) Special regulations 1944 (S.R. and O. 1944) No. 739.
Woodwork Machines Regulations 1974 (S.I. 1974 No. 903). They revoke the Woodworking Machinery Regulations 1922, 2. Woodworking Machinery 1927, 3. Woodworking (Amendment of Scope) Special Regulations 1945.
Lead Paint Regulations 1927 (S.R. and O. 1927 No. 847).
Ionizing Radiations (Sealed Sources) Regulations 1969 (S.I. 1969 No. 808).
Ionizing Radiations (Unsealed Radiation Substances) Regulations 1968 (S.I. 1968 No. 780).-
Asbestos Regulations 1969 (S.I. 1969 No. 690).
Abrasive Wheels Regulations 1970 (S.I. 1970 No. 535).
Highly Flammable Liquids and Liquefied Petroleum Gases Regulations 1972 (S.I. 1972 No. 917).
Dangerous Occurrences (Notification) Regulations 1947 (S.R. and O. 1947. No. 31).
Offices, Shops and Railway Premises Act, 1963, modified by Exemption Order No. 1. (S.I. 1964 No. 964).
Offices at Building Operations, etc. (First Aid), Regulations 1964 (S.I. 1964 No. 1322).
Petroleum (Consolidation) Act 1928.
Petroleum Spirit (Motor Vehicles) Regulations 1929 (S.R. and O. 1929 No. 952).
Explosives Act 1875, amended by Explosives Act 1923.
Boiler Explosives Act 1882.
Mines and Quarries Act, 1954.
National Insurance Act 1965 (Industrial Injuries).
Employment of Women, Young Persons and Children Act, 1920.
Employment of Women and Young Persons Act 1936.
Employers Liability (Defective Equipment) Act 1969.
Employers Liability (Compulsory Insurance) Act 1969.
Protection of Eyes Act 1975.
Brand, R. E. *Falsework and Access Scaffolds in Tubular Steel,* McGraw-Hill, Maidenhead, 1975.

22. Commercial occupations

Dr G. R. Kershaw

Medical Officer,
Rolls Royce Ltd,
Coventry

and W. G. Symons

Formerly H.M. Superintending Inspector of Factories

Originally, almost all legislation aimed at protecting persons at work was seen as emergency legislation, directed at exceptional hazards or abuses. On this basis, mining and manufacturing were considered to be 'high risk' areas, calling for special safety laws which were not needed elsewhere. In the days of steam and water power this distinction had some justification. The development of electric power, light-weight prime movers and new synthetic materials has now created a situation in which mechanical and chemical hazards, formerly considered to be exclusively 'industrial', are now to be found in almost all forms of employment. Along with this has come a growing belief that employed persons generally should be legally entitled to basic standards of safety and to good environmental conditions whatever they are doing.

In 1949 the Gowers Report was published on *Health, welfare and safety in non-industrial employment,* examining some areas not covered at that time by health and safety legislation.[1]

Shortly before this, a Treasury study had been made on *Working conditions in the Civil Service,* concentrating especially on office work.[2] A later study of conditions in offices was made on behalf of the Corporation of London by A. W. W. Robinson.[3] At this period, office work attracted particular attention, partly because it was one of the larger sections of employment not covered by health and safety legislation, and partly because of the development of large office blocks used by commercial undertakings and civil government. There was no evidence at that time that offices had an especially high accident rate, but there were indications that comfort and general environmental conditions needed attention in places.

The Offices, Shops and Railway Premises Act, 1963 was largely in response to this concern. Seen from the perspective of subsequent developments, it was an oddly illogical piece of legislation. It grouped together very diverse types of workplace. It filled some large gaps in the pattern of health and safety legislation, but it left many other and equally important areas of employment uncovered. It did not, therefore, meet the desire for comprehensive protection of employed persons generally. On the other hand, it could not be maintained that the workplaces it covered contained exceptional hazards compared with those not covered. Indeed, there is little doubt that offices are inherently safer places of work than many other places which were, at that time, completely outside the scope of safety legislation and inspection. The Health and Safety at Work, etc., Act, 1974, is now in process of removing some of these anomalies.

In spite of its shortcomings, the O.S.R.P. Act has provided a period of practical experiment in the application of health and safety provisions to non-industrial employment. It has brought rather over 8 million workpeople under safety legislation. These comprise about 5 million in offices, $2\frac{1}{4}$ million in wholesale and retail establishments and $\frac{3}{4}$ million in catering. These figures are, however, very rough, in view of doubts as to definition and the large proportion of self-employed persons in the distributive trades. The statutory provisions for reporting accidents are the same as under the Factories Act—i.e. that any accident causing over three days off work should be reported. Table 22.1 is an analysis of accidents so reported during 1973.[4] Unfortunately the standard of reporting of accidents under the O.S.R.P. Act appears to be low. The Annual Report for 1972 gives grounds for believing that, in the distributive trades, only about 25 per cent of the reportable accidents are in fact reported, and statistics for offices are equally uncertain.[5] Since the standard of reporting is likely to be uneven as between different kinds of premises, figures of 'reported accidents' give little guidance for detailed comparisons.

If we assume the figure of a 25 per cent standard of reporting across the field, it seems likely that reportable accidents are running at a level of about 20 000 per annum for offices, 40 000 for wholesale and retail shops and 10 000 for catering establishments. This suggests incidence rates of about 4 in offices, 16 for shops and 12 for catering establishments, compared with 31 for factories. These estimates are very rough, but suggest that accident rates are significantly lower in offices than in most other forms of employment. Accident rates in shops and catering establishments, however, seem to be not far different from those in parts of manufacturing industry. Little more can be deduced with any certainty from these figures. These estimates are, however, broadly in line with some sample surveys in individual establishments.

One feature of these figures is that the proportion of accidents of various types remains remarkably constant from year to year. Table 22.2, which is

Table 22.1   Reported accidents (1973) under O.S.R.P. Act

| | Machinery | Transport | Falls of persons | | Fires and Explosions | Electricity | Hand tools | Handling goods | Other | Total |
			on level	level to level						
Offices	154	105	1 003	1 425	30	20	57	857	1 497	5 148
Retail shops	446	240	807	1 161	18	18	690	1 569	1 494	6 433
Wholesale depts. and warehouses	214	483	326	411	4	5	60	1 000	607	3 220
Catering and canteens	105	19	655	310	66	4	118	608	635	2 520
Railway Premises and fuel depts.	13	71	42	49	2	2	9	12	111	411
Total	932	918	2 833	3 366	120	49	934	4 146	4 344	17 742

based on 1973 figures, is therefore representative.[4] The main feature is that 'falls of persons' remains much the largest group, at nearly 40 per cent. Machinery accidents remain at about 5 per cent well over half of them are due to lifting and conveying machinery, and to food slicers.

Table 22.2   Distribution by type of accident (1973)

Type of accident	Number	Percentage
Falls of persons	6 199	34·9
Handling goods	4 146	23·4
Stepping on or striking against	1 639	9·2
Struck by falling object	1 169	6·6
Use of hand tools	934	5·3
Machinery	932	5·3
Transport	918	5·1
Fires and explosions	120	0·7
Electrical	49	0·3
Others	1 636	9·2
Total	17 742	100·0

Persons falling

About 40 per cent of reported accidents under the O.S.R.P. Act are due to persons falling, compared with 15 per cent in factories. The preponderance of this cause seems to be due to there being proportionately less accidents due to other causes, rather than to a very high incidence of falls per head. Nevertheless, this cause is important enough to receive attention.

Starting from first principles, it is not too trite to state that equilibrium is readily disturbed when kinetic and gravitational forces are combined, hence falls are more likely to occur during movement than when stationary, the likelihood increasing with the speed; they are more likely when there is a change of direction of movement, and most likely of all when movement is in the vertical rather than the horizontal plane. Therefore premises which are at ground level, or which are confined to one floor, have an inherent advantage, and more so if the nature of the work and the arrangement of sections is such that necessary journeys are short, straight, and unhurried.

If stairs are necessary, then two short flights connected by a half landing are preferable to curved or spiral staircases because, although the change of direction is more abrupt, the others have segmental treads.

It is absolutely essential that the rise of each step should be the same, and stairs with open spaces between the treads are to be deplored. Refinements include conspicuous nosings, and abrasive or corrugated strips in the treads.

A hand rail should be provided on the free side at least; this should stand free of adjacent surfaces and be of a circumference small enough to be encircled by the grasp. A banister rail should be at least 1 m from the floor and, if the intervening space is not completely filled, a toe-board and intermediate fencing should be provided.

Falls from ladders, in this context, occur mainly in stores and stockrooms, but it should not be forgotten that cleaning and maintenance staff are also at risk, and in both situations the victim may be working in isolation. Falls from one level to another are likely to be due to climbing on chairs, tables, or piles of boxes, etc., but the cellar trap-door, and in warehouses the goods-hoist and loading-bay, are particular examples for which adequate guard rails are necessary.

The majority of falls occur on the level. Not a few are due to obstructions such as parcels on the floor, the drawer of a filing cabinet left open, trailing telephone cables and loops of string. Obstruction due to design, as opposed to behaviour, may occur in narrow passages, corners and doorways, and in such places as washrooms and lift-landings which are subject to temporary congestion.

Floor coverings have their own hazards. Linoleum tends to spread, and some newer materials to shrink, so that it is sometimes necessary for them to be trimmed after a short interval to avoid crests or gaps at the edges. Fitted carpets are preferable to rugs, but both carpets and linoleum tend to become worn or frayed near doorways in time, and such patches are sometimes covered by mats which in turn curl at the edges and become further stumbling-blocks; worn areas should be cut out and replaced. Where a mat is provided at the entrance to a building it should be recessed to the level of the floor.

Worn treads and floorboards, chipped concrete, and tiles which are loose or cracked, should be repaired immediately. Floors which are wet or greasy from spillages, as is frequent in kitchens, are a special risk, and these should be wiped up at once; otherwise, the cleaning of floors should be undertaken out of working hours.

Ideally, floors should be level, durable, and not slippery. Some durable materials are noisy, but the main trouble is that surfaces and shoe-materials which are suitable for one purpose may not be so for another, and that the two may be incompatible with each other. Surfaces which are suitable for office floors may not be so for passages, stairs, and entrance foyers; leather heels on terazzo can be just as unstable as rubber heels on wet wood. The answer to this problem has yet to be discovered, but it must be recognized that any change of surface is a source of accidents, and that there is an advantage in providing the same surface throughout a building.

There remains about a quarter of all accidents caused by persons falling and for which no reason is advanced; there is no doubt that many of these are caused by shoes, especially women's shoes. The Minister's Reports show that report-able accidents due to persons falling, both on the level and on stairs, are much commoner among women than men, especially in the case of young persons, and that they form a much higher proportion of all reportable accidents in those commercial premises (offices, kitchens, and shops) where women predominate.

Because office staff, and to a lesser extent shop assistants, comprise a high proportion of young women it is understandable that many of them wear shoes that are fashionable rather than practical; indeed, of some of them this may be expected. There is no doubt that stiletto heels have been responsible for many accidents.

Kitchen staff usually wear shoes with flat heels, but these are often very old, worn, and distorted; moreover, the staff themselves are sometimes of an age when corns and bunions are troublesome, and may be selected for virtues which do not include physical agility. They are often in a hurry, and carrying objects which are in themselves hazardous and may be quite heavy. In canteens, as opposed to small restaurants, it should be possible for them to work to a time-table.

In addition to haste and congestion as contributory causes, inattention through distraction, conversation, or absentmindedness sometimes plays a part; from the medical point of view physical defects of the vision or locomotor sys-tem, early disseminated sclerosis, and inebriation, are occasionally relevant.

Finally, the importance of good illumination must be stressed, especially where elderly people are concerned; there are those who consider even the standards of illumination recommended by the trade[6] to be on the low side, and a body of opinion which regrets that the Act did not at once define minimum requirements.

Kitchens

Kitchens have all the ingredients of accidents, and unofficial figures in some undertakings confirm that catering and kitchen staffs can, in some instances, have an incidence of lost-time accidents substantially higher than the average for manufacturing industry. The main causes are falls, often on slippery floors, handling materials and equipment, burns and scalds, and cuts from hand tools and machinery.

Kitchen floors are particularly likely to be wet or greasy from spillages and these should be wiped up at once; condensation is another factor, and extraction hoods require a properly drained sill to prevent them overflowing. Floors should be impervious and easily swilled, and may be laid with a slight fall towards a channel or gully, but if so the latter should be well away from working positions and the grille must be kept in place.

Many accidents occur when handling articles, which is understandable when so many of the objects in kitchens are hot, sharp, or fragile. Picking up handfuls of cutlery or implements, especially when they are grasped unseen from a drawer or washing-up sink, is a common cause of cuts, and so are the crates, boxes and tins, glass and china, which must be handled frequently; opened tins, cracked glass and china, should be discarded ruthlessly and immediately, and fragments swept up with a brush and pan; otherwise a second injury nearly always occurs.

The first stroke with a knife or saw is more easily controlled when it is made towards the body, and in such a way that the softer tissue is cut before the harder; in the case of knives, the left forefinger should be placed on top of the blade, instead of alongside it, if used as a guide. When cutting horizontally the stroke should be from one side to the other, and when inserting a skewer the point should be directed away from the body; in both cases the left hand should be placed flat on top of the material with the thumb well retracted. Carving forks should, of course, be fitted with a guard and this should be raised. Sharp or pointed instruments are best kept out of the way when not in use, and may either be hung up on hooks or stored in a compartment separate from the others. Quite deep lacerations are common from the use of graters, shredders, scrapers, etc., and fixed rotary tin-openers are safer than the manual ones.

A significant number of accidents in kitchens are due to power-driven machinery, and these include some causing severe injury. Accidents from slicing machines are the most frequent, especially when they are being cleaned, but many designs have recently been improved.[7,8] The blade should be as nearly as possible enclosed by a fixed guard, and the thickness gauge complementary in shape; an endpiece pusher should be provided, and a stop to prevent the carriage making contact with the blade. Guarded chutes and interlocks are refinements on some machines, and one has a canister to guard the cutting edge of the blade while it is being cleaned.

Machinery for mincing, mixing, and beating, should have deep and guarded hoppers, and be equipped with interlocks which are effective on all moving parts.

The deepest burns are caused by the hottest processes such as frying, baking, and roasting, while the most extensive ones are commonly due to boiling or steaming; splashes from hot fat may be multiple, and deep-fryers sometimes catch fire when they are too full or spilled. It is not always appreciated that a dry oven-cloth is a better insulator than a wet one.

Projecting pan-handles are seldom a problem in canteens, where most of the pans are large and two-handled, though heavy to carry. Even the most expert staff may need to be reminded to lift the side of the lid first, rather than the nearest part, and to stand behind the oven door when opening it. Leaks from steam pipes can cause bad scalds if not attended to at once, and the taps of urns and boilers should be fitted with drip-tins to prevent the feet being scalded.

There are certain foodstuffs which are known to provoke skin lesions in sensitive persons[9] and some cleaning materials are quite strong alkalis.

The subject of food hygiene is not considered here, but it is pertinent to mention that puncture wounds are particularly liable to become infected unless treated with a hot fomentation in the first instance. These are often from wire or skewers, but the bones of pork and rabbit are said to have particularly sharp spicules. There are unlikely to be many wooden splinters these days as wooden spoons, mallets, and chopping boards are about the only wooden implements now used, but the handles of implements and brushes may be wooden and these should be discarded if they become worn.

Offices

The frequency of accidents reported from offices is low, but the figures are too uncertain to provide reliable evidence. Records of some individual undertakings suggest that, while accidents to clerical workers in offices are relatively rare, a significant number of office accidents involve cleaners and caretakers. These include falls from ladders, etc., while cleaning walls and attending to windows and fittings, and handling injuries when moving furniture and equipment.

There are two quite distinct types of office premises. On one hand there is the old-fashioned office in an adapted dwelling house, where the floorboards are irregular and the linoleum frayed, and the staircase dark, narrow, and with worn treads; on the other hand there is the large office block of recent construction, designed for the purpose and incorporating every refinement for safety. In fact there may well be more space and light available in older offices, and although the arrangement of furniture and equipment may not be convenient, it is likely to be of a simple nature and the movement of personnel small, whereas the high cost of modern offices may mean that they are relatively more crowded; large or multiple offices mean more walking about, and may raise difficulty in natural illumination, and sometimes the appearance of shining floors and an impressive staircase may be overvalued.

There is a third type of office which is very common; this is the small cubicle partitioned off a larger room or workshop. These tend to be noisy and ill-ventilated, but offer few hazards unless cluttered up with filing cabinets or permeated with volatile liquids.

Machinery does not at present cause many accidents in offices, but, with the increasing use of machinery, this is likely to become more important because, although many machines are satisfactory, the potential hazards are serious. At present the greatest influence for safety lies in the provision of machinery which is inherently safe, quite as much as by training and discipline in their use, and so managements have particular reason to inform themselves on this subject or to seek expert advice.

The hazards are mechanical,[10] physical,[11] and chemical, the latter arising from duplicating and copying machines.[12]

While the majority of guillotines are operated by hand they may conveniently be considered here as machinery. Small guillotines can be satisfactorily guarded by means of a fixed and preferably transparent guard allowing only a slit aperture on one side; where a continuous feed of paper is required this cannot be stabilized at the free end, however, and a properly guarded power-driven guillotine is essential.

One of the commonest injuries from machinery is an in-running nip at the feeder of such machines as are used for slitting, franking or addressing envelopes. The upper roller of the pair should be spring-loaded so that it rises if the fingers are inserted, but some of them are hard or so abrasive that it is better to insist on a fixed guard which makes access impossible. Larger apertures are to be found on some reproducing machines, where the rollers are smooth and the intake speed is slow; these are not so safe as they appear, because it is quite easy for a finger ring to slip through and get jammed beyond the rollers, preventing withdrawal.

No part of the mechanism should be exposed and spindle ends should not project beyond the casing, especially those with locking screws. The casing, which should be properly earthed if made of conducting material, is removable in some machines to clean the components, replenish the solutions, or abstract loose leaves of paper; if so, it must be impossible to remove it while any part of the machine is in motion under power, and unless this is so hazardous parts such as gear trains, drives, fans, etc., must be completely guarded, as should any 'live' or hot components irrespective of an interlocking device. Electrostatic copiers may employ high voltages for the corona discharge.

Some machines enclose reservoirs in the form of shallow trays for their chemical solutions, and if the latter are electrolytic it is important that they should not corrode metal parts or contact 'live' parts by spillage. Highly volatile solutions present other hazards, including the possibility of fire and explosion through sparking or contact with hot surfaces which can be specially dangerous when paper fluff has been allowed to accumulate.

Some copiers incorporate infra-red heaters, and in many heat is generated by the powerful illumination required; even those in which a cooling fan is provided may get quite hot, and it is unwise to place any organic material on the casing and most important to ensure that the vents are not obstructed.

Another physical hazard is ultra-violet radiation, from which the operator should be screened. First degree burns and arc-eye have been encountered, and not only in large-scale photography where carbon arcs or mercury discharge tubes may be exposed. Operators under treatment by photosensitizing drugs are at special risk.

Associated with the above is the production of small quantities of ozone and nitrogen peroxide, but there are many other systems which require ample ventilation because they use volatile liquids such as alcohols or hydrocarbon solvents either as transfer fluids or cleaners. The atmosphere can become heavily contaminated under ordinary running conditions if the room is small, hot, and ill-ventilated, but a common cause is the storage of these fluids, in containers which are not stoppered, in cupboards near radiators or heating pipes. Cleaning cloths, too, may be soaked in fluids and then hung on a radiator to dry. Carbon tetrachloride is sometimes used by tracers to remove pencil marks.

Ammonia is used as a developer in a very common diazo process for the reproduction of drawings. This requires local extraction, and clear instructions on the action to be taken in emergency, ranging from dilution of small contaminations to the evacuation of the room in the case of large spillages.

The oil-based inks used in duplicators are harmless, and few of the other chemicals used offer a direct hazard to the skin apart from the defatting action of spirit duplicators, and solvent blanket washes in offset printing, for example, though care should be taken to protect the eyes when filling reservoirs from stock containers. Amino compounds used as developers in orthodox photography and in the diffusion process are sensitizers, and from time to time sensitization occurs to the accelerators or antioxidants in rubbers and to miscellaneous dyes.[9]

In general, all hazards from these processes can be avoided if the operator understands the nature of the materials and is instructed in their safe-handling; unfortunately the labels of proprietary preparations are seldom informative and often reticent or euphemistic. Only small quantities of fluids should be used at a time; the containers should be properly labelled, kept stoppered, and preferably stored elsewhere in cool surroundings. Cleaning cloths should be as small as possible, and discarded in an outdoor dustbin after use; contaminations of the clothes and skin should be avoided, and a skin cleanser used to remove stains rather than the solvent provided to clean the equipment.

Other premises

Railway premises, although coming within the scope of the O.S.R.P. Act, form a specialized group which lies outside the scope of this chapter. Shops and wholesale establishments are an important and very diversified group of workplaces. The commonest are the many ordinary one-purpose shops which line every shopping street, each with perhaps two or three assistants and possibly with the proprietor working in the shop as self-employed. They also include large departmental stores, supermarkets, street kiosks and wholesale depots with heavy lifting and handling equipment. A great variety of work goes on in these places, and accident hazards are diverse. The most important appear to be manual and mechanical handling and transport, and the use of hand tools. Machinery accidents are less common than in industry, but appear to be more frequent than in offices. Risks from toxic and corrosive materials are not unknown.

For reasons mentioned earlier, accident statistics relating to these forms of employment are not as reliable and complete as might be wished, and there is certainly need for more detailed study. Environmental conditions may also need attention in some places. Departmental stores and supermarkets can be crowded at certain times. Shops partially open to the weather may be cold or wet. In food shops especially, control of environmental conditions may have to be related to the needs of food hygiene. The recent Health and Safety at Work Act has established the principle that, in considering responsibilities for safety, attention has to be given to the safety of non-employees as well as employees. This may be of special relevance in departmental stores and supermarkets, where large numbers of the public may be present and subject to the same risks as employees.

Conclusions

Experience of working the O.S.R.P. Act has shown that many of the hazards which have long been familiar to those concerned with industrial safety also exist in non-industrial employment, although often in different forms and calling for rather different preventive methods. Even where, as in office work, accident risks are relatively low, there are often environmental conditions needing attention. The Annual Report for 1973, for example, draws attention to the difficulties of controlling temperatures in modern buildings with large areas of glass, sometimes resulting in employees suffering from heat exhaustion.[4] Even where there is no direct injury to health, such conditions impair morale and productivity. This is one of the matters where conditions of health and safety need attention at building design stage. In the catering industry, accident risks are higher but fairly clearly defined, calling for attention to design of equipment and good training. Accident hazards in shops and in the distributive trades generally can be quite significant, but are more diversified and need further study.

One disappointing feature of the working of the O.S.R.P. Act has been that the system of accident reporting and investigation has not worked so effectively as under the Factories Act. In part, this has doubtless been due to the involvement of a large number of small units, who have been slow to realize their duty to report accidents. The failure is serious, however, since the process of assembling and using accident experience is essential in shaping accident prevention policy.

At the present time, the Health and Safety at Work Act is bringing additional non-industrial employment within the range of health and safety legislation. One of the urgent problems facing the Health and Safety Executive will be how best to assemble information on accidents and other such matters from the diffuse field of non-industrial employment. Different methods may be needed from those appropriate for factories.

A related question is how advice about occupational hazards and their control is best made available to large numbers of small undertakings in the non-industrial field, such as shopkeepers and employers with small offices. Such undertakings cannot be expected to equip themselves with the specialized safety organization which is becoming more common in large commercial undertakings and in public authorities. They will need outside advice, either from official or trade sources.

References

1. Home Office: Scottish Home Department. *Health, welfare and safety in non-industrial employment.* H.M.S.O. Cmnd. 7664, 1949.
2. H.M. Treasury. *Working conditions in the civil service.* H.M.S.O. Code No. 63-104, 1947.
3. Robinson, A. W. W. *Working in the city.* Corporation of London, 1962.
4. *Offices Shops and Railway Premises Act*; Annual Report of Sec. of State for 1973. H.M.S.O., London.
5. *Offices Shops and Railway Premises Act; Annual Report of Sec. of State for 1972,* para. 72. H.M.S.O., London.
6. Illuminating Engineering Society. *I.E.S. code for the lighting of building interiors,* 1949 and revised edns.
7. Royal Society for the Prevention of Accidents. Food slicing with safety. *Brit. J. Comm. Saf.* 1966, **1**, 4, 46.
8. Ministry of Labour, H.M. Factory Inspectorate (now Department of Employment and Productivity). Guarding of food slicing machines. *Accidents.* 1967, **72**, 24.
9. Kipling, M. D. The hazards of occupational skin diseases. *Brit. J. Comm. Saf.* 1966, **1**, 2, 14.
10. Beak, K. L. The safety of business machines. 2. Mechanical and other hazards. *Brit. J. Comm. Saf.* 1966, **1**, 4, 42.
11. Beak, K. L. The safety of business machines. 1. Electrical hazards. *Brit. J. Comm. Saf.* 1966, **1**, 3, 30.
12. Greenburg, L. Health hazards in the modern office. *J. Am. Soc. Saf. Engrs.* 1965, **10**, 15.
13. *Offices Shops and Railway Premises Act;* Annual Report of Sec. of State for 1975, para. 34–35.

23. Ergonomics

Dr A. H. Hands

Chief Medical Officer,
British Leyland U.K. Ltd,
Leyland Truck & Bus

The word 'ergonomics' means literally 'the principles of work'. It is a word used in Britain to describe the study of that which fits a machine or process to its operator, 'a machine being made with man in mind'. It aims to enable man to work with optimum physical and mental comfort and to use his special senses with best effect. In the U.S.A. the term 'human engineering' is used. More recently the term 'human factors engineering' has been introduced. This is a natural consequence and is concerned with the relationship of engineering with psychology.

The concept of human engineering began during the First World War but continued in earnest as ergonomics by a group of scientists during the Second World War. These men were concerned with developing the required and complex weapon systems to match the human skills needed for their operation. This progress has been carried on into civil life and has now become an important aspect of any engineering and development project. Because of its very nature ergonomics is, and has been, the fundamental support in the awakening of an environmental consciousness. Thus it promotes greater efficiency, by establishing good and safe working practices.

The compound skills of various disciplines are required in order fully to understand and develop the subject. These include:

(a) anthropometry, or functional anatomy,
(b) physiology,
(c) psychology,
(d) engineering.

The anthropometrist supplies data of anatomy and human measurements. The physiologist is concerned with the calorific requirements of work and the functioning of the body, including the reception of stimuli, their processing and the effective action taken. The psychologist is concerned with much that appertains to industrial life and safe working. Human factors engineering is possibly the most important part of future development in this field. It is sometimes described as engineering psychology. The engineer must collate the information provided by the above and build accordingly.

The industrial doctor has a part to play since he is professionally trained in the first three of these disciplines and is now employed with the fourth. The safety engineer has to put into practice much of what is provided by the above.

Anthropometry

This is the study of human measurements (shape and size) and should include the range of joint movements. The unit of measurement is known as a percentile. As the term suggests, the table is from 0–100. A person who is 80 percentile in height is taller than 80 per cent of the measured group and shorter than 20 per cent.

As a simple illustration of design procedure, a doorway may be designed so that at least a 95 percentile person may pass through—but a control switch to be reached from a fixed position would be so placed that a 5 percentile person can operate it. The rare or 'freak' cases lie at the extremes of the percentile range and cannot normally be catered for. Most design engineers try to meet the requirement of the 5–95 percentile range.

Body sizes and anthropometric data are very important in machine design. Much data has been produced in the U.S.A. and there are several authorities on the subject. Proper use of anthropometric data eliminates awkward body positions and therefore inaccurate control. No two human beings are identical and the designer must allow for as wide a range of build as is mechanically possible. There are also variations in groups, and in view of this it is wise, if possible, to select the group for which you are intending to manufacture and make your own measurements. If this is not possible the comprehensive tables in the literature should suffice.

The use of a manikin is now of value. A mechanical figure can be adjusted with a range from a 10 percentile to a 90 percentile male, and allows adequate scope for the design engineer with regard to leg room, seat width, and height of viewing. However, it does not include a method of measuring reach. The latter is of increasing importance, for example a vehicle driver must be able to reach his controls and switches while at the same time remaining strapped by his safety belt. Legislation in certain countries now requires this.

The 'average man' constitutes a minority of the population. The term 'average' is not, therefore, used in this context since it would constitute only the 50 percentile group omitting 0–49 and 51–100 percentile.

It is important that sex and race are considered when designing a piece of equipment. For example, the Japanese as a race are shorter than the Negro. The Sikh who is the professional driver in India is obliged, by his religion, to wear a turban. This increases his overall height and must be allowed for.

Comfort angles are concerned with the optimum joint position which is usually neutral, i.e., neither extended, flexed, nor rotated. This allows operation with a minimum of fatigue.

Physiology

This deals with environmental conditions and their effect upon the operator. Man and machine form a system which is becoming ever more complex.

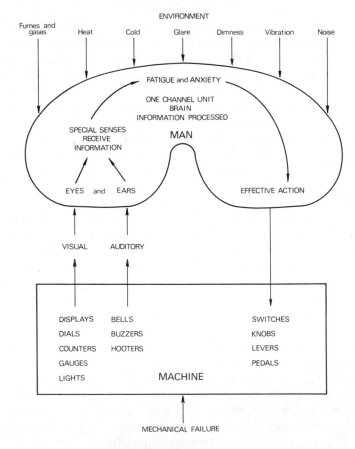

Man and machine have a relationship which is complementary since each requires the other to fulfil a function satisfactorily. There are certain things which man can do better and others better performed by machine.

Man is flexible with an ability to programme and organize his work. He can cope with the unforeseen occurrence. Although having a memory of limited capacity and permanence, he has power of recall and, what is probably more important, he can exercise judgement. On the other hand a machine can store more information with great reliability, and is faster and more accurate in sorting out data. It can work indefinitely at high speed without fatigue. A machine can generate or deliver considerable amounts of energy (power) and exercise control much better than man, who is relatively slow, weak and physically limited, whilst producing variable performance with liability to error.

The five special senses of man are often highly developed over a wide range and are usually more effective than mechanical sensing devices. The aim is to facilitate optimum reception by the operator with efficient processing and appropriate action. Any physical or mental condition which interrupts this flow must be eliminated or reduced.

Factors which interrupt the flow are as follows.

(a) Design anomalies which result in a physical inability to reach controls. Often there is failure in standardization of controls which results in uncertainty, with an operator becoming confused.

(b) Difficulty in manipulation and control causes unnecessary demand for physical effort.

(c) Poor presentation, and distribution, of visual data as, for example, may occur in a road vehicle with wide pillars or badly positioned driving mirrors.

(d) Environmental factors which affect the operator. These include noise, heat, cold, fumes and gases. Discomfort, headache, and fatigue are frequent effects with consequent loss of efficiency.

All efforts must be made to ensure that displays are clear and positive. Colour can provide a most useful contribution to a visual display. Best presentation of scaled measurement comes from rotational indicators against well-spaced marking on a well-contrasted background plus adequate illumination of scale. When undue physical effort is demanded of an operator, servo mechanisms should be incorporated in the machine. This has happened with steering and braking in heavy road vehicles where, also, automatic gear change is now commonplace.

Temperature control of environment is a fairly simple matter, but an air-flow without draught can be a problem. Fresh moving air is stimulating and helps an operator in his task.

Design features

A purpose of ergonomics is to facilitate the immediate and correct comprehension of information and the identification and manipulation of controls.

This purpose will necessitate the best design displays dials and controls.

Vision is the special sense which is most easily overloaded, and this problem is best illustrated by the presentation of display on the flight deck of a large aircraft. It is impossible for the flight deck crew simultaneously to check every visual signal and dial. Sometimes it is necessary to reinforce the visual display, say of a variation which goes past a limiting condition, by means of an alarm to attract the attention of the operator through another of his senses. Most frequently an audible signal is used, such as a buzzer. Dials should be arranged so that the indicators fall into pattern when at normal to enable an operator to scan a bank of dials without undue strain. Whenever the abnormal occurs the indicator of the relevant dial will change and so alter the pattern, a fact which is relatively easy to observe.

There are three types of indicator.

(a) A check reading indicator which, when demanded, gives information as to the state of equipment. It can also confirm as to functioning of a mechanism when in use. This is demonstrated by the direction flasher indicator in an automobile.

(b) A qualitative indicator will demonstrate a dynamic measurement without exact precision. The ideal example is that of a temperature gauge which records—cold—normal—hot. The exact temperature is not required.

(c) A quantitive indicator will require precision. In this context the speedometer is an example. A quantitative indicator requires more frequent attention than the other two.

A quantitative indicator should never be used if a lower-order one will do. The use of check reading and qualitative indicators presents little problem in design. However, there are several varieties of dial in quantitive indicators. These include the vertical and horizontal linear types, also the semicircular, round and open-window rotational types. In the latter the pointer is fixed and the scale moves. In the remainder the pointer moves. Experiment has shown that the open-window dial is the one which produces least errors in reading although the round dial has probably as much merit. It is shown that 1 cm spacing markings on a dial are best suited to accurate reading.

Design of controls relative to indicator types

Check reading indicator, single action control.

    e.g., on–off, as electric light switch.

Qualitative indicator, a multiple-choice action control.

    e.g., push-button station selection, as with radio or T.V. channel control.

Quantitative indicator, continuous control within parameters.

    e.g., automobile steering wheel during vehicle movement. The control which is most suited to the operation must be used.

One must consider the 'degree of expectancy' which is inherent in most people. By employing a customary movement an operator expects a certain result. For example, if a steering wheel is turned in a clockwise fashion one expects to turn to the right. With a clockwise turn of a knob one might expect increased output or *vice versa* for an anticlockwise turn. When a switch is pressed down one expects lights to go on and *vice versa*. This is normal expectation. Often, however, this is not so and accidents result. The operator does exactly the opposite to what is required. A big advance in ergonomic design of a commercial vehicle is made when ease of entry and egress are facilitated. Nowadays it is possible to walk right into a commercial vehicle without having to perform the acrobatic feats that were essential until quite recently.

There is the case of the two-car family. Unless the cars are identical in make, model and fitted extras it is certain that indicator and control lay out will be different. Hence, the family member occasionally taking over the less familiar vehicle may find himself confused in the event of an emergency and thereby temporarily not in total control. Although lasting for only a second or two the outcome is an accident. Thus a fan is switched on instead of lights, an incorrect gear shift is attempted, or perhaps through tiredness the loss of oil pressure is overlooked and there is engine seizure. In the latter situation a combination of visual and audio signals could save nasty consequences.

Ergonomics may contribute to vehicle design leading to standard controls and more rational arrays of indicators. Also, it may lead to the introduction of devices to offset the consequence of a sheer habit, such as attempting to drive on the left of the road in right-hand side traffic flows.

A person has a limited mental capacity which can be overloaded with extraneous worries. The strain of modern living together with a complex operation can produce a state wherein the operator will no longer respond to warning information. The so-called inexplicable failure of a routine operation is often found on further investigation to be due to mental overload. We all experience the 'Now look what you have made me do!' syndrome. When a person suffers from mental overload he becomes worried and distracted and in consquence commits actions which he would not normally do. In other words he becomes what is described as 'accident prone'. Such a case is usually treated with amusement. However, it is a very real condition and is likely to produce consequences far more serious than dropping china in the kitchen.

It is necessary to have more intelligent operators for the more complicated modern equipment. It follows that a person of low intelligence or one who has been accustomed to a simple machine for many years will not be capable of operating the more complex and costly machines of today. The saying 'it is difficult to teach an old dog new tricks' remains a truism. If it is attempted, failure of the operator and damage to the costly machines will often result.

It is no longer simply a question of moving knobs and levers. It is essential for the man to understand the machine and its functions and for him to play his part in the system. There are optimum hours of work and rest and there is legislation in most countries to ensure this. There is a personal limit to the attention anyone can offer and after which his concentration fails. It is, therefore, essential that rest is taken before reaching this particular condition. Operating comfort is an ideal which most people seek. Man is adaptable and able to cope with many awkward and uncomfortable situations; he will make his own adaptations to equipment to serve his own particular needs. It is, however, because of his very adaptability that he may become a hindrance to the work of the designer since the latter is frequently not made aware of the difficulties being experienced by the operator.

The work study expert employs those techniques devised and evaluated by the ergonomists. The B.S.I. has defined work study as a 'generic term for those techniques particularly Method Study and Work Measurement which are used in the examination of human work in all its contexts and which lead systematically to the investigation of all factors which affect the efficiency and economy of the situation being reviewed in order to effect improvement'. It helps to produce the best from both men and material engaged on a working programme and in being concerned with the efficient use of man power and machine it also promotes safe working.

It is necessary, when engaged in work study, to break down an operation into its component parts and to examine each part carefully. It is a combined study of 'how best?' and 'how quick?'. What is the best way of doing a particular job, and what is the optimum time? With regard to the latter it must be remembered that rest periods are essential, and higher production can often be obtained by ensuring that these rest periods are of the right length. These periods are physiological in requirement and are advised as a result of long experience in these matters by work study experts.

It is important to remember that the rate of recovery from fatigue is greatest in the first few minutes of rest. Therefore, frequent short rests are better than fewer longer breaks.

There are two aspects of ergonomics:

(a) to design equipment to meet physical standards of the operator; by considering the 5–95 percentile group, most problems of a physical nature can be overcome;
(b) to understand and plan for the mental capacity and problems of human behaviour and reaction which are often unpredictable but, when faced with emergency action, are those of habit; it is this second problem which is a contributory factor to many accidents.

Sometimes a machine design or quality is at fault, but more usually it is human error. The reason for machine design being at fault is that too often there is a failure in standardization but this again leads to human error in manipulation.

This chapter has freely used motor vehicles in examples, but it will be easily recognized that the use of ergonomics appertains equally well throughout industry. Few but the largest companies will employ experts in the various disciplines I have described as being necessary for the study and development of ergonomics.

However, there is a comprehensive bibliography available of both British and American origin and the intelligent use of this literature by industrial doctors working together with safety engineers and hygienists should prove of value. Significant improvements have been attained by many manufacturers of equipment, and as long as this continues there will be a resulting improvement in safe working.

Bibliography

1. McCormick, E. J. *Human factors engineering*. 3rd edn. McGraw-Hill, 1970.
2. Morgan, C. T. *Introduction to psychology*. 5th edn. McGraw-Hill, 1975.
3. Murrell, K. F. H. *Ergonomics—man in his working environment*. Chapman and Hall, 1965.
4. Edholm, O. G. *The biology of work*. World University Library (Member of the Weidenfeld Publishers Group), 1967.
5. Grattan, E. and Jeffcoate, G. O. Medical factors and road accidents. *Brit. Med. J.* 1968, **1**, pages 75–79.
6. McGowan, J. Ergonomics and the driver. *Commercial vehicles—engineering and operation*. Institution of Mechanical Engineers, 1967.

24. Atmospheric conditions

F. E. Grandidge

Ventilation Engineer,
H. H. Robertson (U.K.) Ltd

The composition of pure air is approximately $20\frac{1}{2}$ per cent oxygen, and 79 per cent nitrogen by volume. Small percentages of carbon dioxide, argon, neon, helium, and water vapour are also present. Purity of the outside air is fairly constant in different localities. In large towns the carbon dioxide content is likely to be higher than in country districts, but it rarely exceeds 0·4 per cent or four parts in 10 000.

The amount of dust or soot in the atmosphere in industrial towns and cities may call for air cleaning. Major cities are taking numerous tests of air to find the quantity of pollution in the atmosphere, and we are very soon likely to reach a situation where cars are banned from all large towns in the world, due to their emission of excessive carbon monoxide into the atmosphere.

In occupied buildings the atmosphere is affected by the bodily functions of the occupants and by their activities. Carbon dioxide and water vapour are given off in the air exhaled from the lungs. Bacteria are discharged during breathing, sneezing, or coughing. The body gives off organic impurities to an extent varying with personal cleanliness. Where smoking occurs, or where exposed flames are present, the product of combustion will cause contamination. Further pollution arises where industrial processes release fumes, gas, or dust.

Although oxygen is the vital component of air, its proportion can be considerably reduced without distressing effects. A candle flame will be extinguished when the percentage drops to 17 per cent, but life can be supported by an oxygen content of only 13 per cent, providing the air is kept moving and the body is at rest.

It is now universally accepted by competent authorities that the chemical changes undergone by air due to respiration in normal environments, have little effect on human well-being. The factors which are of importance are those which control the loss of heat from the skin, and this loss of heat rate will depend on the combined influence of three distinct atmospheric factors: temperature, humidity, and air movement.

Comfort is essentially a personal assessment of the conditions in which we live. Sensations of comfort vary with our state of health, vitality, age, and personal habits. Despite the fact that comfort is a personal and subjective state, a surprising degree of unanimity is found whenever a group of individuals is asked to assess the comfort of a given atmosphere. During analysis of the factors determining comfort, psychologists have learned that the temperature, radiation, humidity, and amount of air movement are chiefly responsible for the production of the sense of comfort. Much has been learned concerning the relations and relative importance of these factors. It is also recognized that in respect of measured air temperature, air movement, humidity, and heat radiation, there are definite maximum and minimum limits to the ability of the human body to adapt itself to achieve comfort.

If the surrounding conditions are such that the heat sources exceed the losses, then the temperature of the body rises, a person feels hot and, therefore, uncomfortable. Naturally if the heat losses are greater than the sources, a person feels cold.

Every building needs some form of ventilation to remove one or more of the following.

Bacteria. Respiratory diseases and other complaints are spread by air-borne infection. Tests have demonstrated that sneezing and coughing spread bacteria-laden droplets over considerable distance. Infectious droplets remain floating in the air.

Heat loss from bodies. The human body continuously produces heat which it derives from the consumption of food. The amount of heat produced increases with exertion, but the body cannot store heat. The inner temperature of the body must remain constant at about 37°C. Even a small rise above the normal temperature may be dangerous. The body must dissipate heat as fast as it produces it, which it does by physiological processes which regulate heat loss from the skin. In cool surroundings the heat loss is fairly rapid but in a warm environment the body may gain heat from its surroundings. It must then get rid of not only the heat it generates but also the heat it gains. Air movement helps it do so by accelerating heat loss from the skin.

Table 24.1 Heat losses from the human body under different conditions (watts)

Persons seated at rest	115
Engaged on very light work	140
Light bench work (Industrial)	235
Heavy work (Industrial)	440

Body odours. Opinions vary about the unpleasantness of these odours. It is sometimes argued that many people like a 'good fug'. This has led to low standards of ventilation being tolerated in the interests of economy, particularly where heating costs may present problems. It must be remembered, however, that body odours arise from organic substances in the air, which increase where personal hygiene is deficient. It is reasonable to assume that such odours represent undesirable pollution.

Evaporation. The body gives off varying amounts of water in the form of perspiration according to its needs to get rid of heat. The quantity may be as much as 76 mg/s or even more. Perspiration, which is mopped off the face or soaks into clothing, contributes nothing to body heat loss. It is the evaporation of perspiration from the skin which takes away body heat. The rate of heat loss is governed by the rapidity of evaporation, which in turn depends on the capacity of the surrounding air to carry off moisture, that is the relative humidity of the air.

Industrial fumes and gases. Most of the fumes and gases created by industrial processes are injurious to health. Generally the concentration which can be allowed in the atmosphere is very small, and it is necessary to prevent contamination of this kind being freely released into occupied space. It should be extracted at source where possible.

Solar radiation. Solar heat transmitted through windows, walls, and roofs raises interior temperatures considerably. Roof surface temperatures of 60°C have been recorded in Britain. Part of this heat will be given off into the air, part will penetrate into the building. The amount of solar heat absorbed by building surfaces varies with their colour—dark coloured surfaces absorb more heat than those of lighter colour. The flow of heat through walls and roofs depends on their type and thickness. Glass is transparent so that it allows almost the full intensity to pass inside instantaneously. Up to 950 W will pass through each m^2 of glass in the roof of a building due to solar radiation.

Until recently, ventilation has been accidental rather than planned; the need for ventilation was not understood. The ill effects resulting from the lack of it were not realized, and the reasons for air being stale were completely unappreciated.

The history of scientifically planned natural ventilation is of less than 60 years duration. Prior to 1924 it was rare to find constructive thought given to the ventilation of workshops, offices, or business premises generally. It was only where conditions were appalling that attempts were made to remove the fumes, gases, and heat by means of such unsatisfactory devices as jack roofs and louvres.

The more alert managements of well-organized concerns realized that provision of medical supervision, personnel management or welfare staff, physical ease in working conditions and well managed heating, ventilation, and lighting, represented a wise outlay in business.

It is a fact that only the most far-sighted employers of labour have realized that to ventilate a factory does more than provide for the comfort of the work people. Employers are inclined to be sceptical of the benefits to be derived in hard cash for what should amount to no more than common humanity. Comfort surely pays a bonus in contented personnel and increased efficiency.

Ventilation implies fresh air supply, the removal of contaminants and heat, and air motion for cooling, freshening and counteracting discomfort due to humidity. In estimating the amount of ventilation and air motion required, the following factors must be taken into account: the size of the room or the building, number and type of occupants and their activities, heat gains from equipment and solar radiation, relative humidity, outside air temperatures, and range of temperature.

The rate of ventilation can be worked out from these factors, although this is not always necessary. In certain cases it is sufficient to follow recommended rates of air renewal for buildings where no fume, heat, or moisture is present, i.e., warehouses. Most people think in terms of air changes/h without consideration of the height of the building. This can be a vague and variable unit of measurement. If the same number of changes were provided in buildings of variable height, the amount of ventilation provided in each would vary as the height varied. When using air change units in calculations, allowances must be made for the height of the building or compartment to be ventilated.

In steelworks and glassworks, where large quantities of excess heat and fume are liberated into the atmosphere to the detriment of the worker, great attention must be paid to the design of the ventilation system. The more information that can be obtained about the size, location, and type of plant at the initial design stage of the building itself, the more efficient the ventilation system is likely to be. It is a fact that on large schemes the design and size of the building is, to a large extent, governed by the type of ventilation.

Underestimation of ventilation requirements can have serious consequences, apart from discomfort to the workers. Corrosive fumes held in the building will attack the building structure with disastrous results.

An investigation into the excessive summer heat conditions in a furnace building revealed an intolerable temperature at floor level of 66°C with an outside temperature of 20°C, a rise of 46°C. An engineered natural ventilation scheme was proposed and installed and the resultant temperatures recorded were 43°C at floor level with 29°C outside giving a temperature rise of only 14°C as against the 46°C rise previously recorded.

Some recent installations carried out in this country include the ventilation over four large electric furnaces, each of approximately 100 ton capacity where the spread of fume into adjoining bays had been a constant problem. After installation the spread of fume was non-existent, the ventilators efficiently removing over 5·66 million m³ of air/h.

Fig. 24.1   Large ventilators sited over electric furnaces to handle the extreme heat and fume problems of a steelmaking process.

An engineered ventilation scheme was also designed for a bottle making plant turning out approximately 3 million bottles/day. The ventilator is removing 35 million watts and over 1·4 million m³ of air/h. With an outside temperature of 8°C the temperature around the bottle making machines is 19°C.

25. Colour and environment

I. Stancliffe

Head of Colour and Design Service
Crown Decorative Products Ltd

The importance of environment in making for human happiness and efficiency is gaining wider recognition in industrial, academic, and governmental circles. Environmental studies are now established in some universities and the Ministry of Technology is also making a valuable contribution in its work on light, colour, heat, and acoustics. While these environmental studies are proceeding and widening our knowledge there is already a good deal of sound information and experience available today and the purpose of this chapter is briefly to examine, assess, and formulate the requirements of a safe and pleasant industrial environment and to discuss practical means of attainment.

A rational approach

It is now possible to create and control the quality of environment we require by suitable enclosure of space, provision of light and heat, and control of noise levels, although cost may often preclude the ideal being attained. Whatever compromise may be called for on grounds of economy, however, an adequate supply of light (see also page 315) must be the first provision since light is the basic requirement for seeing. Light is also the all-important element in creating the right environment—we all know the tonic effect of sunlight after a spell of dull, overcast weather, and similarly the feeling of uplift we get in passing from a dark unkempt place into a bright newly-painted workshop with light-coloured walls. The light wall colour is almost as important as the light source which it reflects, since black and dark colours *absorb* light and tend to create a gloomy and depressing environment.

The most important component of light, however, is *colour*, for when colour is used correctly it can provide not only a cheerful environment but it can also aid visibility, direct or focus attention where required, and give visual warning of danger. However, to realize the full possibilities of applying colour in industry as an aid to safety we must regard it first and foremost as a means of communication rather than of decoration. This means that we must select a colour to do a particular job—not merely because we happen to like it. Selecting colours for a factory or workshop involves the happiness and safety of a number of people, and also efficient working conditions leading, ultimately, to increased output.

The term 'decoration', in the sense that one applies to the home, the café, or the cinema, should not be applied in the factory, except perhaps in the canteen or the rest rooms. In painting a factory or workshop, colours should be chosen not merely for their appearance value or cosmetic effect. The rational approach is to select colours primarily for their inherent *functional* value and to apply them for a specific purpose as, for example, to reflect light without glare; to improve visibility by defining shapes clearly; to 'highlight' working areas, to 'spotlight' danger zones and hazards; to 'signal' fire and first-aid stations and appliances; to 'identify' service pipes, conduits, and so on.

Used rationally in this way colour must inevitably create a better and safer environment in which there are likely to be fewer accidents, and reduced absenteeism and labour drift. The provision of good working conditions and a regard for the welfare of personnel improves worker–management relations, and progressive managements find it worthwhile to provide these conditions.

The attributes of colour

Before we can use colours rationally or functionally we need to know something of the functions we may reasonably expect them to perform. Colour is such an

---

Fig. 25.1

(a) Yellow, Orange, and Red have high visibility and great visual impact and are therefore selected for the most urgent warnings—FIRE, DANGER, STOP, CAUTION, etc. Greens and Blues recede and are more appropriate for less urgent signals.

(b) Black and Yellow (B.S. 08 E 51) diagonal banding for marking dangerous hazards.

(c) The B.S. Symbol for First Aid is a Green Cross. The Red Cross is better known and has greater impact, especially when shown on a White circle.

(d) B.S. SAFETY COLOURS AND SYMBOLS. These distinctive shapes are recommended to ensure recognition by those with colour defective vision.

(e) STOP, DANGER, FIRE, etc.

(f) CAUTION.

(g) SAFETY.

(h) Bandsaw painted Yellow inside safety guard to warn of danger when open.

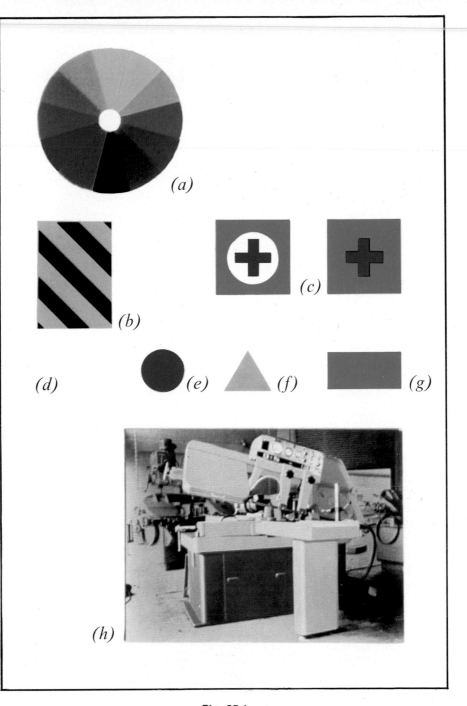

(a)

(b)

(c)

(d)

(e)

(f)

(g)

(h)

Fig. 25.1

everyday experience to those with normal vision, and is provided in nature in such abundance, that we tend to take it for granted. The fact is, however, that colours have various characteristics or attributes which make them appropriate in some situations but not in others. In the factory these attributes must be considered carefully in relation to specific requirements if the colours are to function properly. Some colours are warm, others are cool; some colours advance, others recede; there are light colours of high visibility, and dark colours with comparatively low visibility. There are focal colours and background colours: masculine colours and feminine colours; colours which reflect light and appear enlarging; dark colours which absorb light and have an enclosing effect.

When these attributes are recognized and correctly categorized it is a fairly simple matter to determine when and how they might be employed in a rational manner to produce a desired environment. It is appropriate, for instance, to use warm colours in cold, sunless rooms and cool colours in rooms where warm conditions prevail. Similarly, it is appropriate to use pale pastel colours having strong feminine appeal in workshops where only women are employed, whereas the same colours would be quite out of place in a steel works—nothing could be more inappropriate than, say, a large power press painted pink!

Colours have a natural tonal progression from light to dark, yellow being the lightest colour nearest to white, orange next, then scarlet red, crimson and purple, down to violet, the darkest colour nearest to black. After violet, and now getting progressively lighter, comes blue, then, blue-green, green, yellow-green and yellow the lightest again. See colour circle, Fig. 25.1(a). This progression is known as the 'natural order of colour' which appears consistently in nature. Yellow, orange, and red are the colours of highest visibility; they are also the warmest colours and tend to *advance*. For this reason they are the most powerful, focal, or signal colours we have at our disposal. Yellow-green is also a useful high-visibility colour, but since it inclines towards the cooler receding colours it has less powerful impact than orange or red.

Whilst yellow and orange have greater visibility than red, since their luminosity and light reflection value is higher, red is nevertheless the more powerful and exciting colour and is undoubtedly the colour with the greatest visual impact. From time immemorial it has symbolized fire with its dual implication of warmth and danger. Red is also gay and stimulating but, as with all stimulants, excess can be tiring and distracting. It should therefore be used with due reticence and discretion.

Green and blue are the very opposite to red, being cool or cold and *receding*. Green suggests calm serenity and is therefore appropriate for use in quiet rest rooms. It suggests peace of mind and safety. Blue suggests the immensity of space, the coldness of ice and steel. It is essentially a *static* colour in contrast with red, orange, and yellow which suggest warmth, light, and movement.

Purple and violet are rich regal colours associated with status and special occasions. They are deeply calm and reflective but also depressing to some people with their suggestion of mourning. Paler tints produce lilac and lavender which are essentially feminine in appeal.

So far we have been discussing the basic 'primary' colours and their characteristic attributes. Each of these primary colours or hues is capable of infinite variation by lightening with white or darkening with black, or by intermixing to produce intermediate hues, so modifying their attributes according to the degree of variation in colour. Pastel pink for example is pleasant, cheerful, and suggestive of 'well being', but it does not have the impact and excitement of the parent red from which it derives. Nevertheless it can be said that the main attributes of the parent or primary colours are retained to some degree in all the tints, shades, and subtle hues which derive from them and these attributes will affect the environment of an interior in proportion to the amount of the particular colour used. Thus an interior painted a pale yellow or primrose tint will always look warm and sunny compared with a similar room painted a cool ice-blue.

Colour classification

Colour names are not the best guide to colour identity. A more exact basis is required and this is provided in the Munsell system, which has been adopted by the British Standard Institution as a basis for B.S. 4800:1972, a range of 88 colours including black and white. The Munsell atlas classifies colours systematically in three dimensions referred to as hue, value (lightness), and chroma (saturation):

> *hue* refers to the basic colour scale, i.e., red, blue, green, etc.;
> *value* (lightness) refers to the tonal scale, i.e., the lightness or darkness of a colour, so that a high value colour is a pale tint whilst a colour of low value is a dark shade;
> *chroma* (saturation) refers to the purity scale or strength of colour which can vary from the purest colour available (high chroma) to almost neutral grey (low chroma).

Munsell notation describes each colour exactly in these three dimensions, the hue reference given first followed by value and lastly chroma; thus 7·5 R 4-5/16 (a strong pure red), or 2·5 GY 6/2 (a pale grey-green). In practice, however, it is usual to specify only the B.S. number which in the cases of the two colours mentioned, are 04 E 53 and 12 B 21 respectively. The 88 colours in the B.S. 4800 range are shown on colour cards which all leading paint manufacturers now supply and reference to one of these cards will make the above notation clear.

In the tonal scale a *tint* is a colour reduced or made lighter with white, whilst a *shade* is a colour made darker with black. Colours in *harmony* are those lying close together in the colour circle, e.g., yellow, orange, and red; or yellow, green, and blue. *Contrast* or complementary colours are those lying opposite in the colour circle, e.g., red and green; orange and blue; yellow and violet, etc.

When juxtaposed, contrast colours greatly intensify each other. For instance when red and green are placed together, the red looks redder and the green looks greener, but the effect can be clashing and hurtful in large areas especially when the colours are equal in tone value. When a small patch of bright red is placed on a large area of dark green, however, the red is greatly enhanced and stands out vividly from its background. In this case there is contrast of value as well as hue and this is especially important where visibility is concerned. Contrasts of tone value are the most significant in design and legibility.

Whilst black and white provide the maximum contrast in tone value, black and yellow provide even greater visual impact because the eye is naturally attracted to the brightest and most colourful object in view (hence the black bands painted diagonally on a yellow ground to mark obstructions and dangerous hazards, see Fig. 25.1(b)).

Colour and visibility

Many industrial interiors resemble a vast confused expanse of grey monotony, ill-lit, and devoid of colour. Noise greatly adds to the confusion and danger when visibility is poor. This gloomy picture can be completely transformed by light and colour, but in effecting this transformation there is always a danger of overdoing things by well-meaning but misguided efforts to produce a bright and colourful environment. Overenthusiastic misapplication of colour can result in a factory looking like a fairground, distracting and visually chaotic. In a large workshop, warning colours can be reduced to insignificance or completely lost in the complex mass of machinery and structural steel, particularly if stanchions, girders, roof trusses and the like are painted in strong bright colours. Colour misapplied in this way is not only distracting but can be a positive danger in high-lighting irrelevant details at the expense of more important dangerous hazards.

What is required, of course, is visual clarity and the sparing use of colour is often the more effective way of achieving this. By careful selection and placing, colour can create 'visual targets' to guide the eye to critical features, danger zones, and hazards. It is a matter of getting the priorities right.

Adequate light is the first priority, for without light we cannot see. The second priority will vary with different industries. In some it is the danger of fire or explosion in which case it is imperative that adequate warning is always visible as a constant reminder and safeguard, and that fire-fighting appliances and escape routes are prominently signposted in case of emergency. Other priorities will depend on the degree of hazard but in any occupation where the accident risk is serious the indication of first-aid posts should have priority.

The problem in any large factory is to ensure that all emergency warnings, danger hazards, and so on, are visible and easily seen. They should be quite distinctive in colour and stand out clearly and unmistakably from their surroundings. Nothing of minor importance should be allowed to compete for attention and no similar colour should be used elsewhere which might lead to confusion in an emergency. Warning colours should be of maximum brilliance and everything else should be subordinate or 'played down' in order to enhance the effect of these warning focal colours.

This means, in practice, that colours for general surfaces—roof areas, ceilings, and walls, including structural elements such as stanchions, girders, and roof trusses—should all be of fairly weak chroma to ensure that they cannot possibly be confused with the more powerful focal or signal colours which should be of the strongest chroma available.

In selecting colours for roof areas, ceilings, and walls it should also be borne in mind that these surfaces provide the background against which any pipe identification colours will be seen and, since the latter will be in various positive colours of pronounced chroma, the background should be unobtrusive, as light as possible and fairly neutral in order to enhance and not clash with the pipe colours. The ideal background, of course, is white or near-white (B.S. 08 B 15, 10 B 15, or 00 A 01). Seen against such a background, pipework often provides quite a colourful element which can look attractive in itself.

Industrial environment will vary according to the nature of the job. A machine room in a clothing factory, or a well-lit transistor assembly shop where accident risk is minimal is a vastly different environment to, say, a steel rolling mill, or a locomotive assembly shop, where overhead cranes and other hazards are always present. Where risk is light, one can afford to give more thought to the cosmetic rather than the purely functional aspect of colour (especially where women are employed), but where risk is high then function should be the first and foremost consideration.

The roof area or ceiling is usually the largest visible area seen from the shop floor since walls are often hidden by bulky plant. These large roof areas can be coloured yellow, e.g., B.S. 10 C 35 to provide a warm sunny environment, or alternatively pale blue such as B.S. 18 E 51 to provide a cool aspect; such colours look best when the trusses or ceiling beams and the walls are painted white. Coloured ceilings admittedly involve some loss in light reflection but it must be remembered that white roof beams and walls compensate for the lower value of the ceiling colour and overall light reflection therefore remains high.

Floors should also be regarded as important areas of light reflection and colours should not be lower than say value 5 or 6 whenever possible. As a rough guide the colour of dry new concrete is approximately value 6. The Illuminating Engineering Society's Code recommend the following light reflection factors:

Ceilings and roof areas	70 per cent or more
Upper walls and lower walls	not less than 40 per cent
Floors	between 20 and 40 per cent

Whilst high light reflection is particularly desirable on ceilings and roof areas it is not the only consideration, otherwise everything would be painted white. What is even more important, however, is the environmental effect of colour and in the examples mentioned above any slight loss in light reflection would be far outweighed by the gain in visual satisfaction.

Red is sometimes used on large areas of ceiling or walls, possibly to suggest warmth and good cheer, and whilst this may be acceptable in circulation areas, despatch departments and loading bays which are often draughty places, much careful consideration is required before using it elsewhere in the factory. As mentioned earlier, red is our most powerful colour and should be strictly reserved for the most urgent hazards, i.e., fire warnings, fire doors, and fire appliances, etc. The danger is that when red is used in a large area merely as a 'decoration' it will lose its impact value as a warning or signal colour because one becomes gradually conditioned to it and eventually ceases to take any notice of it.

The basic requirements for a good safe working environment then are, first, good lighting; second, large areas of light tints of high reflectivity on ceilings and walls; and third, structural elements (stanchions, girders, etc.) in tones of lower value and weak chroma. These three basics will provide an ideal background against which any focal or signal colours will 'tell out' distinctly, marking instantly the location of fire and first-aid posts, moving cranes, trucks, and other hazards.

In painting ceilings and walls a matt finish is the most desirable for visual comfort and the even dispersion of light, but it requires more frequent maintenance than a gloss or semi-gloss finish. High gloss gives maximum protection but also gives troublesome glare from specular reflection of light sources. The best compromise is therefore a semigloss, satin, or eggshell finish which gives good light dispersion without undue specular reflection.

Safety code and signal colours

To ensure that warnings are clearly visible, the brightest and strongest colours should be used as focal or signal colours to mark and delineate all dangerous hazards, to give visual warning, to direct attention to exits and escape routes, and so on. In addition to these more urgent functions, there are subsidiary tasks in coding services and pipe identification. Ideally each of these tasks should have its own special distinctive colour, but in spite of there being almost a thousand colours in the Munsell Atlas there are really only about ten basic hues which are easily recognizable, the rest being intermediate hues with infinite variations in value and chroma.

The problem is to select the most powerful and distinctive colours for the most urgent functions. We have seen already that yellow, orange, and red have the greatest visibility and that, of these colours, red has the most immediate and powerful impact. For this reason it is always used for the most urgent tasks, i.e., for STOP in road and rail signalling; for DANGER warnings; and for FIRE notices and equipment.

Since fire and explosion represent the greatest and most urgent hazard, these should have the most efficient precautions where risk is high as for instance in places such as gas and petrochemical installations. Here the warnings must be conspicuous and always visible, night and day, but since even the brightest normal red pigment loses its identity and appears dark brown in certain types of mercury lighting, it will not do. This is where the brilliant fluorescent pigments are unequalled for use as warning colours, both day and night, since they retain their high visibility under all types of lighting. The location of fire extinguishers and appliances should be marked with a large circle painted on the wall in fluorescent red. Signboards warning against fire or explosion should also be painted fluorescent red and lettered in white for maximum visibility.

Moving objects such as overhead cranes, fork lift trucks, and so on, are next on our list of priorities, and the position of these should be indicated in a high visibility orange-yellow, e.g., B.S. 08 E 51; likewise the marking of dangerous obstructions, unfenced openings and traps. The painting of black and yellow diagonal bands can make large obstructions even more conspicuous but this can be overdone, especially on small items which are best painted in one colour only—bright yellow. Dangerous hazards such as rollers, cogs, pulleys, guillotines, and power saws should also be clearly indicated with B.S. 08 E 51 yellow; electrical switchgear and *inside* fuse boxes should be similarly marked.

The B.S.I. has published B.S. 2929 *Safety colours for use in industry* and B.S. 1710 *Specification for identification of pipelines* and when these colours are adopted they should be used consistently and reserved exclusively for their respective purposes in order to avoid confusion in an emergency.

Each warning colour should also have its own distinctive shape or symbol to ensure recognition by anyone with colour-defective vision. This is not always easy to apply in practice but in general the warning colour can be painted in the prescribed shape, or the symbol itself can appear in black or white on the warning colour.

Pipe identification

The painting of pipes, conduits, and ducts in distinctive colours has obvious advantages from the point of view of identifying services, maintenance and safety and each factory should have its own identification code, based primarily on B.S. 1710 but possibly amended where required to suit specific needs which may not be covered in B.S. 1710 (see Table 25.1). The requirements in any colour identification or safety codes are that there should be sufficient contrast and difference in hue and value to make the colours easily recognizable in any light, making due allowance for darkening or fading of colours due to aging of the paint film. This is quite a tall order and rather than risk confusion with colours

Table 25.1 Colours for pipe identification

Pipe contents	B.S. 1710:1971		Recommended B.S. 4800 colour	
	Ground colour	Colour bands	Ground colour	Colour band
WATER				
Drinking	Green	Blue	12 D 45	18 E 53
Cooling	Green	White	12 D 45	White
Boiler feed	Green	Crimson/White/Crimson	12 D 45	04 D 45/White/04 D 45
Condensate	Green	Crimson/Em. Green/Crimson	12 D 45	04 D 45/14 E 53/04 D 45
Chilled	Green	White/Em. Green/White	12 D 45	White/14 E 53/White
Central heating				
< 100°C	Green	Blue/Crimson/Blue	12 D 45	18 E 53/04 D 45/18 E 53
> 100°C	Green	Crimson/Blue/Crimson	12 D 45	04 D 45/18 E 53/04 D 45
Cold, down sluice	Green	White/Blue/White	12 D 45	White/18 E 53/White
Hot water supply	Green	White/Crimson/White	12 D 45	White/04 D 45/White
Hydraulic power	Green	Salmon pink	12 D 45	04 C 33
Sea, river, untreated	Green	—	12 D 45	—
Fire extinguishing	Green	Safety red	12 D 45	04 E 53
COMPRESSED AIR	Light blue	—	20 E 51	—
VACUUM	Light blue	White	20 E 51	White
STEAM	Silver grey	—	10 A 03	—
DRAINAGE	Black	—	Black	—
ELECTRICAL	Orange	—	06 E 51	—
TOWN GAS				
Manufactured	Yellow ochre	Emerald green	08 C 35	14 E 53
Natural	Yellow ochre	—	08 C 35	—
OILS				
Diesel fuel	Brown	White	06 C 39	White
Furnace fuel	Brown	Emerald green	06 C 39	14 E 53
Lubricating	Brown	Salmon pink	06 C 39	04 C 33
Hydraulic power	Brown	Crimson	06 C 39	04 D 45
Transformer	Brown	—	06 C 39	—
ACIDS and ALKALIS	Violet	—	22 C 37	—

which are too close in value or hue it is much better to mark clearly with a system of colour bands or lettering; these identity codes are now readily available in coloured adhesive tapes.

The question as to whether the entire length of ducts, service pipes and conduits should be painted in their respective B.S. code colours, or only in bands at intervals or junctions, depends largely on the layout of the pipework; an untidy clutter of pipes and conduits, for instance, can be 'lost' by painting in the wall colour and marking in code colours only at the source and termination of the supply and at important intervals or junctions. On the other hand a neat and tidy installation is best painted throughout in the appropriate code colours, but only if the wall or background against which it is seen is painted white or a very light neutral tint that will not clash with the code colours. In this way the bright code colours, chosen entirely for functional reasons, can help to provide an attractive, even gay, environment, whilst providing recognition and identity of services which can be clearly seen from the shop floor. There is also less chance of error when pipes are painted along their whole length, particularly when the pipes go through walls from one place to another.

Painting of plant and machinery

The treatment of plant and machinery is of great importance and can make a significant contribution to safety. Large bulky plant in dark colours absorbs too much light and creates dark working areas on the shadow side. Plant should therefore be painted as light as possible consistent with practical maintenance.

No amount of cleaning will improve the appearance of dark, shabby paintwork, but when machinery is painted a light or medium tone in appropriate colours the operative is given an incentive to take a pride in keeping it clean. The choice of colours for machinery will depend largely on the general interior scheme and also on the colour of the materials being processed. 'Neutral' colours of low chroma are generally recommended for the framework or body of the machine. A small range of colours having Munsell values around 5 are generally suitable for most purposes, i.e., a range of greyed hues; these are sufficiently neutralized to allow the more positive code or signal colours to be clearly visible against the background of the machine. This means, in practice, that machine colours should ideally be selected from greyed hues having Munsell values of between 4 and 6 and of low chroma, e.g., B.S. 10 A 07, or inclining slightly to green, blue, yellow, etc., as desired (as, for example, B.S. 12 B 21, 18 B 21, or 10 B 21 respectively). A second colour may be used to distinguish minor elements of large machines in which case a lighter or darker tone of the main colour may be used; thus B.S. 10 A 11 may be used with 10 A 07; 12 B 25 with 12 B 21; 18 B 25 with 18 B 21; and 10 B 25 with 10 B 21. Such two-tone schemes always look well and do not conflict with safety code colours which might be used on dangerous moving parts, emergency stop buttons, etc.

Machine guard covers should be painted externally in the main or secondary colour of the machine, but the dangerous working parts *inside* the machine guards should be painted in bright yellow (and here again, fluorescent colours are ideal for maximum impact) so that these are instantly visible when the guards are removed. The insides of electrical switchgear and fuse boxes, etc., should also be in brightest orange for the same reason; the outside casing may be painted in the wall colour on which they are fixed, or in a light neutral grey, e.g., B.S. 00 A 05.

The use of high-gloss finishes on plant and machinery can give troublesome glare due to specular reflection of light sources and in order to overcome this trouble semi-gloss or eggshell finishes, formulated to withstand the rigorous conditions to which machinery is usually subjected, are recommended.

Colour and lighting

Although industrial lighting is dealt with more fully elsewhere, we feel that a special word on colour and lighting is necessary here. Adequate lighting in any workplace must be given first priority, and this is especially important in occupations where noise level is high and one has to depend on seeing rather than hearing approaching danger.

However, light itself can become a hazard if it is so placed as to cause blinding glare, or if it is the kind of light (e.g., certain types of mercury or sodium) which prevents instant recognition of warning colours. Light 'flicker' of fluorescent tubes and discharge lamps can also be a hazard and, if not corrected, the strobo-scopic effect may cause revolving parts of machinery to appear stationary and so lead to accidents. These lighting defects may be overcome or minimized by certain wiring techniques and the advice of an electrical consultant should be sought.

With regard to the possible effects of lighting on the appearance of colours this can result in some cases in complete loss of colour identity and it is most important that lighting should always be considered at the time of colour selection, especially where safety is involved. Ideally, colours should be selected in the kind of lighting in which they are to be used.

As far as colour recognition is concerned tungsten filament and any of the white fluorescent tubes are satisfactory since they do not cause any major falsification in the appearance of colours. The worst offenders are undoubtedly sodium and uncorrected mercury, both of which are used mainly for street lighting but are also used in many factory installations for reasons of economy.

Sodium lighting is orange-yellow in colour and since it is monochromatic it makes all light colours appear yellowish in hue, whilst all dark colours appear in tones varying from yellow-grey to black. Bright red appears almost black, but fluorescent red appears light yellow; thus while losing its colour identity it still *retains its high visibility*.

Uncorrected mercury (MB/U) is a bluish-violet light with a two-line spectrum of yellow-green and blue-violet. It thus favours yellows, greens, blues, and violets but debases most other colours. Bright reds appear dark brown or black, but fortunately fluorescent red still retains its colour identity due to activation by the ultra-violet element in the mercury. For this reason only fluorescent red or orange can be used with safety as warning colours in uncorrected mercury lighting.

The colour rendering of mercury is greatly improved by blending with either tungsten or fluorescent (designated MBT and MBF/U respectively). When thus 'corrected' all colours illumined by it retain their essential identity and it is therefore safe to use with warning and signal colours. It is, however, a cold, hard, brilliant light, much less pleasant environmentally than tungsten filament or fluorescent.

In any lighting installation it is desirable that a fair amount of light is thrown upwards to illuminate the roof or ceiling area, but in far too many factories the light fittings have solid metal dispersive reflectors which concentrate all light downwards, leaving the upper roof area virtually 'blacked-out'. This creates a dark 'tunnel' effect in which the bright light source is seen like car headlights against a virtually black background, giving rise to acute glare through excessive brightness contrast. Under these conditions any colour applied to roof areas for environmental reasons cannot be enjoyed as it can during daylight hours since it is invisible at night time. By installing *pierced* metal or translucent plastic reflectors which allow some 20 per cent of the light to filter upwards, the distribution of light is greatly improved; the oppressive 'tunnel' effect disappears and the better spread of light produces quite a dramatic environmental improvement, with consequent gain in working efficiency and safety.

In painting welding shops it is desirable to avoid paints of high ultra-violet reflectance. Most matt or eggshell wall paints nowadays are based on titanium dioxide which is effective in absorbing ultra-violet reflectance and therefore satisfactory in this respect. In welding booths the walls or screens in the immediate vicinity should be of low value and warm hue, e.g., B.S. 10 B 21 to minimize ultra-violet reflectance and glare, but walls further away from the welding operation may be lighter, e.g., 10 A 03, 10 B 17, or 12 B 17.

Colour planning

It will be seen from the foregoing that the planning of an industrial colour scheme calls for specialized knowledge of colour and its practical application in the industrial field. There is really no room for the amateur approach, nor any need either since there is no lack of expert advice so readily available from leading paint manufacturers.

Once a comprehensive colour plan is prepared for a factory it can be put into phased operation, department by department, whenever painting becomes necessary or possible. This obviates piecemeal application and ensures a consistent approach throughout the whole plant. However, to ensure complete success in implementing the plan it requires the fullest co-operation of all from the boardroom to the shop floor and this can only be achieved when the objects of the scheme are communicated to those involved—foremen, heads of departments, and management. The effort in striving to achieve a pleasant and safe working environment can show dividends in attracting and retaining labour, in better 'housekeeping', fewer accidents, greater efficiency and, ultimately, increased production.

Summary

The following is a summary of the salient requirements for a safe and pleasant industrial environment.

(a) An adequate provision of light—both natural and artificial—free from glare, with good distribution above and below the light source which must also possess good colour rendering properties (sodium and uncorrected mercury to be avoided at all costs). Colours should be selected under the type of lighting to be used.

(b) A rational approach to selection and application of colour for functional reasons rather than for cosmetic effect. Colours of high value and low chroma recommended for ceilings and walls to provide unobtrusive background with good light reflection, against which focal and safety colours are clearly visible. Floors also require to be as light as practicable to assist overall light reflection.

(c) Safety code colours of highest visibility and chroma, e.g., yellow (B.S. 08 E 51) for marking hazards, obstructions and moving objects (cranes, trucks, etc.). Colour with greatest visual impact (fluorescent red) for warnings of fire and explosion hazards and for marking location of fire doors, appliances, etc. Safety code colours should also have a distinctive shape or symbol to ensure recognition by those with colour-defective vision.

(d) Colours for plant and machinery in medium tone value and low chroma to provide non-distracting appearance against which signal colours for emergency stops, etc., may be clearly seen. Safety guards painted outside in the general machine colour, but brightest yellow *inside* to warn when a guard is removed. Where an additional colour is desired to distinguish minor parts of large plant, a two-tone effect is recommended, the second colour being a lighter or darker tone of the general machine colour.

(e) Colours for factory interiors, plant, safety codes and pipe identification are generally selected from B.S. 4800, this being the British Standard 1972 paint colour range generally available from principal paint manufacturers. Semi-gloss paint finishes are generally recommended for ceilings and walls and also for working areas on machine tools to minimize glare from specular reflection.

26. Lighting for safety

K. H. Ragsdale

*County Surveyor's Department,
Lancashire County Council*

Lighting is something which we take for granted. Yet in this supposedly technological age, much of the electric lighting installed over the shop floor and offices
of industry can be described as a handicap rather than an aid to safety.

Without light there can be no seeing and no safety. On the other hand, light
itself can be dangerous if misapplied, as any motorist who has suffered the
approaching car with undipped headlights will verify. Good lighting is lighting
which promotes safety and industrial efficiency; it is lighting which is designed
specifically to suit the requirements of the job, the employees, and the location.
It should be kept abreast of all the other improvements that are continually
made within a factory to increase efficiency and productivity.

There is evidence to believe that poor lighting contributed directly or
indirectly to a significant proportion of the falls which form such a high proportion of the total number of industrial accidents. Whilst many accidents are due
to lapses on the part of employees (possibly because they neglect to observe the
safety procedures laid down by management), employees have little control
over the standard of lighting provided for their work. This is a management
responsibility which must be faced up to if lighting is to fulfil its important role
in industry.

Lighting faults

There are a number of lighting faults which all too often are found on the shop
floor and in the factory office. The most significant are as follows.

Insufficient light

Darkness can often conceal real danger and semi-darkness is little better. It can lead to misinterpretation of visual information because the position, shape, or speed of an object is misjudged if the illuminance provided is insufficient for the task that is being attempted; in total darkness we would probably not attempt the task.

Fig. 26.1 In any working interior good lighting should promote efficiency, safety, and good housekeeping and, in partnership with the decorations, create a pleasant and acceptable working environment.

Although the eye can function over a very wide range of brightness values, there is a limit below which a job cannot be performed safely. The illuminance levels recommended by the Illuminating Engineering Society (I.E.S.) indicate the levels currently acceptable for good work (see Table 26.1).

It has also to be remembered that eyes take time to adapt to abrupt variations in the intensity of lighting, especially where the change is from a high level of illuminance (say outside on a bright sunny day) to a lower one (perhaps a poorly lit staircase just inside a building) and we are not able to discern detail until the adjustment is complete.

Shadows

Shadows result if luminaires (lighting fittings) are too widely spaced in relation to their mounting height, or if they are in the wrong position. Unless the presence of bulky obstructions obviously requires precautions to be taken, shadows will generally be unobjectionable if the spacing/height ratio of the luminaires does not exceed the maximum recommended by the manufacturer. This is normally between $1\frac{1}{2}:1$ and $1:1$, depending on the type of luminaire concerned.

An example of badly positioned lighting is found in every situation where, in his normal working position, the operative is between the task he is trying to see and the main source of light provided for that task, and so works in his own shadow.

Table 26.1 Recommended standard service illuminance values for different classes of visual task.

Task group	Type of task or interior	Standard Service illuminance (lux)	Limiting glare index
—	Storage areas and plant rooms with no continuous work	150	—
Rough work	Rough machining and assembly	300	25
Routine work	Offices, control rooms, medium machining and assembly	500	22
Demanding work	Deep-plant drawing or business machine offices. Inspection of medium machining	750	19
Fine work	Colour discrimination, textile processing, fine machining and assembly	1000	19
Very fine work	Hand engraving, inspection of fine machining or assembly	1500	16
Minute work	Inspection of very fine assembly	3000	19

Shadow must on no account be allowed on staircases which in themselves are hazardous enough, without the added danger of shadow concealing a step edge or giving the appearance of an additional one. Luminaires should be provided at the top and bottom of each flight of steps and should be used continuously if there is inadequate natural lighting during daylight hours. The stair lighting must be switched on at dusk.

The annual cost of running two 100 W lamps throughout the total working hours of a single shift factory will be about £5, a small price to pay for preventing the possibility of a fatal accident.

Glare

Glare is experienced in three different forms.

Disability glare. This is the visually paralysing effect caused by bright, bare lamps directly in the line of sight. This kind of glare is seldom experienced in working interiors because most bright lamps, e.g., filament and mercury vapour, are usually at least partially screened by some kind of fitting.

Discomfort glare. This is caused mainly by too much contrast of brightness between an object and its background and is a common by-product of poorly designed lighting. Since discomfort glare does not usually cause any immediate adverse reaction no-one is unduly concerned, but, over a period of time, discomfort glare can cause eyestrain, headaches, and fatigue; this, in turn, can be a contributory cause of an accident. In an existing installation which is already producing discomfort glare, a number of remedial steps can be taken. These include:

(a) changing to luminaires which adequately screen the lamp at all normal viewing angles;

(b) keeping the luminaires as high as practicable; the optical controls incorporated in many of the luminaires for high pressure sodium (HPS) and mercury discharge (MBF/U) lamps permit reasonable flexibility in choice of mounting height in contrast to the restrictions inherent in the old-fashioned type of enamelled open industrial reflector. Tubular fluorescent (MCF/U) lamps have a relatively low surface brightness and are usually acceptable at any height provided always that the installation is competently engineered;*

(c) using light coloured decorations on the walls and ceiling so that the backgrounds are as bright as possible; it is usually desirable for the luminaires to emit a reasonable proportion of upward light, approximately 7–17 per cent of the lamp light output, to help brighten the ceiling (as from the slots in the top of a fluorescent trough reflector)— the object is to reduce the brightness contrast between the luminaires and the background against which they are viewed;

(d) ensuring that the orientation of fluorescent luminaires is the most suitable from the point of view of glare reduction; most types give less glare if they are mounted parallel to the main direction of view in an interior.

Reflected glare. This is the reflection of bright light sources in shiny or wet 'work surfaces' such as glass or plated metal. Even a modest amount of glint can almost entirely conceal the detail in or behind the object which is glinting and where there is likely to be trouble of this kind great care is necessary either to use light sources of low brightness, or to arrange the geometry of the instal-

---

* Tubular fluorescent and colour corrected mercury reflector lamps (MCFR/U and MBFR/U) are versions of the ordinary lamps and have internal reflecting layers on the upper lamp surfaces, so that a high proportion of downward light is transmitted from the underside irrespective of the dust build-up on the top of the lamps. The undersides of these lamps are obviously extremely bright and particular attention has to be paid to lamp screening. It is difficult to design schemes using these lamps to meet I.E.S. glare recommendations, and advice on minimum mounting heights should be sought from the lamp manufacturer concerned.

lation so that there is no glint at the particular viewing direction, or both. (A great deal of difficulty would be saved if the use of shiny materials were avoided whenever possible and it is worth noting that micrometers, for example, invariably have a matt finish nowadays.)

Although more a hazard than a fault, it is appropriate to mention that mercury (MBF/U and MBFR/U) and sodium discharge lamps take some minutes to reach full light output after being switched on and also require time to relight if there is a sudden interruption to the electrical supply. In the interests of safety, there is a need for a pilot scheme of lighting that will provide a limited amount of 'instant' light, from say fluorescent or filament lamps, where the main installation is of mercury or sodium lamps. Although it is not obligatory, it is common sense to install an emergency lighting circuit in any area where a power failure would result in total darkness, such as in a windowless factory. The circuit could be arranged to control the pilot lighting referred to above and could do this either entirely independently of the mains supply or automatically in the event of mains failure.

The lighting faults discussed here are those which are particularly important from the safety aspect. There are, however, other aspects of interior lighting which bear on productivity. The type, colour, and direction of the lighting can aid or hinder quick and accurate perception of the details of the work task. The I.E.S. Code gives guidance on these matters and their importance must not be neglected by anyone concerned with business efficiency in its widest sense.

Standards

As mentioned earlier, the I.E.S. Code (1973) gives a schedule of standard service illuminances in steps which range from 100 to 3 000 lux* depending upon the task involved. The values are related to visual performance and visual preference and also take into account other factors such as the influence of daylight in the interior, recent improvements in lamp efficacies and practical experience. The recommended values are neither minima nor optima but represent good current practice. The 'standard' illuminance is the mean illuminance throughout the life of the lighting system and averaged over the relevant area, which may be the whole area of the interior or the area of the visual task and its immediate surround.

Where the task contrasts are abnormally low, or where the consequences of wrong perception may be serious (e.g. where expensive materials are used or where it is particularly important to reject faulty materials), the 'standard'

---

* The unit of illuminance:

$$1 \text{ lux} = 1 \text{ lumen per square metre (lm/m}^2)$$
$$= 0 \cdot 093 \text{ lumen per square foot (lm/ft}^2)$$
$$\text{or } 10 \cdot 76 \text{ lux} = 1 \text{ lm/ft}^2.$$

service illuminance must be increased by one or more steps to take account of such of these factors as may be present. Also, where the majority of workers are above, say, 50 years of age, or where protective goggles must be worn, or in windowless working interiors where the standard service illuminance for the task involved would normally be not less than 500 lux, it would be advantageous to increase the illuminance by at least 50 per cent.

Considerations of status or prestige, or the desire to be in the forefront of progress, may also prompt the use of levels higher than the recommended standard service illuminance. It is true in many fields that best practice is always, and understandably, ahead of official codes. It is certainly well in advance of any legal minima that regulations on factory or office conditions may require.

Most of the recommendations in the Code, therefore, are directly related to the difficulty of the visual task. However, in those interiors where the task is relatively simple or undemanding, another consideration appears; that of meeting the general psychological need for bright surroundings. To meet this need the Code recommends that the general level for these interiors should be not less than 150 lux if prolonged activity occurs. It does not apply to store-rooms, corridors, or other areas that are visited briefly or passed through.

It will be seen from Table 26.1 that the Code also makes recommendations for the limitation of glare (from general lighting installations). A full explanation of the calculations involved is not attempted here; suffice it to say that the limiting glare indices given in Table 26.1 represent design values which, if adhered to, will mean that glare has been kept within tolerable and reasonable limits.

Legislation

The mandatory requirements for the general run of both industrial and commercial premises are mostly either outdated or couched in terms too vague to be satisfactory. 'Lighting shall be sufficient and suitable . . . free from objectionable glare', etc., mean different things to different people. In general, H.M. Factory Inspectorate and those Public Health Inspectors responsible to local authorities for the enforcement of the Offices, Shops and Railway Premises Act regard the I.E.S. Code as defining the kind of lighting which is good for efficiency and productivity as well as for safety, health, and welfare. Because they are concerned only with the health and safety aspects, however, they will probably only insist on sufficient light to cater adequately for these.

The position as regards offices has been partially clarified by an explanatory booklet which the Department of Employment and Productivity issued during 1969.

Lighting installations up to I.E.S. Standards are regarded as both 'sufficient' and 'suitable' in the legal sense, unless other specific regulations apply.

Good lighting

Good lighting is not achieved by haphazard methods; it requires forethought and sound planning. Obviously, some simple, clear, and reliable guidance would be a great help to any user and this is readily obtainable from Electricity Boards, members of the Electrical Contractors' Association and the manufacturers of lamps and lighting equipment. These organizations will willingly survey and report on existing installations and prepare new schemes without charge or obligation.

Equally, it is desirable for the user to be able to do some basic checking for himself and, to this end, a pocket lightmeter is an essential item of equipment for measuring existing illuminance levels. Such an instrument is relatively inexpensive (approx. £15) and simple to use, and the information it gives cannot be obtained in any other way. Readings can be taken at all important working points and places of potential danger, such as stairways, as the first step in appraising the effectiveness and suitability of the existing lighting. At the same time as the illuminance values are being recorded, a number of visual checks should be made:

Are all the lamps adequately shielded?
Is there enough upward light on the ceiling?
Are the walls well illuminated?
Is the decoration helping the lighting?
Is there reflected glare from important work surfaces?
Are there dark corners and/or deceptive shadows on gangways and elsewhere?
How much light is being lost through side windows?
Are all the lamps and fittings clean and in reasonably good condition?

A thorough investigation along these lines will reveal defects in all but the most exceptional lighting installations. Many faults will be remediable without much trouble, but serious defects may show up which are more difficult to correct. The lighting of many factories will be found to have fallen behind their development in other respects. Modern plant and new processes may have been incorporated which impose more stringent seeing conditions, or the greater employment of girls and women may raise the question as to whether more attention should be paid to the appearance of the workshop. In these circumstances there may well be a temptation to adopt various expedients in order to avoid complete re-lighting and in this connection considerable care should be exercised.

If, for instance, it should be decided to substitute fluorescent or discharge luminaires for filament luminaires, the size of the operation involved justifies consulting an expert since the full benefit of modern lighting equipment cannot be realized if it is constrained in an obsolete or unsuitable layout.

Luminaires are sometimes lowered in order to increase the illuminance of areas beneath. This expedient is not recommended since the increase obtained below the luminaire must be accompanied by a corresponding reduction in illuminance between luminaires. The use of this device is a clear indication either that local lighting is required or that the lighting generally needs over-hauling.

The fact is that artificial lighting is less flexible than is commonly supposed. The idea that overhead lighting can be improved or added to at a later date overlooks the difficulties involved (unless previously allowed for in the design) and sooner or later many factories will face the necessity for new lighting installations. The success of such installations will depend not only on the skill of the lighting engineer, but on the decisions which have to be taken by the works management and executives and on the requirements which they give to the lighting engineer or contractor.

Bibliography

The I.E.S. Code for Interior Lighting. The Illuminating Engineering Society, 1973.
Interior Lighting Design. The Lighting Industry Federation and the Electricity Council (4th. edn., 1973).
Leaflets and booklets on lighting for Offices and for various Industries (Printing, Papermaking, Clothing, etc.) are available from the Electricity Council or from district and headquarters offices of Electricity Boards.

27. Glass as an aid to safety

H. F. Rigby

Technical Advisory Service,
Flat Glass Division,
Pilkington Brothers Ltd

Glass is generally accepted as being a brittle material, which, when broken, has dangerous razor-like edges; that it can be considered as an aid to safety may then appear contradictory, but glass can, and does, play an important part in the prevention of accidents. This is particularly so when the unique properties of certain types of glass are taken into account at an early stage in the design of equipment, machinery, doors and screens, low level and roof glazing and similar applications. Knowledge of the properties of the full range of glasses available and the use of the correct type of glass are essential to safety considerations.

Glass may be regarded as a material which, of itself or combined with other materials, adds to the provision of safety, and can do so in a number of ways:

1. for environmental control, by

    (a) providing clear vision where required;
    (b) providing weather barrier or other protection whilst permitting light to pass;
    (c) reducing uncomfortable glare;
    (d) insulating against loss of or excess of heat;
    (e) insulating against disturbing noise;

2. for protection, by

    (a) being resistant to heat and/or fire;
    (b) being resistant to impact;
    (c) providing a measure of safety if broken, for example by 'dicing' into comparatively harmless pieces.

Some terms in common use for glasses which have special qualities are 'toughened', 'laminated', 'heat resistant', 'fire resistant', 'insulating'—all appropriate in safety considerations.

Environmental control

Buildings have windows for a variety of reasons including the use of available natural light, the provision of view and means of ventilation. Adequate standards of daylight help in the prevention of accidents. Sky and reflected glare, often sources of trouble, can be overcome by the use of solar control glass of heat absorbing and/or reflecting types; whilst at the same time allowing the use of larger glazed areas to provide adequate light levels. The provision of a view helps in the sense of being in communication with the outside of a building or even the other side of a partition within a building and consequently becomes a relief from monotony and task fatigue. In such ways glass has become accepted as a means of helping towards safety. Again in the forms of solar control and double glazing, it adds to the comfort and well-being of people by keeping down excessive heat gains in summer and reducing heat loss in winter. The results of these features are greater efficiency, with minds more alert to the task, and safety consciousness.

Noise, more appropriately din when it has a nuisance value, can be a most disturbing influence in terms of safety, particularly should the intruding din be capable of damping down or even making inaudible those noises from machines which could spell danger. Double window systems, in which panes of glass are glazed 50 mm or more apart and in which sound absorbent material is positioned, can obviously lessen the possibilities of accidents.

Immediate protection

Fire resistance

Glass is a non-combustible material and will not, therefore, contribute to fire or directly help fire to spread. However, it is not considered to be an effective fire resistant material unless of certain types and glazed in approved ways. Even then the total area of glazing may be restricted by regulations and codes of practice because, whilst the test performance criteria of integrity and stability may be satisfied, that of insulation will not be. Consequently the extent of glazing in doors and screens forming escape routes will be limited to ensure that people are not exposed to unbearable radiation. Consideration must be given also to the possibility of radiant heat through glass causing ignition of flammable materials.

Apart from small vision panels of unwired glass in fire resisting doors, the glass not less than 6 mm thick and of area not exceeding 0.065 m^2, the only types of glazing which are suitable are:

1. 6 mm wired glass for half hour and one hour resistance, when panes not exceeding 1·2 m² may be permitted in conjunction with non-combustible channelling or metal frames which have been designed to avoid distortion. Where timber beads are used, the pane size may have to be reduced, and unless the beads are specially treated or capped, half hour integrity will not be attained.
2. Copper light glazing in unwired glass not less than 6 mm thick, individual panes not exceeding 0·015 m². Panels of copper lights should not exceed 0·4 m², but composite panels may be assembled using metal dividing bars provided that the metal is itself suitable for half hour or one hour resistance as appropriate to the situation.
3. Hollow glass blocks built *in situ* and having mortar joints can provide half hour and one hour fire resistance as regards stability and integrity.
4. Unwired glass qualifies for P60 designation under external roof test conditions provided that the glass is not less than 4 mm thick, does not exceed 760 mm wide and is glazed in timber bars (minimum dimension of section 38 mm with rebates not less than 12·7 mm) or in metal frames.

Official fire tests reports, available from the glass manufacturers, can be an aid towards design of fire resisting glazing and may also be useful when discussing fire problems and seeking approval of schemes with local authorities.

Heat resistance

This property of glass is different from fire resistance in that the glass itself is capable of withstanding thermal shock and/or elevated temperatures for longer periods than ordinary annealed glass. Two types in general use in industry are soda–lime–silica (the composition of window glass), which is toughened to endow the glass with high thermal and mechanical strength, and borosilicate, which, by composition is heat resisting and can be toughened to meet more stringent conditions.

Typical applications are sight glasses for pressure vessels; radiant heat screens, protection being directly for people or indirectly by protecting other glass such as heat reflecting which may not of itself withstand extreme heat; screens against welding spatter and flying scale from hot metals.

Laboratories

Glass is necessary for fume cupboards, glove boxes and the like, so that experiments may be observed, but many experiments can result in explosion and it is advisable to use toughened glass which, as well as being able to withstand thermal shocks up to 250°C, can withstand higher loading due to its increased strength. For situations where violent explosions may occur, double windows of toughened glass or laminated toughened glass should be considered.

Hot acids and alkalis can attack the surface of glass. Chemical attack may be prevented by the use of mica protection to the glass. In other cases, two panes of glass, glazed surface to surface or as separated panes in a double window system may be the answer, but as conditions and requirements vary considerably there is no clear cut answer to all such problems, and it is strongly recommended that the manufacturers' advice is sought at the design stage. Double windows may be designed to restrict the area of glass subject to impact, whilst still providing adequate vision, even allowing the glass on the observer's side to be of greater size to permit angle viewing. The system of angle viewing may be obtained too by the use of mirrors for situations where direct viewing is not possible.

Machinery guards

Toughened glass is an ideal material for machinery guards, affording protection for the operator and enabling good vision of the work being undertaken. Depending on the nature of the work and hazards involved in machine failure, the glass can be designed to give the necessary protection.

Glazed doors and low-level panels

Glass is frequently required in doors and at low levels, up to about 1 m from floor level, to permit light entry and to give unobstructed view of corridors and workshops below the observer's level. Particularly in swing doors where the chances of accidents are high due to impact, the glass should be toughened. It will withstand a high degree of rough handling but, if it is broken, will dice and should not result in serious injury. As a safety type glass it is also recommended for situations where a difference in levels is necessary and a transparent barrier permits unobstructed viewing and at the same time a guard against accidents by falling people or equipment, tools and the like. For greater detail and for appropriate applications of toughened, wired, laminated and annealed glass the relevant British Standard Codes of Practice should be consulted.

Roof glazing

Wired glass is commonly used for roof glazing, because should the glass be broken there is the safety aspect that the wire tends to hold the glass together and prevent it falling into the building. However, wired glass cannot withstand sudden intense heat which can occur in steel works and dye works for example; in such cases toughened glass should be glazed.

Other applications

In addition to the above rather general applications where glass is used for reasons of safety, there are many others:

miners' cap lamp glasses,
welders' goggle glasses,
angled mirrors at blind corners for viewing oncoming traffic,
furnace observation glasses giving protection against eye damage,
diffuse reflection glass for meter dials,
automatically opening doors of toughened glass,

to mention but a few.

Reading about glass for safety purposes, one becomes more and more aware of the important part it has to play in providing means to permit immediate recognition of danger and by relieving eye strain, preventing defective sight and hearing, preventing personal injury and property damage and in consequence its role in keeping with the requirements of the Health and Safety at Work, etc. Act 1974.

A selection of relevant British Standard Specifications and Codes of Practice

B.S. 229:1957. *Flameproof enclosure of electrical apparatus.*
B.S. 476: *Fire tests on building materials and structures:*
  Part 3:1958: *External fire exposure roof tests;*
  Part 7:1971: *Surface spread of flame test for materials;*
  Part 8:1972: *Test methods and criteria for the fire resistance of elements of building construction.*
B.S. 857:1967: *Safety glass for land transport.*
B.S. 3463:1962: *Observation and gauge glasses for pressure vessels.*
C.P.3 (Chapter I). *Lighting Part 1:1964 Daylighting.*
C.P. 145. *Glazing Systems Part 1:1969. Patent Glazing.*
C.P.152:1972: *Glazing and fixing of glass for buildings.*
C.P.153. *Windows and rooflights:*
  Part 3:1972: *Sound insulation;*
  Part 4:1972: *Fire hazards associated with glazing in buildings.*

28. Noise

Professor G. R. C. Atherley

Dept. of Safety and Hygiene,
The University of
Aston in Birmingham

Noise is an important concern for advisers in safety and hygiene. But acoustics—the science which underlies noise—is a very large subject. It embraces engineering, physics, and physiology. An adviser in safety and hygiene is concerned with matters besides noise. He does not need command of the whole of acoustical knowledge, but he does need detailed knowledge of selected topics chosen from the field of acoustics because of their importance in safety and hygiene.

The purpose of this chapter is to identify the key knowledge which the safety and hygiene adviser needs in order to fulfil these tasks. It is not possible within one chapter to give all the knowledge which is necessary. Instead, the chapter shows where the key knowledge can be found and gives explanations to facilitate the reader's task in assimilating the knowledge.

Degree and nature of risk

Excessive exposure to noise causes injury to hearing. This is the source of danger with which the safety and hygiene adviser is principally concerned. And this danger is the sole topic for this chapter. However, safety and hygiene advisers will be aware that noise has other undesirable effects on people. In particular, it causes annoyance and, on occasion, it interferes with the understanding of speech. If a safety and hygiene adviser is drawn into questions of annoyance from noise he may find the following references helpful:

Land Compensation Act 1973
Planning and Noise, Department of the Environment Circular 1973
Control of Pollution Act 1974
British Standard 4142:1975.

Apart from these, publications from the Noise Advisory Council available from time to time from H.M.S.O. are a useful source of information.

Knowledge about occupational deafness and its relation with noise has advanced rapidly over the past decade. It is now possible to specify quite closely the risk presented by virtually any noise likely to be encountered in industry. The risk can be specified in various ways, as we shall see later. All of these are well within the scope of the safety and hygiene adviser. They enable him to provide management with strategic information which is essential for managers to plan the allocation of resources in order to fulfil their obligations under the Health and Safety at Work Act 1974.

Agent–response relations

Knowledge has advanced to the point where the agent–response relations— sometimes called dose–response relations—can be specified quantitatively for noise in relation to occupational deafness. An understanding of these quantitative relationships is essential for an understanding of the measurement of noise, the assessment of risk from noise, the specification of control, and the investigation of incidents involving previous exposure to noise.

Obligations; codes, limits, and standards

At the time of writing (July 1975) the only statutory obligation relating to noise is to be found in the Woodworking Machinery Regulations 1974. These require that where woodworking machines are used in factories and where any person is likely to be exposed continuously to a sound level of 90 dB (A) or to an equivalent or greater exposure to sound, the noise shall be reduced as far as is reasonably practicable, and suitable hearing protection should be provided. The hearing protectors shall be 'maintained, and shall be used by the person for whom they are provided'.

Section 2 of the Health and Safety at work Act lays down general obligations for an employer to provide a place of work which is safe and healthy. There can be no doubt that noise is unhealthy. Therefore, on this reasoning, a noisy factory represents a breach of Section 2. The Health and Safety Executive appear to take the same view; improvement notices have been served on employers because of noisy factories.

Section 6 of the Act contains general duties for designers, manufacturers, importers, and suppliers of equipment and machinery. It is likely that this Section will be used to achieve noise reduction at the design stage in new machinery. The safety and hygiene adviser needs to be familiar with these provisions when the purchase of new machinery is being contemplated. Noise specifications for new machinery are important in connection with Section 6. It is known that the Health and Safety Executive is working on a code of practice for machinery designers. At the time of writing this is yet to be published.

Under Section 16 of the Health and Safety at work Act, the Health and Safety Commission has powers to approve codes of practice. In 1972 the Department of Employment issued the code of practice for reducing the exposure of employed persons to noise. It is likely that this document will provide the basis for a code of practice to be approved by the Commission. The approved code may well differ in detail from the original code, but the principles are unlikely to change. The significance of approval for a code of practice has been dealt with elsewhere in this book. It is sufficient to say here that an approved code can be taken into account by an inspector when he frames an improvement or a prohibition notice.

Codes, limits, and standards are issued by other bodies besides the Health and Safety Commission. The British Occupational Hygiene Society in 1971 issued the Hygiene Standard for wide-band noise. This follows the same basic principles as the Department of Employment's code, although there are certain points of difference.

Codes, etc. are also issued outside Britain. American standards in particular often find their way into British industry. In the case of noise, however, it should be noted that American standards differ in important essentials from their British counterparts. Although the American standards make instructive reading, great care should be taken not to be misled by differences in the two systems. In particular, certain imported noise-measuring equipment may conform to American rather than British principles, and its use in Britain would therefore be misleading.

Compliance

The nature and extent of obligations imposed on employers and employees by the Health and Safety at Work Act are discussed in Chapters 37–39. Compliance with these obligations generally and specifically in relation to noise (in so far as these are legally enforceable) is a matter for the Health and Safety Executive. Ignorance of the law is no excuse. An important role for the safety and hygiene adviser is to ensure that all concerned are fully aware of the obligations imposed upon them by the Act. When once management are aware of their obligations they will require technical advice on how these should be fulfilled. The safety and hygiene adviser should be able to give them the technical advice they need or point out where it can be found. He should also be able to help management plan and co-ordinate the response.

In planning a course of action there is now an agreed scale of priorities. The Department of Employment's code makes clear that the first action to be considered is control over noise. This is a safe-place strategy as outlined in Chapter 1. The provision of personal hearing protection—a safe-person strategy—is seen as a second line of defence only to be used when the first line of defence—that is, noise control—has been fully explored.

The safety and hygiene adviser can see clearly where he should be attempting to stimulate compliance. First, he needs to stimulate management to take all necessary steps against noise, and second he needs to stimulate work people to accept the need for personal hearing protectors. Section 2 of the Health and Safety at work Act provides for the involvement of work people. Although no regulations have yet been issued under this section it is nevertheless clear that the Health and Safety Commission intends that the spirit of this requirement should be fulfilled. The safety and hygiene adviser should consider carefully for the factory for which he is responsible the implications of employees' involvement.

There are difficult issues involved. The allocation of resources for noise control is traditionally a matter of management prerogative. On the other hand the work people should want to be assured that all possible action is being taken against noise before they willingly accept the need for hearing protectors. The safety and hygiene adviser should take up a neutral and technical position in which he can speak freely to both management and workers.

Previous experience and investigation

Safety and hygiene advisers build up considerable experience in persuading management and workers to take action which neither perceive as their first priority. It is important that this experience is channelled into the handling of noise problems. Industry has certainly made considerable progress in tackling noise, but there is a good way to go before work people's exposure is universally within the Department of Employment's limits. A safety and hygiene adviser knows from previous experience that expenditure on safety is axed all too readily in times of economy. The way forward is generally through clear-cut and well-defined requirements which leave no one in any doubt about what is required. Under these circumstances, management accepts the need for allocating resources to safety on a continuing basis. It is therefore vitally important that best possible value for money is obtained for expenditure on safety hardware. Noise control for some industries represents an opportunity for almost unlimited expenditure. A key role for the safety and hygiene adviser is to ensure that solutions which are applied to noise problems represent good value for money. But the safety and hygiene adviser cannot be expected to have expert knowledge in all aspects of acoustical engineering—this point has already been made. However, one of his skills is to know where to get reliable advice on technical matters. This is especially important in questions of noise control.

In many organizations the safety and hygiene adviser becomes involved in questions of compensation. It has been argued that safety and hygiene advisers should not be involved in investigation of incidents and accidents which have resulted in disease and injuries arising from work. I do not take this view. The safety and hygiene adviser needs to know what is happening in his company. In particular, he needs to know what is going wrong. He may be the only person with the necessary skills to investigate accidents and exposures. If he leaves these investigations to other people there is a risk that they will not be done adequately and that he will be cut off from an important source of information. Obviously, it would be undesirable for the adviser to spend his whole time carrying out investigations. In circumstances where the burden from these promises to be great, his role should be to generally superintend investigations and to act as technical adviser.

At present there are increasing numbers of employer's liability claims for occupational deafness. These inevitably require investigation of previous noise exposure, consideration of preventive action taken at various times in the past, and related matters. It is likely that these claims will continue for some years to come. Safety and hygiene advisers should, as far as possible, involve themselves in the investigations of these claims. But their role should be that of technical adviser and supplier of technical information. As long as the role is confined to these two considerations there will be no need to take sides in any legal arguments. The safety and hygiene adviser cannot give purely legal advice—unless of course he is legally qualified. But he should be able to discuss with management and workers in layman's terms the technical questions which arise in connection with claims for compensation. He should understand procedures and he should be able to direct workers to appropriate advice.

Occupational deafness has recently been prescribed under the National Insurance Industrial Injuries Scheme (1974). The safety and hygiene adviser should understand how this scheme works and be prepared to explain it in general terms to management and workers.

Essentials of acoustics and noise measurement

Equipment

All measurements of noise in connection with occupational deafness centre on one system: the A-weighted decibel, abbreviated to dB(A). A wide range of instruments are available which measure dB(A) directly. More detailed measurement is necessary only for certain specific purposes. The safety and hygiene adviser requires a good instrument measuring in dB(A) as his basic equipment. In organizations where there are several different noise problems or a single problem presenting special difficulty, other instrumentation may be necessary. Octave-band analysis is necessary in the selection of hearing protection. Noise average meters, sometimes called noise dose meters, may be needed where noise exposure is highly irregular.

Acoustical laboratories will be equipped with precision tape recorders, acoustical analysers, and various other items of highly specialized equipment. None of this equipment is essential for day-to-day use by safety and hygiene advisers. Of all the specialized equipment, a precision tape recorder is probably the most useful.

Measurement

Sound-level meters give a measure of the rate of flow of acoustical energy. The A-weighting refers to a filter which removes a proportion of the low frequencies (and also certain high frequencies). This filter is included on two grounds: it mimics to some extent the characteristics of the human ear, and dB(A) has been found empirically to be the most serviceable choice of measuring scale for purposes connected with occupational deafness.

The key research is that reported by Burns and Robinson (1970). They showed that the degree and extent of injury to hearing caused by noise was related to the 'dose' of A-weighted sound energy. Dose is represented by the cumulative A-weighted sound energy. In general terms the cumulative dose at the end of Y years is given by

$$E(A) = \text{sound flux} \times Y$$

where $E(A)$ is the noise dose.

Sound level meters measure sound flux. But the units in which the measurements are expressed are decibels, which are logarithmic. It is therefore necessary to use logarithms in order to calculate the noise dose when a sound-level meter is providing the basic data. The expression given above needs modification:

$$E_A = L_A + 10 \log_{10} (Y/Y_0)$$

The noise dose, E_A, in logarithmic form is called noise immission level and measured in decibels; L_A is the reading in decibels from the sound-level meter; and Y represents the years of exposure, and Y_0 is one year. Table 28.1 gives values of the expression $10 \log_{10} Y/Y_0$ for various values of Y.

The Hygiene Standard for wide-band noise settled on a limit of 105 dB for the noise immission level over a working lifetime. For many occupational hygiene purposes a working lifetime has a notional value of 30 years. From Table 28.1 we find that $Y = 30$ corresponds to $10 \log_{10} (Y/Y_0) = 14 \cdot 8$. We can substitute 14·8 in the expression given above and we find that L_A corresponds to 90·2 dB(A) for a noise immission level of 105 dB for 30 years.

This information can be stated in another way: 90·2 dB(A) experienced habitually for a working lifetime of 30 years gives the dose of noise corresponding to the upper acceptable limit for noise dose.

The Department of Employment's code of practice for noise (1972) uses 90 dB(A) as the upper limit for habitual exposure to noise. It can be seen that the two standards correspond with each other, even though they are couched in different terms.

Table 28.1   Values of $10 \log_{10} (Y/Y_0)$

Y (years)	$10 \log_{10} (Y/Y_0)$	Y (years)	$10 \log_{10} (Y/Y_0)$
2	3	22	13·4
3	4·8	23	13·6
4	6	24	13·8
5	7	25	13·9
6	7·8	26	14·1
7	8·5	27	14·3
8	9	28	14·5
9	9·5	29	14·6
10	10	30	14·8
11	10·4	31	14·9
12	10·8	32	15
13	11·1	33	15·2
14	11·5	34	15·3
15	11·8	35	15·4
16	12	36	15·6
17	12·3	37	15·7
18	12·6	38	15·8
19	12·8	39	15·9
20	13	40	16
21	13·2	41	16·1

Noise immission level is an important concept in two respects: As we have seen, codes and standards for safeguarding work people against occupational deafness aim essentially to restrict noise immission level at the end of an average working lifetime; and knowledge of noise immission level can be used to estimate the degree and extent of occupational deafness in individuals as well as populations. The latter use of noise immission level is described below.

The measurement of L_A

In circumstances where the noise level is steady and the pattern of exposure is unbroken, the measurement of L_A is straightforward. A single reading in dB(A) on a sound-level meter suffices for the calculation of noise immission level. In circumstances where the noise is intermittent or irregular, or the exposure pattern is irregular, single measurements in dB(A) will not suffice. For the purposes of explanation it is convenient to consider separately the two aspects of non-steady noise exposure.

First, let us take a case where noise is steady but the exposure to it occurs in a broken pattern. Electricity generating stations present this sort of problem. Under load conditions, the generating sets produce steady noise. A plant attendant's work might take him away from the generating set for periods of time during his ordinary working day. The noise level of the generating set can be measured, but this would not by itself give an accurate estimate of the plant attendant's exposure; L_A could not be used directly for the calculation of the plant attendant's noise immission level. Allowance has to be made for the time in which he is away from the noise.

Under circumstances such as these, use is made of equivalent-continuous sound level L_{eq}. The Hygiene Standard defines this as the level of continuous noise in dB(A) which, in the course of an 8-hour working day, would cause the same sound energy to be received as that due to the actual noise over a typical day. Equivalent continuous sound level is found from the following expression:

$$L_{eq} = L_A + 10 \log_{10} (T/T_d),$$

where L_A is the sound level in dB(A), T is the number of hours of exposure per day to L_A, and T_d is the number of working hours in a day (normally taken to be eight). Table 28.2 gives values of $10 \log_{10} (T/T_d)$ for various values of T, using the standard value of $T_d = 8$.

Table 28.2   Values of $10 \log_{10} (T/T_d)$

T (hours)	$10 \log_{10} (T/T_d)$	T (hours)	$10 \log_{10} (T/T_d)$
0·25	−15·0	6·5	−1·0
0·5	−12·0	7·0	−0·6
1·0	−9·0	7·5	−0·3
2·0	−6·0	8·0	0
2·5	−5·0	8·5	+0·3
3·0	−4·3	9·0	+0·5
3·5	−3·6	10·0	+0·9
4·0	−3·0	10·5	+1·2
4·5	−2·5	11·0	+1·4
5·0	−2·0	11·5	+1·6
5·5	−1·7	12·0	+1·8
6·0	−1·2		

Let us return, by way of example, to the plant attendant in the generating station. Let us suppose that L_A for the generating station is 92 dB(A) and that the plant attendant spends four hours in the vicinity of the generating set. His total working day is eight hours and he spends the remaining four hours in a quiet control cabin. Table 28.2 shows that the value of $10 \log_{10} T/T_d$ is −3 dB. Substitution of −3 dB in the expression for the calculation of equivalent continuous sound level gives a value of 89 dB(A).

From the example, we can see that the generating set's noise exceeded the limit of 90 dB(A). But the plant attendant's exposure did not exceed the Code's recommended limit of an equivalent continuous sound level of 90 dB(A). Circumstances similar to these are assessed by means of readings taken with a sound-level meter used in conjunction with details of exposure time. The assessments are easy to make and are used to show whether an individual or a group of individuals exceed the Code's limit for exposure. Provided that the noise is reasonably steady, a straightforward sound-level meter may be used.

The second type of irregular exposure pattern which has to be considered is that where the noise is very irregular in level. An example is drop forging, where the hammer blows produce very loud impact sound. In between the impacts there is continuous noise from furnaces and certain other kinds of machinery. A sound-level meter used in these circumstances shows rapid movements of the needle. Some meters have a facility which can be switched in to slow down the meter response to rapidly changing sound. But neither fast nor slow settings give an adequate measure of all impact and impulse noise. For a reliable assessment, a noise average meter or a noise dose meter must be used. These instruments give a direct reading of equivalent continuous sound level. Provided that the reading is typical of a day's exposure, it can be compared directly with the Code's limit for exposure. These instruments are essential for measuring impact or impulse noise, and their use is wholly acceptable for the purposes of the Code. However, care should be taken to find out whether the instrument in use conforms to the industrial or precisions specification (see below).

Essential principles

The concept of noise immission level was first put forward by Robinson (1968). It was derived from the observations by Burns and Robinson (1970) from their studies of the hearing of work people exposed to industrial noise largely of the continuous type. In 1971, Atherley and Martin reported a study which showed that the concept of noise immission level could also be successfully applied to drop forging. They were able to work out the equivalent-continuous sound level for drop forging noise (the study was made before noise average meters became available). They measured the hearing of men employed on drop forges and showed that the hearing levels of the men fitted closely values predicted by means of Robinson's equation when the drop forging equivalent-continuous sound levels were used. In 1974 Atherley showed that concept of noise immission level could also be applied to noises with a rapid impact rate, such as pneumatic fettling, hammering, and riveting. Coles, Rice *et al.* (1974) have suggested that the noise immission concept can also be applied to impulse noise such as that produced by gunfire. Overall, therefore, it can be assumed that almost all noises encountered in industry can be assessed by means of the methods outlined above. The only exceptions are single explosive sounds of very high intensity and very short duration. At present, there is insufficient evidence about this kind of exposure, though there is no evidence to suggest that it *cannot* be treated in exactly the same way as all other noises.

Grades of measuring instrument

Appendix 2 of the Code of Practice draws attention to the two grades of sound-level meter. Precision grade conforms to the specification in B.S. 4197; the industrial grade conforms to B.S. 3489. In the former specification the tolerance must not exceed ± 1 dB in the mid-frequency range. For the industrial grade, on the other hand, it must not exceed ± 3 dB. For most practical situations industrial grade instruments suffice. However, where measurements reveal a marginal situation, it is clearly preferable for assessments to be based on precision grade readings.

The assessment of noise problems

In order to assess the risk from noise, three categories of information are required:

(a) noise levels from plant and machinery;
(b) the pattern of exposure of all persons liable to be affected by the noise;
(c) the numbers of people involved in the various exposure patterns.

This information can be used to assess the degree of risk in various parts of a factory, to select appropriate control measures, and to judge the value for money of the control which is proposed.

It is important to keep records. Forms are a convenient means for doing this. Appendix 8 of the Department of Employment's Code gives a specimen which, with certain modification, can be adapted for general use. The essential point is that noise levels should be accompanied by sufficient information to identify the source of the noise. Equivalent-continuous sound level, on the other hand, relates specifically to the exposure of people. A single source of noise may be responsible for the noise exposure of several groups of people, each of whom has a different exposure pattern. Record keeping needs to reflect these factors separately and in sufficient detail for the various calculations to be made.

As a first step it is useful to list all plant and machinery which produce noise levels in excess of 80 dB(A). The recorded information should show the type and pattern of machine, etc., and the actual noise levels produced by that machine under various conditions of operation. The next step is to survey the exposure of the work people concerned. From this information, the equivalent-continuous sound level is calculated on a representative basis. If the work patterns in a factory are so varied that no two jobs are alike then it is necessary to calculate the equivalent-continuous sound level for every single job. On the other hand, if the exposure pattern is similar for large numbers of people, then it is only necessary to calculate a typical value for equivalent-continuous sound level. The next step is to show the numbers of people at the various values of equivalent continuous sound level.

The information recorded so far is sufficient for a preliminary assessment of a noise problem. It gives an indication of the extent of risk. This can be calculated in greater detail by means of information given in the Hygiene Standard or B.S. 5330:1976.

For the purposes of the following discussion, I shall assume that the safety and hygiene adviser has identified a noise problem in that a material number of people are exposed to equivalent-continuous sound levels in excess of the Code's limit. At this stage management need to be brought into the picture. There is a choice before them: should noise control be instituted, or can a safe-person strategy be relied upon? Normally the choice turns on the question of feasibility of noise control, which itself turns on the question of cost. At this stage it is useful to have a preliminary appraisal made by a competent acoustical engineer who is asked to say whether noise control is a practicable proposition for the machinery, etc., in question. He should be asked to say on engineering grounds whether it is reasonable to expect noise control to alleviate the problem completely, partially, or not at all. If the answer is completely or partially, then he should be asked to provide a rough guide to the cost of the noise control. The estimates need only be rough. Information at this stage is needed for management to make broad policy decisions. Of course, in making the policy decision, management may well require further and detailed information. Obtaining and collating this information may or may not be a responsibility for the safety and hygiene adviser. Often, these will be matters for the chief engineer's department.

It is worth pointing out at this stage that the Health and Safety at work Act is qualified in many parts by the well-known legal phrase 'so far as is reasonably practicable'. The legal interpretation placed on this phrase allows cost to be weighed against risk. (For a fuller discussion of this interpretation see the introductory sections to any of the recent editions of Redgrave's Factories Acts.) The cost and risk data are thus important from a legal as well as a managerial point of view.

Acoustical engineering is a relatively new subject. There are a number of reputable consultants and an increasing number of people with the relevant academic qualifications in the field. The safety and hygiene adviser should establish close working contacts with someone whose judgement and expertise can be relied upon. I. Sharland, *Woods Practical Guide to Noise Control* (1972) is useful additional reading.

Where circumstances rule out noise control as an immediate solution, noise exposure control has to be considered. The simplest method is to reorganize work patterns so that no individual's equivalent-continuous sound level exceeds 90 dB(A). The principles involved have already been demonstrated in the calculations for the generating plant attendant's exposure set out previously. Table 28.2 can be used to work out number of hours exposure per day which must not be exceeded in order to keep within a 90 dB(A) limit for equivalent-continuous sound level. In practice, there are few circumstances where this method of exposure control can be successfully and reliably applied. The inherent weakness of the arrangement is that it relies on patterns of exposure being adhered to at all times. This represents a constraint on employment pattern which neither management nor workpeople are likely to welcome as a permanent arrangement. In particular, it creates difficulties for overtime working.

Hearing protectors represent a much more practicable and therefore common means of exercising control over noise exposure. In chapter 1, the distinction was drawn between safe-place and safe-person strategies. Readers should refer to that chapter for a discussion of the essential differences between the two classes of strategy, and the essential weaknesses of safe-person strategies.

At this stage we return briefly to record keeping. The entry for each source shows the noise level associated with it. It should now also show the control measures which are planned: noise control, exposure pattern control, hearing protectors, or any combination of these. There is now a record of managerial decisions. These need to be given authority by whatever means authority is given within the organization. The record can then be used as a check list for implementation. It should show a progressive fall in the numbers of people exposed to harmful noise.

Where noise control has been opted for, the crucial test is whether noise levels drop. Thus, the record should show noise levels taken at regular intervals. The fall in decibels is the sole measure of progress. The safety and hygiene adviser should monitor this progress, and where the pace of change is unsatisfactory he should bring matters to the attention of the appropriate manager.

Where exposure pattern control has been opted for, the safety and hygiene adviser needs to check from time to time that the exposure pattern is in fact being controlled. Where the record shows that hearing protectors are to be relied upon, the safety and hygiene adviser will be concerned with the adequacy of the hearing protection. Assessment of adequacy is normally a responsibility for the safety and hygiene adviser, because he is likely to be the only person with the necessary understanding of the principles and issues involved.

Hearing protectors

The range of hearing protectors now commercially available is very wide. The safety journals carry advertisements from suppliers from whom manufacturers' literature can be obtained. In practice, the selection of hearing protection resolves itself into a choice between ear muffs of one kind or another and one or other of the various patterns of ear insert. In making the choice, the safety and hygiene adviser needs to consider acoustical and non-acoustical factors.

The key purpose for hearing protection is to reduce the noise immission level of the wearer. Under special circumstances, hearing protectors have other purposes. They may be used by marksmen to prevent them from flinching from the noise of other marksmen's weapons, especially in competitions. Welders may use hearing protectors to safeguard their ears against the entry of sparks from their welding.

Reduction in noise immission level by hearing protectors determines the protection which they afford against noise. Else (1973) has set out the principles involved. It has been widely understood for a long time that the protection afforded by hearing protectors depends on their attenuation. Else pointed out that there was another, equally important, factor to be considered. This is the percentage of time during which the hearing protection is worn.

He started by assuming for the purposes of his calculation that we could find hearing protectors with infinite attenuation; that is, all sound is kept out. With this assumption, he showed that the maximum protection in dB(A) for protectors could be calculated from the following expression:

$$\text{maximum protection in dB(A)} = 10 \log_{10} \frac{100}{[100 - \text{percentage of time worn}]}$$

He used the expression to construct the graph shown in Fig. 28.1. The graph shows the maximum protection in dB(A) plotted against percentage of time worn, for hearing protectors with infinite attenuation. From the graph we can see that protection worn for 50 per cent of the exposure time confers no more than 3 dB(A) of protection. The implication is that there is no point in choosing hearing protectors which are acoustically excellent if the wearers are not going to wear them for much of the exposure time. From the graph it can also be seen that even if the wearer wears the protection for as much as 99·9 per cent of the exposure time the maximum protection which can be achieved is 30 dB(A), even though the hearing protectors might have an attenuation far in excess of this figure.

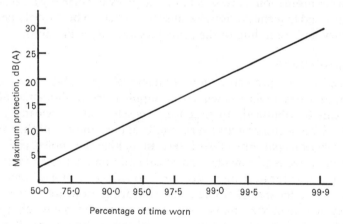

Fig. 28.1 (Reproduced from 'A note on the protection afforded by hearing protectors' by D. Else, *Ann. Occup. Hyg.*, **16**, 1973)

The lesson to be drawn from the graph is that when hearing protectors are opted for, very great attention should be attached to the problem of getting people to wear them. This problem is one in which the safety and hygiene adviser should be able to offer specialist advice. Certainly, it is unlikely that management and work people will be aware of the crucial influence of percentage of time worn unless the safety and hygiene adviser points this out to them. In circumstances where noise levels are high and hearing protection is being relied upon to provide 20 or more dB(A) of protection, attention must be paid to questions of comfort and acceptability of hearing protectors, because these will materially influence the extent to which people actually wear the protection which is provided.

Percentage of time worn is the first element to be considered in the selection of hearing protection; the second element is the attenuation which can be expected from the protectors which are to be supplied. The crucial information is the reduction in dB(A) which the hearing protectors are theoretically capable of providing. The figure to be aimed at is found by substracting 90 dB(A) from the equivalent-continuous sound level for the individual worker or group of workers for whom protection is provided. It is worth re-emphasizing that equivalent-continuous sound level is deduced from information about sound level and exposure pattern. Any changes in exposure pattern are likely to affect equivalent-continuous sound level, and a reconsideration of the necessary protection may be called for. If the safety and hygiene adviser cannot be sure of the constancy of exposure pattern or of the reliability of review procedures, he might be well advised to use machinery and plant noise levels for his calculations rather than equivalent-continuous sound level. In practice it may be prudent to make both sets of calculations.

The protection afforded by hearing protectors in specific circumstances requires calculation. The manufacturers of the hearing protection cannot give all the necessary information because the degree of protection afforded depends on the spectrum of the noise which they are to be used in, as well as the attentuation of the hearing protectors. The manufacturer provides the necessary information about attenuation, but the safety and hygiene adviser has to obtain the other information for himself. Unfortunately, the straightforward sound-level meter reading solely in dB(A) is not adequate for this purpose. An instrument which provides octave-band analysis is required. The decision to purchase such an instrument should be based on an estimate of the likely work load. An alternative is to hire instruments. All acoustical consultants should be able to carry out octave-band analysis. Because the procedure is straightforward, there is no reason why this should cost a lot of money. On the other hand, the safety and hygiene adviser may find it more convenient to have his own equipment at hand.

The method for making the calculations is set out in Appendix 4 of the Department of Employment Code. There is no purpose in repeating what is set out there, and the reader is referred to Appendix 4 for the detail of the calculations. However, it may be useful if the principles are set out briefly.

A manufacturer of the ear muffs will provide information about the attenuation. He should provide two sets of figures: the mean attenuation at various points in the frequency range, and the standard deviation of those data. The first step in the calculations is to arrive at the *assumed protection* by subtracting the standard deviation from the mean attenuation. A specimen calculation is given in the Code showing how this is done.

The next step is to subtract figures for the assumed protection from the octave-band data relating to the noise in question. The result, in effect, is the transmission loss of sound through the structure of the hearing protector. The result of this stage of the calculation gives us the spectrum of noise at the wearer's ear. The next step is to convert this spectrum to dB(A). This is done by subtracting the A-weighting correction on which the dB(A) scale is standardized. The relevant data are given in the Code. There are various means of converting octave-band levels to A-weighted level.

The A-weighted level arrived at by whatever means represents the level inside the hearing protector. This level is then subtracted from the equivalent-continuous sound level, giving the maximum *theoretical protection* for the specific hearing protectors in the specific noise, on the assumption that they are going to be worn for 100 per cent of the exposure time. Clearly, the equivalent-continuous sound level at the ear of the wearer must be less than 90 dB(A) in order to conform to the requirements of the Code. Where the theoretical maximum protection is insufficient, another type of protector must be chosen and the calculations should be repeated. Sometimes calculations show that the hearing protectors are conferring considerably more protection than is actually required to bring the A-weighted sound level below 90 dB(A) at the wearer's ears. Although it is generally desirable for as much noise as possible to be excluded, there are disadvantages in over-protection.

As pointed out by Atherley and Noble (1970) and confirmed by Else (1976), there may be danger for some workers if the hearing protection excludes too much sound such as warning sounds or cues from machinery which is not functioning correctly. In work shops with travelling cranes, for example, men depend on their directional hearing to perceive the travel of the crane towards them. And in certain work shops there are additional audible warnings such as hooters for warning of the crane's presence. The safety and hygiene adviser should survey carefully the auditory communication involved in a job before hearing protection is provided. Where there is danger from noise as well as dependence on sound for communication, hearing protection needs to be chosen judiciously, so that there is a sufficient degree of protection and no more. The shutting out of useful sound is sometimes given by workers as a reason for their rejecting hearing protectors, along with discomfort, sweating, unsightliness, and general inconvenience. Such complaints are often justified. Careful ground work is essential if they are to be avoided.

Planning a hearing protection project

Under the Health and Safety at work Act, there are general duties for employers, management, and employees which are applicable in respect of hearing protectors. Section 3 of the Code of Practice specifies certain duties. In general, duties can be summarized as follows.

Employers should provide a place of work which is free from noise as far as this can be achieved at reasonably expense. Where this cannot be achieved on grounds of cost or technical feasibility, employers have a duty to provide suitable hearing protectors. In circumstances where hearing protectors are to be provided, *management* has a duty to identify places where hearing protectors are required, control entry into such areas, ensure that suitable hearing protectors are provided and used, ensure that people concerned understand the care and use of hearing protectors, and ensure that hearing protection is a continuing concern.

In circumstances where hearing protectors have been selected, workpeople have a duty to use the hearing protectors, not to enter areas where hearing protectors are required (unless authorized by management), not to wilfully damage or misuse hearing protectors, and to report any loss or damage of hearing protectors.

Section 2 of the Health and Safety at work Act requires employers to provide a written safety policy. As part of the policy it is necessary to identify any special dangers to which the workers are exposed. Noise would presumably come into this category. Therefore, the safety policy is an excellent chance to spell out for all concerned their duties relating to noise. Allocation of these duties should be a matter of agreement following detailed discussion. A suitably constituted safety committee is an ideal forum for this kind of discussion. From the safety adviser's point of view, the purpose of the discussions should be to identify clearly the obligations falling on the people involved. Discussions about allocation of responsibilities should take place against a background of everybody recognizing the need for the responsibilities to be fulfilled. The safety and hygiene adviser's function is to explain the issues involved and identify the various ways in which the problems can be tackled. Many of the questions which arise in connection with hearing protectors closely parallel those arising with other forms of personal protection. Here is an opportunity for the safety and hygiene adviser to draw upon his previous experience.

We have seen that the selection of hearing protectors rests in part on acoustical considerations. But there is also the crucial question of whether the protection is actually worn. Because of this it is usually wise to allow the work people or their representatives to take part in the selection process. Before they can contribute effectively to this, however, they will require some explanation of various questions involved. In too many factories, the purchase of hearing protectors is left to the buying department. There, the predominant influence is normally economics, and this may not lead to the most appropriate choice of hearing protector.

When once the pattern of hearing protectors has been decided upon, attention is given to the routine procedures needed to ensure the continuing adequacy of the hearing protection project (this is what the Woodworking Machinery Regulations mean by 'maintained'). The project needs to provide for the regular cleaning, maintenance, replacement, and reordering of hearing protectors. This part of the project needs to be managed effectively and should be a specific responsibility for some designated person. With a sizeable work force, a hearing protector project can involve a considerable work load and adequate resources need to be allocated for this. The detail of arrangements is very much a matter for local decision. Experience shows that first-line supervisors normally have too many pressures on them for them to be able to shoulder the whole of the burden of a hearing-protectors project. However, first-line supervisors are key people in maintaining discipline in regard to hearing protectors. This duty needs to be spelt out carefully in the safety policy so that supervisors and workers are aware that it is the supervisor's job to ensure that hearing protectors are actually used.

The general obligation for management to ensure that work people have the necessary knowledge about hearing protectors can be seen as part of the general legal duty for managers to ensure that work people are sufficiently knowledgeable about the dangers to which they are exposed and the steps that are necessary to protect them from those dangers. The safety policy is an important vehicle for this kind of information. It will often require supplementation by lectures, posters, films, and other forms of communication. These will be successful if the various preliminary stages have been fully attended to. If the work force has been adequately consulted at all stages, has participated in certain aspects of the decision making, and has accepted certain responsibilities, it is to be expected that safety information of an instructional kind will be well received. On the other hand, if the communication material is presented without the preliminaries, then there is a prospect of its not succeeding in its object.

If the matters outlined in the previous paragraphs were attended to thoroughly, there would be no grounds for anxiety about liability for employer's liability claims. However, it must be acknowledged that many managements are haunted by the prospect of a successful claim. An additional factor is that few people besides lawyers have a complete grasp of the duties which the common law places on employers; and few lawyers know sufficient about noise to understand how the general duties should be interpreted with respect to noise. In some firms, for example, it is the practice to require men to sign for their hearing protectors. Obviously a system of record keeping is useful irrespective of any legal considerations, but care should be taken to avoid any suggestion that a man's signature is being obtained for legal purposes and for possible future use against him. The question of record keeping and whether signatures are required should be tackled in consultation with employees' representatives. In this way, misunderstandings can be avoided and the rights of the individual can be safeguarded. Generally speaking, it is undesirable to have a system which depends upon people signing for their acceptance of risk the control of which is in the hands of employers and management—irrespective of any legal questions which these 'bloodchits' may raise.

Audiometry, as a matter of routine for work people exposed to noise, is a matter of great medical controversy. The case for it has been put by Pelmear (1973) and against it by Atherley et al. (1973).

A purpose given for routine audiometry is that it safeguards employers against employers' liability claims. Whether audiometry is useful for this purpose is open to question. However, it is worth stressing that the Health and Safety at Work Act is concerned with safeguarding employees and not with safeguarding employers against employees' claims. Many health and safety advisers see their objective as the fulfilment of the Act, and for them the employer protection aspects of audiometry are irrelevant.

Another purpose given for audiometry is that it persuades more people to use hearing protectors. If the procedures discussed in this chapter are followed, it should not be necessary to use scarce and expensive medical resources to achieve an educational objective for health and safety training.

Conclusion

The prevention of occupational deafness is an important problem in many factories. It comes within the general scope of the Health and Safety at Work Act. Employers, management and work people all have obligations under the Act. In order to fulfil these in regard to noise they will need good advice which the safety and hygiene adviser should be able to provide. Above all, he will stress the safe-place strategies.

References

1. Health and Safety at Work, etc. Act 1974 (see also Chapter 37). H.M.S.O., London.
2. Statutory Instrument 903, 1974. H.M.S.O., London.
3. Department of Employment. *Code of Practice for Reducing the Exposure of Employed Persons to Noise.* H.M.S.O., London, 1972.
4. British Occupational Hygiene Society (1971). *Hygiene Standard for Wide-Band Noise.* Pergamon, Oxford.
5. Statutory Instrument 1414, 1974. H.M.S.O., London.
6. Robinson, D. W. *The Relationship between Hearing Loss and Noise Exposure.* N.P.L. Aero. Rept AC 32. The National Physical Laboratory, London, 1968.
7. Burns, W. and Robinson, D. W. *Hearing and Noise in Industry.* H.M.S.O., London, 1970.
8. Atherley, G. R. C. and Martin, A. M. Equivalent continuous noise level as a measure of injury from impact and impulse noise. *Annals of Occupational Hygiene* **14**, 11–28, 1971.

9. Coles, R. R. A., Rice, C. G., and Martin, A. M. Noise-induced hearing loss from impulse noise: present status. *Proceedings of the International Congress on Noise as a Public Health Problem*, Dubrovnik, 1973.

10. British Standard 5330:1976. *Method for Estimating the Risk of Hearing Handicap due to Noise Exposure*. British Standards Institution, London, 1976.

11. Sharland, I. *Woods Practical Guide to Noise Control*, Woods, Colchester, 1972.

12. Else, D. A note on the protection afforded by hearing protectors—implications of the energy principle. *Annals of Occupational Hygiene*, **16**, 81–83, 1973.

13. Atherley, G. R. C. and Noble, W. G. Effect of ear-defenders (ear-muffs) on the localization of sound. *British Journal of Industrial Medicine*, **27**, 260–265, 1970.

14. Else, D. (1976). Ph.D Thesis, University of Aston in Birmingham.

15. Pelmear, P. L. Hearing conservation. *Journal of the Society of Occupational Medicine* **23**, 22–26, 1973.

16. Atherley, G. R. C., Duncan, J. A., and Williamson, K. S. The value of audiometry in industry. *Journal of the Society of Occupational Medicine*, **23**, 22–26, 1973.

17. British Occupational Hygiene Society, *Hygiene Standard for Impact Noise*. To be published.

29. Eye protection

G. C. E. Barker

General Manager,
Protector Safety Products (U.K.) Ltd

Nobody knows the exact cost of eye accidents in industry because the only accidents discussed and the only statistics compiled centre on those that resulted in lost time. In 1965 one company of roughly 2 000 employees calculated the approximate cost to them of the effects of eye injuries and arrived at a sum of £17 000 for 1 year. Wastage of this nature: stoppage of work, visits to first-aid rooms, fouled-up jobs needing to be scrapped, can be found in every industrial installation in the country. Yet most eye injuries can be prevented by foresight, planning, and the provision of the right kind of training and the right kind of equipment. The slightest injury is more expensive than the eye protector that prevents it.

The importance of eye safety is highlighted by the vulnerability of the eye, although, to counteract this, nature has provided the eye with a number of built-in protective devices. The bone structure acts as a shield against large objects. The muscles around the eye act as shock-absorbers against blows. Eyebrows prevent moisture running directly into the eyes. Eyelids and lashes act like safety-curtains, closing rapidly with a reflex action to trap small particles or insects, or to shut out sudden glaring light. An extra long optic nerve allows for some displacement without permanent damage. And, after all this, if something should penetrate the defences, the tear-ducts do their best to wash away the offending foreign body.

All these natural defences, although efficient enough for natural out-door surroundings, are inadequate in man-made environments where chemicals, radiation, and fast-flying particles are common hazards.

Hazards and injuries

There is hardly an industrial process that does not produce an eye hazard of some sort or another. Apart from the obvious risks, there are many which are not so apparent and, as in any form of accident prevention, it is essential to develop the ability to foresee potential hazards. The time-study man, the girl from the office taking a short cut through the spot-welding shop, the labourer in the carpenter's shop mis-hitting a nail—all can be at risk. The safety man's job is to evaluate these risks and set up a programme of sensible precautions.

We cannot here go into all the details of all the possible causes and consequences of eye injuries. Some of them are minor: the most common injury is simply reported as 'something in the eye'. This can range from a speck of dust, easily removed by competent first-aid room personnel—to something really serious, requiring hospital treatment.

Eye injuries generally can be classified either as burns, lacerations, or contusions. Contusions arise from heavy blows which, if they are hard enough, may displace the lens or retina or even rupture the entire eyeball. Lacerations can be caused by any sharp piece of material, even the edge of a sheet of notepaper. Depending on the cause, pain does not always accompany such an injury and there are many recorded instances where the sheer speed and sharpness of flying objects has allowed them to penetrate completely without the injured men being aware that damage has been done. Also, if it happens that the particle is of iron or copper, there is a serious risk of permanent damage because of the toxic effects that such metals have. Burns, whether thermal or chemical, can produce grave consequences. Alkalis, for example, tend to soak into the tissue and once the process has begun cannot easily be stopped. In the case of any chemical splash, urgent action is absolutely vital. Flush with clean water, and obey the golden rule—*keep it going*. There's no such thing as too much water, and too little might make all the difference between saving and losing an eye.

Radiations can be extremely dangerous, with results ranging from irritation to death, depending on the type of radiation and exposure. Most common sources of radiation in industry are from welding operations but, perhaps more particularly in laboratories and special installations, some forms of powerful infra-red or ionizing radiations could be met.

Protective equipment

The law today already goes some way in its statutory requirements for eye safety devices and its scope will increase. All safety-minded people understand the need for goggles and shields, but personal protective equipment has one serious weakness. It does nothing to eliminate or reduce the hazard. It simply offers a defence, and any failure of this defence, whether due to inadequacy or defectiveness or just that it was not being used at a critical moment, means an immediate exposure to the hazard. As with all other industrial safety activities, the provision and wearing of protective equipment should not be considered as the ultimate in accident prevention. Every effort must be made to eliminate the hazard at its source. Dangerous machinery should be isolated or enclosed; guards, shields, or filters should be integral parts of all equipment wherever possible. Eye protectors have an important part to play and their use should never be neglected, regardless of how apparently slight seems the risk.

In considering the various types and designs of protective equipment with which to guard against eye hazards, it is useful to consider goggles or frames as separate components, adding any of the many types of safety lenses and filters so as to permit an almost infinite permutation of alternatives. While it is possible to build an eye protector to protect against any risk, there is no one single type that will protect against all the hazards at once.

Safety glasses

It may be possible to build an eye protector for every conceivable hazard; but the one big (and often unsolved) problem lies in the reluctance of many operatives to wear eye safety equipment continuously. It is unreasonable to expect a man who spends long hours at his job without a break to wear goggles that are uncomfortable, that hurt, or in some other way make his working life a misery. Also, from the standpoint of simple industrial economics, it is unsound practice to spend money on equipment of any kind if it is not going to be used properly.

Modern safety glasses are made in a variety of styles and colours with many different types of sideshields to protect against lateral hazard. They can meet perhaps 80–90 per cent of industry's needs. Those of us who have to use personal corrective spectacles know that we can put them on as soon as we get up in the morning, and wear them all day. This same high level of comfort and acceptance applies to properly fitted safety glasses, which is why their use in industry has now reached a high level of acceptability.

Not all so-called safety glasses have the right factors. Many, in fact, are ordinary ophthalmic frames fitted with toughened lenses. Some are offered for sale with 'snap-in' lenses and 'slide-on' sideshields, presumably in an attempt to provide ease of maintenance and adaptability. The trouble is, they can 'snap-out' and 'slide-off' with equal ease. Therefore some thought and care must be exercised when making a choice.

Ideally, safety glasses need to be specially designed and engineered for the task they need to perform. Whether they are made of plastic, metal, or even some other material will depend on their application. Lens retention properties should be such that it is impossible to propel a lens towards the eye, even after a heavy blow. The best safety lens in the world is only as effective as the frame that holds it, and it should be built to withstand the heavy wear and tear that the glasses will probably suffer during their working lives.

Fig. 29.1 Typical safety glasses.

Safety glasses, like ordinary sighted spectacles, are very much personal articles. No one appliance can be expected to fit every face. The shape and size of the nose, the width of the temple, the level of the ears and their distance from the front of the safety glasses, all contribute to the final fit. Therefore individual fitting is essential—it means that the glasses will be worn, and not left lying in the pocket or tool cupboard.

Lenses can cover a very wide range of applications and they can be made to personal prescriptions, so avoiding the use of a cover goggle which is often awkward and adds a second surface to catch reflections.

Cup goggles

Although beginning to lose popularity, cup goggles are versatile and, generally speaking, are less expensive than safety glasses. They have the advantage of fairly simple fitting properties, and screw-rim lens retainers allow easy replacement of lenses or filters. On the other hand, the cups are usually made of some hard material which can cause discomfort after a long period, and adjustment of the distance between lenses is sometimes difficult and can cause obstruction to central vision. In a few modern designs, cup goggles can be worn over ordinary personal spectacles. Depending on the type, cup goggles can be used in work areas where flying particles or gas-welding glare and radiation form the principal hazards.

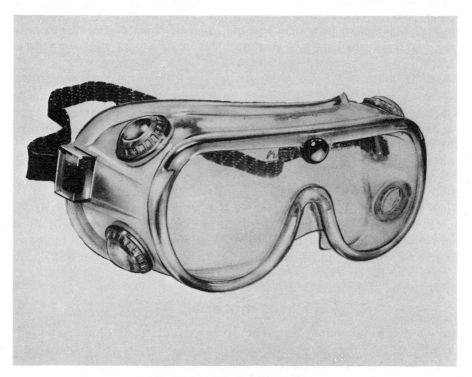

Fig. 29.2   Typical splash goggles.

Wide vision goggles

Sometimes known as 'one-piece' or 'box-type' goggles, these can be made fairly inexpensively, are light in weight, have a good seating above the brows and around the cheeks, and offer little or no obstruction to central vision. According to the design, they can be used for protection against fine dust, fumes, liquids, splashes, radiation as well as impacts from flying particles. Some designs have an adapter in place of the one-piece lens, which will hold regular glass or plastic safety lenses or welding filters.

Faceshields

Although there are variations in design, and usage depends on possible hazards, faceshields generally are intended to give protection from the forehead to the neck including the whole of the facial area. They are used for operations where flying particles could cut the face as well as cause an eye injury, such as in woodworking, metal machining, buffing and polishing, grinding, etc. They are also used in spot welding and during the handling of hot or corrosive materials. The screens attached to headgears have a fairly wide range of applications, some being made of steel mesh or a combination of steel mesh and clear plastic so as to provide a clear window. The more usual screen, however, is plastic which can be either clear or tinted or even aluminized to help reflect glare and heat. At least one manufacturer has built into a face shield a chinguard to prevent the upward splashing of acids or corrosives.

Welding helmets

These are made from fibreglass, fibre or some other composition material which will protect against the radiation and heat from electric welding. The window size is often a matter of custom or personal preference, and in Britain 108 mm × 83 mm is more usual. Other countries use 108 mm × 51 mm, but the choice of window measurements will depend very much on the application.

Visitors' goggles and clip-ons

There is an enormously wide variety of both clip-ons and visitors' goggles or spectacles from which to choose. These can range from substantial metal-framed ones to those simply made from shaped pieces of plastic. Although their uses in the hard, tough, conditions of industry are limited, they have their value in a number of circumstances.

Lens materials

If one were to single out any part of an eye protector as being more important than the rest, it would probably be the lenses. Many different materials are used, each having its own particular merits, and the final decision on a choice of lens material must depend on the hazard.

Glass

Glass lenses, either thermally toughened or laminated, are the most common in use. Both types will give a fairly high degree of resistance to flying particles but, broadly speaking, thermally toughened glass will best withstand impacts from high-mass, low-velocity particles. The toughening process means that should the lens break then there are no sharp edges or slivers of glass to cause damage to the eye. Further, if a toughened lens is securely clamped into a safety frame the chances of broken pieces leaving the housing after being broken are considerably reduced.

Glass safety lenses can be clear or tinted or specially made to personal prescription, and they are versatile enough to provide protection over a wide range of risks from fast-flying particles to radiation. They can be either flat (made from sheet glass) or curved, the latter being preferred because the domed effect gives more certain deflection of flying particles, greater strength, a closer fit to the eye, and reduced reflections.

Degrees in quality of lenses are not always obvious and it is possible for some to have imperfections and distortions which are not easily detected but which, over a long period, can cause acute discomfort or headaches. Many countries have standards which specify the minimum optical requirements of safety lenses, and certification or conformity to these standards is a good guide.

Plastics

During recent years many different types of plastic materials have been tried or introduced. *Cellulose-acetate* is probably the most commonly used, having the advantage of being inexpensive, and easily shaped and formed. If it is thick enough it gives a reasonable impact resistance and, like most plastics, throws-off spatter quite well. It can be tinted to almost any colour or shade so as to protect against glare. The major disadvantage of cellulose-acetate is that it scratches and abrades quite easily and some types lack the optical fidelity so necessary if the eye protector is to be used over long periods. *Acrylics* (such as 'Perspex') are often used for lenses to fit safety glasses, and they have the advantage of being easily worked. *Polycarbonate*, which has a really excellent impact resistance, is a plastic which has been used effectively where very high speed impacts are to be expected. At the moment of writing, it is the only one single lens material that will meet the impact test set out in British Standard 2092, although many others in combination will do so. Polycarbonate has the disadvantage of being soft and easily abraded and is subject to early discoloration. A fairly recent development has enabled polycarbonate lenses to be treated with a hard surface coating which adds considerably to the abrasion resistance. Available under a number of brand names, this kind of lens might well be worth looking for if impact protection is a requirement. *CR-39* (*allyl-diglycol-carbonate*) in its basic form is crystal clear, like glass, although it can be tinted for use as an antiglare lens. It can be made with a high degree of optical perfection and is easily worked to personal prescriptions. Because CR-39 has to be shaped by casting, as opposed to moulding or extrusion, it is more usual for lenses of this material to be made for cup goggles or safety glasses. It is one of the most abrasive-resistant plastics, stands up well to molten metals and hot solids, and in the proper thickness has strength enough to resist most of the impacts experienced in industry. These qualities make CR-39 suitable for many operations where all these hazards may be present, such as in foundries. In thin sheet form, it is ideal for use as cover plates for welding goggles and shields because of its anti-spatter properties. During the last few years, hydrophilic coatings, which prevent the fogging or misting up of goggle lenses, have become available. There are a number of different types of coating, each more suitable than the others in differing circumstances, and manufacturers' data sheets should be studied when considering such lenses.

Radiation filters

Although there are a number of special filter lenses with which to guard against ionizing or laser radiations, the bulk of industry's hazards of this nature are covered by the use of welding filters. In Britain these are made to comply with British Standards Specification 679 and, almost entirely, are made from glass. Additional to welding filters are many types of blue or cobalt filter, each with its special application, which range from the simple 'pot blue' to more sophisticated ones which are density controlled with special colour-dampening effects.

Filter lenses have three main functions:

(a) reduction of radiations to a level at which harmful effects are negligible,
(b) reduction of visible light to a comfortable level,
(c) mechanical protection of the eyes from molten metal and flying particles.

Reduction of radiation depends on the ability of the lens to absorb particular wavelengths and strengths. It is the chemical content rather than the apparent density that prevents the passage of harmful radiation and, for example, a filter devised to give protection against radiation generated by a carbon dioxide laser is perfectly clear without any indication of tint. Yet, because the chemical content is exactly right, this filter will protect against continuous power from this laser in excess of 500 W at 10 600 Å.

The reduction of unwanted visible light depends on the process and the intensity of luminosity. For example, glass workers have problems arising from the yellow glare or sodium flare that appears when hot flame is applied to glass, and which obscures working surfaces. This is done by using lenses which contain certain rare earth elements. On the whole, the amount of reduction of visible light that is considered necessary is something that has been established by custom and usage for each individual process.

The use of filters that have been toughened and/or protected by cover plates has largely overcome the mechanical hazards associated with radiation, but the answer to each problem must depend on the type and severity of any likely impact. It is enough that the reader appreciates that the risk exists.

Choosing eye protection

Manufacturers have designed and produced such a variety of goggles, safety glasses, screens, and shields, that the choice can sometimes seem bewildering. Experienced safety men will, by using their knowledge and expertise, have little difficulty in determining what is required. However, if there is doubt or if it is felt that there are special circumstances to take into account, ask for the advice and suggestions of manufacturers.

Well-intentioned—but often misguided—managements sometimes authorize a general issue of one style of goggle, regardless of the differences in the sort of job that people do. Everybody in the works receives the same thing and management thinks the problem is solved. This is rarely true. Different departments have different risks; even people in the same section may require variations of one basic theme. Of course, there is also this all important business of adding comfort and good looks to protection so as to ensure proper use of the equipment.

In selecting goggles or safety glasses, bear in mind that today's choice may become your standard for some time to come. You are paying for safety. The first thing that you need to be assured about is that what you are buying is going to do the task of protecting the wearer. Next, study all the aspects of fitting and comfort. All protection may involve some sort of inconvenience but it should not be tolerated in such measure that it operates against acceptability. Modern designs of eye protectors have very nearly eliminated discomfort and 'horse-collar' appearances which for so long have been associated with goggles, and there should be no difficulty in finding a suitable protector that will be used.

The best method of determining the right sort of protection for each job is to study and conclude which lens is best suited to the hazard, and then make sure that the chosen lens is properly fitted to a frame that will not only complete the required protective values, but will also meet the comfort and cosmetic needs. It is good sense to begin by making a list of all job operations together with the protective equipment agreed as proper for each. Unless there are good reasons for it, requests for 'fancy' or 'dressy' types of safety glasses should be rejected.

Once a plan has been formulated, have it endorsed by senior management. One such plan is for a general issue of safety glasses to everyone, with additional protection in the form of faceshields or cover-goggles of the flexible mask type where operators have more hazardous jobs. Even welders can benefit from safety glasses with lightly tinted lenses so as to protect against flash during those moments when a welding shield is raised or removed. Constantly follow up the programme and check its efficiency, and don't be afraid to admit an error and make a reappraisal if it seems necessary. The Factories Acts already list the types of work where eye safety is a legal necessity, but always try to be ahead of minimum statutory requirements.

Experience has taught that the best sort of eye safety programme is one where ideas have been successfully sold throughout the entire works. The use of posters, company newspapers, bulletins, films, displays, and talks, has helped many companies overcome the initial hurdle of introducing eye protection as a routine standard practice. When the need for eye protection is properly recognized and the consequences of ignoring it appreciated, workmen themselves will see that the regulations are obeyed.

An important factor which is fundamental to any eye safety programme is to provide for the proper fitting and servicing of eye protectors. The value of proper adjustments for each individual protector cannot be overstressed, and the records of legal decisions in many compensation claims testify to the importance in law of this aspect. In fact, this one detail can prove the dividing line between success and failure. Disinfection, replacement of worn or faulty parts, regular inspections, storage facilities, cleaning, and demisting, are all considerations that should be included.

Visitors must also be provided for, and that includes the chairman and managing director who, by wearing protective equipment, can set the right example.

30. Skin care

W. C. McCreanney
Century Oils Ltd

In every year, 900 000 working days are lost to industry due to skin troubles. Workers might be expected to look after their skin; they do not, but when something goes wrong with it, there is no greater cause of lowered morale.

In modern industry, apart from the familiar causes of skin troubles, every day brings new dangers. The risk of dermatitis has moved into practically every area of industrial operations and today resins, coolants, solvents, and chemicals present a growing challenge to the people who try to control costly skin ailments. The simplest definition of occupational dermatitis is 'any inflammation of the skin due to exposure to an irritant at work'.

When reviewing the hazards to which the skin is exposed it is important to consider how the skin structure can be affected by work materials, by failure to keep it clean, and by the use of dangerous substances such as solvents, industrial detergents, abrasives, etc. to clean it after work. For clarity the complex structure of the human skin may be simplified into three main layers.

(a) *The epidermis* or outer protective layer, naturally waterproof, consists of layers of dead horny cells which are constantly shed and replaced from the generating layer below. Natural human waxy fat secretions— sebum and keratin—constantly emerge in a clean healthy skin to provide the natural, and finest protection against outside attack.

(b) *The dermis* immediately beneath the epidermis consists of dense fibrous tissue, housing blood vessels, sebaceous glands, hair follicles, nerve endings, and sweat glands. Its uppermost surface is deeply corrugated and fits into the underside of the epidermis rather like stubby fingers into a glove.

(c) *The hyperdermis*, which is really the lower part of the main dermis, is a mass of fat cells and vascular tissue.

The skin is the largest single organ of the body and, far from being a mere covering for our flesh and bones, is a living part of us. In addition to protection against outside irritants and bacteria it regulates our temperature by its reaction to heat or cold; the pores dilate when we are hot, allowing perspiration to come to the surface and evaporate to cool us; when it is cold they close to retain body heat. Perspiration also helps to rid the body of waste matter and, since the normal person produces approximately $\frac{1}{2}$ l of perspiration per day (in high temperatures the rate can be as high as 1 l per day) this function of the skin is important. On the palms of the hands and the soles of the feet there are as many as 460 pores/cm^2 against 60/cm^2 on the legs or back.

The skin also breathes, taking in oxygen and discharging carbon dioxide in the same way as the respiratory system, but at a much slower rate. It is only during this century, although there are clinical data on the skin going back as far as Ramazzini (1633–1714), that the vital role of the skin in assimilation by the body of fats and proteins has been revealed.

Here then is a complex and very important part of human anatomy, exposed daily to the hazards of work in modern industry. Damage to skin accounts for 70 per cent of all the time lost in industry through sickness, and study of an authoritative work on the subject (*Occupational diseases of the skin* by Schwarz, Tulipan, and Birmingham, published by Lea and Febiger, Philadelphia, U.S.A.) shows that the risk, though greater with some materials than others, extends throughout industry.

Medical evidence reveals various factors which may affect the incidence of skin troubles.

(a) It is thought that blonde persons are particularly sensitive.
(b) Hairy skins seem to be more susceptible to oil acne but thick, oily skins withstand solvents better than others.
(c) Women appear to be more sensitive than men, but they tend to look after themselves better.
(d) Persons with a history of skin trouble, or known or suspected allergies, may be easily affected.
(e) The summer produces more cases of dermatitis, possibly because the skin is perspiring and open, but in winter the risk may appear higher because of drying and chapping of the skin.
(f) The increasing use of modern chemicals in the home may have a decided bearing on the incidence of skin troubles.
(g) General health is important, and diet is more important in winter.
(h) Finally, and perhaps most important of all, *personal hygiene*.

The causes of occupational skin troubles usually fall within this list:

(a) mechanical damage due to injury, friction or pressure;
(b) natural plants, wood, micro-organisms (e.g., primula obconica, teak, and fungi);

(c) heat, cold, excessive sunlight (causing photosensitization) especially in industries where filtering materials are deposited on the skin (e.g., pitch and tar); irradiation also comes into this category;

(d) chemicals, especially inorganic alkalis such as caustic soda, lime, soda, caustic potash, cement; inorganic acids such as sulphuric, hydrochloric or hydrofluoric; organic acids: carbolic acid, acetic; oxalic: alkaline substances with a pH value over eleven;

(e) oils and petroleum products, especially 'improved' lubricants and fuels;

(f) various metals, especially non-ferrous; dyes (often due to the measures taken to remove stains from the skin); explosives; plasticizers; hardeners; and various pharmaceutical products.

Inflammation may spring from nervous causes or from psychiatric reasons, and people who are prone to constitutional eczema often react quickly to an irritant.

One of the two main types of dermatitis found in industry is contact dermatitis; an inflammation occurs as a direct result of exposure to an irritant or trauma and this usually exhibits itself on the exposed parts of the body. The other type is sensitization which may take years to build up or, if it is of more rapid onset, may be due to a person's hypersensitivity or allergy to a certain substance. This type can manifest itself elsewhere on the body than on the parts exposed.

This is the problem to be dealt with: a highly complex and vital part of the body is exposed every day to substances which injure it, interrupt its functions, and lead to one of the most distressing of illnesses which, once contracted, usually takes a long time to clear up and often leads to a general lowering of morale. This is a problem where an ounce of prevention is worth a ton of cure. The basic need is to prevent contact or, where that is impracticable, to reduce the period of contact to a minimum.

It is assumed that personnel engaged for a process will have been medically checked for previous skin-disease history and any unusual skin condition such as excessive perspiring or dryness. Personnel employed on any process where there is a skin hazard should be told about it and instructed in the right technique to minimize the risk.

This having been done, every effort should be made to use the least harmful effective materials for the job. The safety routine is then simplified and the following are the precautions to be taken.

(a) *Good housekeeping.* The importance of clean and safe environment and atmosphere cannot be overestimated. A person working in slovenly surroundings may be expected to be careless in his approach to personal hygiene.

(b) *Consultation.* The most effective way of implementing a cleanliness programme is to bring the operatives, through Works Councils or Safety Committees or through their Unions, into discussions about the precautions necessary to minimize risk.

(c) *Protective clothing.* Gloves, masks, and aprons help to reduce contact and are of great value also against physical and mechanical risks to the skin; but the finest protective equipment is worse than useless if it is not kept clean. It was reported that in one factory where an outbreak of oil acne was investigated, for every case involving the exposed forearms there were eight on the thighs and stomach because the overalls were not changed regularly enough.

For persons exposed to primary skin irritants and known dermatitis hazards, barrier creams should be provided, but great care must be exercised to ensure that the correct barrier for the job is employed. It is a waste of time and money to provide a water soluble barrier cream on a job involving water, solvents or even excessive perspiration. A degree of protection can be secured by careful application of the correct barrier substance to *clean* skin. At the very least their use reminds and encourages people to wash and, since the removal of the potential irritant is of prime importance, this is valuable. Beware, however, of giving operatives a false sense of security.

It is now acknowledged that the most powerful factor against occupational skin troubles is the regular removal of all contamination at work breaks and at finishing times. To do this, an efficient and properly formulated skin cleanser is required since it is only in a minority of cases that soap and water (the minimum statutory requirement) is effective in removing contamination from the skin. Efficient cleansers obviate the misuse of solvents, abrasives, domestic detergents, bleaches and scouring powders and thereby eliminate what my 20 years' practical observation shows to be the major cause of skin troubles occurring in industry.

A suitable skin cleanser for industry must be safe, efficient, and quick. The criteria in developing a skin cleanser are as follows:

(a) It must work, quickly and visibly, removing the widest range of substances normally encountered in a day's work. It must cleanse the skin, not only on its surface, but also deep into the pores and hair follicles. It must not depend for its acceptance on that false sense of well-being given by the type of cleanser which leaves a soft feeling due entirely to a thin layer of the very mineral oil it is supposed to remove.

(b) It must be medically safe. Drying out the natural oils cannot be tolerated, nor must it contain any potentially dermatitic substances.

(c) Its pH value must be as near to that of the normal skin as is compatible with efficient cleansing, and have no free alkali or solvent.

(d) It must contain no abrasive to do mechanical damage to the skin.

(e) It must be completely acceptable to those who are to use it, easy to use, dispense and control, and available at a price which will allow it to be made available for general use.

Clean, dry towels for good drying after washing are essential. Hands and other parts left wet are prone to cracking and chapping at all seasons, especially winter, and this condition is often confused with dermatitis.

Having decided on the action to be taken to obviate skin trouble, management must then bring into force a continuous programme of education to ensure that workers receive maximum benefit from it. Posters, where applicable, may be displayed, together with reminder notices provided by reputable suppliers to encourage full use of the precautionary measures provided to eliminate abuse.

Reputable manufacturers of industrial skin care products maintain free technical advisory services with full laboratory facilities. The problems of synthetic resins in industry, particularly those of the epoxy type, are one example of how these free facilities can assist industrial management; promoters of such a service will try hard to help their clients. The problem may be one to which an immediate answer is available because it has occurred previously or, if it is a new one, the answer will be found and this will benefit any company encountering a similar set of circumstances.

Selection, protection, the best personal hygiene, good housekeeping in the factory, and a continuous education programme ... these are the measures which will do most to eliminate skin cases from any industry. Once this has been done the lot of the Safety and Medical Officer will be a happier one since there is no more difficult case to resolve than one of dermatitis.

31. Personal protection and equipment

F. Riddell

Group Safety Engineer,
Reed International Ltd (Northern)

The role of personal protective equipment could be interpreted as being a last line of defence. When the need for personal protective equipment arises, the accident has usually already happened: the casting has already been knocked over, or the chemical has already been splashed.

Having considered the various accident prevention courses open to us, it may still be necessary to provide some kind of barrier between the person at risk and the potential instrument of injury. An instrument of injury may have various characteristics, e.g. sharp, abrasive, corrosive, heavy, hot, irritant, or harmful in ways not immediately apparent as in the case of radio-active substances. It may be frequently encountered, in expected situations, or be quite unexpected and unforeseen. In order to fulfil legal and moral obligations, it is desirable to establish a pattern, or programme, of looking at all aspects of the provision and use of personal protective equipment. A programme should include the following:

1. establishing the need;
2. selection of the equipment;
3. provision and availability of the equipment;
4. ensuring the use of the equipment.

1. Establishing the need

There are a number of methods of obtaining this information available to us, some of which are:

(a) Accident/incident experiences

Every accident or incident, whether or not injury has resulted, should have the one redeeming feature of containing some kind of lesson for us. That lesson may high-light a need to provide and use personal protective equipment. In the case of an injury causing accident this is usually obvious, but in the many more incidents causing slight or no injury, this factor may not be so obvious. One cannot, therefore, stress too highly the value and importance of detailed examination of available information in order to ascertain where the lines of defence are weakest. This information may be in the form of records of surgery or first-aid treatments, in departmental reports from supervision, or in some cases as part of a sophisticated damage control, or total loss control system. Such information may show that over a period there were a number of cases of, for example, chemical splashes into the eyes, or cuts to the hands, prompting questions as to whether suitable equipment was available and/or used.

(b) Safety representatives/safety committee

Protective equipment is an area in which the safety representative can play a very valuable role. Familiarity with shop-floor conditions, and the operations carried out, place him or her in a very favourable position to observe where the dangers lie. An important factor, however, is the effectiveness of the lines of communication. It is vital that in situations in which the risk is an imminent one, or one which frequently occurs, there should be no tendency to delay action until the safety committee next meets.

It must be made quite clear that prompt action through line management channels must be taken, and the safety representative must be encouraged to do this. Action taken can, and should, be later discussed at the safety committee meeting. The examination of accident/incident records as outlined in 1(a), with a view to showing the needs, is a matter that can profitably be discussed at the safety committee meetings.

(c) Safety audits—surveys—sampling

Whilst it is not intended to deal at length with the use of assessment techniques such as safety audits, safety surveys, and safety sampling, it should be mentioned that these are excellent methods of obtaining information. They can be used to cover the whole range of health and safety investigation, or, if desired, can be limited to specific areas, one of which should certainly be that of identifying the needs in respect of personal protective equipment. These methods can be used, not only as fact-finding exercises, but also for providing a form of training for supervisors, charge hands, safety representatives, etc.

(d) Legal requirements

An obligation to provide and maintain personal protective equipment may also be a requirement under existing Regulations and will no doubt be a feature

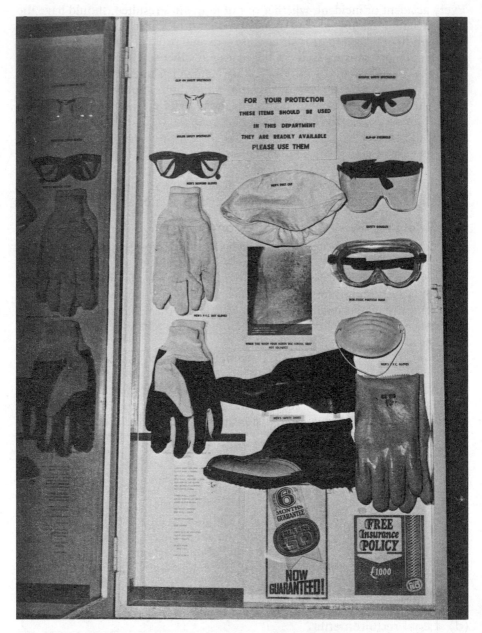

Fig. 31.1 A departmental display showing the items of protective equipment which are available and should be used for work carried out in that department.

under a number of Regulations which will materialize under the Health and Safety at Work, etc., Act. If the recent 'Protection of Eyes Regulations' can be taken as an indication, it appears likely that the requirements under certain Regulations, in addition to the ones under the main Act, will not only place a duty upon the employer to provide and maintain equipment of a prescribed standard, but will also place increased obligations upon the employee to use it. It is important to bear in mind the obligation not only to provide the equipment, but also to maintain it, which could be interpreted to mean to maintain in the sense of keeping in good condition, and also to maintain in the sense of continuing to provide.

(e) Medical department records

A need for protective equipment may arise out of the physical condition of an employee. Some condition may be present which means that a person is either more susceptible to a possible injury-causing situation, or that the effects of an injury would be more severe than it would be for an average person. An example in the first category would be that of a person who is allergic to some substance in the process and, therefore, in need of extra protection. In the second category one could have the case of a person who has effective vision in only one eye, which means that extra care is necessary to protect the one good eye.

Close liaison with the medical and personnel departments is obviously desirable so that information of this type can be made known to the members of management who are responsible for that person during the working day, and the necessary equipment may be provided.

2. Selection of the equipment

The choice of equipment is obviously very important. Its quality, durability, suitability for its purpose, and lack of interference with the user's faculties and movements are factors requiring consideration at the time selection is made. To the employee however, an equally important factor is its appearance or, to be more accurate, his or her appearance when wearing it. Any efforts to obtain co-operation in the use of protective equipment which makes the wearer an object of fun or derision is doomed to failure.

We in this country seem to be particularly sensitive on this point, and a reluctance to wear equipment that would be frequently seen in certain other countries appears to be a national trait.

It becomes evident that, when selecting equipment, not only do we need the advice of manufacturers and safety experts in respect of its suitability, we also need the views of the workforce as to its comfort and acceptability. Perhaps this, too, is an area in which the safety representatives can give valuable assistance in obtaining and communicating the views of the people they represent. There is undoubtedly a greater degree of acceptance and a greater likelihood of usage of the equipment where the people who are required to use it have had some say in the selection of it. All too frequently, if equipment has been presented to the employee with little regard to acceptancy factors, he or she has, in a short time, ceased to use it, a situation which not only leaves the employee exposed to hazard but is also a sheer waste of money. The old Army adage 'Time spent in reconnaissance is never wasted' is equally true in this field.

The manufacturers and suppliers of personal protective equipment can be relied upon for sound advice where needed, and the use of equipment which has met the requirements of the British Standards Institution and consequently displays the kite-mark will ensure quality and reliability.

3. Types of equipment

(a) Respiratory protective equipment

This is a specialized field and is covered as a separate subject within this book (Chapter 32).

(b) Headgear

The most obvious hazard against which headgear is required is that of falling objects, but headgear may be required to protect against heat, chemical splashes, and to prevent the wearer's hair from contact with machinery parts.

For use largely against the risk of falling objects the British Standards Institution specify two standards:

> B.S. 2095—Light duty—requiring a shock absorption test equal to 28 ft lb (5·13 kg m).
> B.S. 2826—Heavy duty—requiring a shock absorption test equal to 40 ft lb (7·3 kg m).

Further information may be obtained from the British Standards Institution, Head Office, 2 Park Street, London W1A 2BS.

Helmets designed to withstand heat and chemicals are obtainable in a variety of materials and types, and the manufacturers and suppliers catalogues contain many examples.

For the less potentially serious, but still painful, situations, for example during work under machine framework where a person may bump his head, bump caps may be worn. These do not have the protective qualities of kite-marked helmets, but they do give a degree of protection against painful knocks, and they tend to be more acceptable to people where the risks are not obviously severe. Probably the most contentious type of headgear is the type intended to prevent the wearer's hair from coming into contact with moving machinery. These take the form of hair nets and caps with or without snoods, and come in a variety of styles. It must be said, however, that desirable though they may be, careful selection from a style/fashion point of view and a good line of sales talk are necessary to persuade people to use them.

(c) Eye protection

This subject is also covered in a separate section of this book (Chapter 29).

(d) Gloves

These are probably the most widely used items of personal protective equipment, not surprisingly, as injuries to hands form a very large proportion of the total injuries received at work. The fashion/appearance factor does not appear to play as large a part in the acceptance by the user of gloves as it does in cases of other items of protective equipment. One can, therefore, concentrate on more important aspects in the choice of the most suitable protection. Factors to be considered in that choice would include:

   (i) hazards to protect against (contacts with sharp abrasive, corrosive, hot, irritating, etc., articles or substances);
   (ii) degree of resistance to contact substances;
   (iii) amount of sensitivity required;
   (iv) area to be protected (fingers, whole hand, wrist, arm).

Traditional glove material is, of course, leather. Gloves in hide, chrome leather, and various other types of leather are available but expensive. The more commonly used material is now PVC, which has a good all-round performance and is obtainable in many patterns and weights, ribbed, smooth, supported and unsupported. Some synthetic materials, Nitrile, Neoprene, etc., are available and can, in the case of certain solvents and hydro-carbons, give a greater degree of resistance.

Terrycloths of heavy and medium weights are useful in handling dry substances or articles which may be hot, sharp, or abrasive. These are usually very comfortable and consequently popular.

Lightweight gloves for use whilst handling smaller components are obtainable in various fabrics and plastics, giving reasonably good sensitivity, and can be inexpensive and disposable.

Some of the leading glove manufacturers publish charts indicating the suitability and degree of resistance of glove materials relative to the different chemicals in use in industry. A very important factor, except in the case of the cheaper disposable types, is the care and handling of the gloves. Certain chemicals will eventually bring about deterioration of even the best material, but that deterioration is accelerated by permitting the chemical to remain on the gloves after use. Rinsing them in clear water and drying naturally will prolong life very considerably.

(e) Body protection

The situations in which protective clothing covering much of the body is required are almost as numerous as the types of garment and kinds of material used and the number of manufacturers and suppliers.

Publications issued by the Royal Society for the Prevention of Accidents, The British Safety Council and other publishers of occupational safety literature contain information on large numbers of sources from which protective clothing can be obtained. Much work has been done by N. T. Freeman and others on the suitability of specific materials for use in a wide range of circumstances and possible contact with a large number of potentially damaging substances.

The types of garments include overalls, duffle coats, aprons, and many other variations, and are selected with a view to giving protection against weather, dirt, chemicals, oils and fats, heat, or contact with general articles which could cause physical damage.

Materials selected may need to be one or more of the following: warm, comfortable, windproof, impervious to dust or liquids, non-static producing, flame resistant, easy to clean and, for use in some circumstances, of high-visibility material.

An interesting fairly recent development is in the use of lightweight clothing made of paper, or other non-woven material which is very cheap and intended for disposal after use. There is much to be said for the hygiene aspect of this type of garment.

(f) Footwear

Earlier in this chapter reference was made to the large number of occupational accidents involving injuries to hands. A very close runner-up must be the accident category in which injuries to the feet occur. There can be no doubt that, were it possible to obtain an accurate figure of the costs involved in injuries of this type throughout this country, the figure would be staggering. This is readily appreciated by a large number of industrial and commercial concerns, as evidenced by the fact that so many of them provide facilities for employees to obtain safety footwear through their own organizational channels. Those who do not yet do so might find it worth considering seriously.

Facilities provided by firms range from arranging for footwear to be purchased by employees and repayments made through wage-deduction schemes—information on such schemes can be found in a booklet entitled *Wage Deduction Schemes*, published by the Golden Shoe Club and obtainable from a number of the leading footwear manufacturers—through various levels of subsidy provided by employers, to, in some cases, the free issue of protective footwear. This is sometimes on a selective basis where the risk is considered to be fairly severe but, in some instances, to all employees.

Very considerable advances have been made in recent years in the design and appearance of safety footwear. The range extends from shoes (frequently referred to as 'executive styles') to Chukka boots, full boots, and rubber boots of the Wellington type. The materials used vary considerably and are selected for such properties as durability, acid resistance, oil resistance, non-conductivity, heat resistance and non-slip. The choice of design and type is such that one can hardly visualize a situation for which suitable footwear cannot be obtained.

There are safety shoes for ladies available, although it must be said that the range here is nowhere near as extensive as in the case of mens' footwear. In defence of the manufacturer one can appreciate the problems of endeavouring to maintain a range for ladies considering the frequent fashion changes.

In addition to safety footwear with steel toecaps for ladies, many manufacturers produce practical types of shoe which are not, strictly speaking, safety footwear in that they do not have steel toecaps. In many cases, where the risk of foot injury is not too great, people feel that the use of these is at least preferable to the wearing of bedroom slippers, peep-toe sandals and other odd types of footwear all too frequently seen on the factory floor. One of the problems associated with persuading employees to obtain and wear safety footwear is that in many instances the boots or shoes must be ordered from the supplier. When they arrive the purchaser finds that they do not fit very well or they pinch somewhere. This sometimes means that they must be returned and the exercise must be repeated, a time-consuming and somewhat frustrating business.

Larger firms may be able to hold a stock of the more popular types and sizes in the 'safety store', but the smaller firm may find this too expensive a method to adopt. Some manufacturers have solved this problem to some extent by having mobile vans which visit factory sites, and employees can try the footwear then and there. Others have set up regional depots to facilitate deliveries. On the whole it is probably true to say that present day safety footwear, in addition to preventing injury, provides very good value for money.

Persuading the employees

All efforts to select and provide suitable protective equipment will be to no avail if, in the end, it is not used, and the final outcome in those circumstances is frustration and annoyance that time, effort, and money have been wasted. In that kind of situation, when it is seen that an employee is not using the item of equipment provided, the temptation is of course to 'fly off the handle'. Before doing that, however, there is one simple, and frequently rewarding question one should put to the employee—'why?' He may have what appears to him to be a perfectly good reason why he is not using the item. It may be uncomfortable, it may not do the job for which it is intended. He may have a boil on the back of his neck, or another personal reason why not. He may, and often will, say a great deal about the suitability and availability of the equipment. There was recently held a seminar at which a talk was being given by an experienced member of H.M. Factory Inspectorate on the subject of the Protection of Eyes Regulations. In the notification literature, delegates had been asked to bring a sample pair of goggles in use at their premises to the seminar. At the beginning of his paper, the Inspector asked the members of the audience to don their goggles and wear them through the 1 to $1\frac{1}{2}$ hours of the presentation of the paper. The result had to be seen to be believed, one just would not have anticipated the amount of fidgeting and adjusting of the goggles that went on after about ten minutes of wear. That practical demonstration taught those delegates more than a torrent of words. The lesson is clear: use care in the selection, involve the worker in the choice, try them yourself and, above all, set an example yourself. If protective equipment is required to be worn in a specific area *that means you too*.

32. Respiratory equipment

S. J. Cheffers

*Formerly Home Sales Manager,
Siebe Gorman and Co. Ltd*

The recent technological advance of modern industry has considerably increased the potential dangers from dusts, fumes, and gases. This, coupled with the growing awareness of the necessity for promoting safe working environments and practices, has resulted in a revitalization of the safety equipment industry.

To appreciate how modern respiratory protective equipment has evolved it is necessary to delve into history. The first recorded method of filtering air was the wetting of the beard by a fireman before entering smoke. It was but a short step from this to the introduction of pads of material held across the nose and mouth to prevent infection by viruses during epidemics.

Early respirators gave only rudimentary protection against gases and due to the necessity for persons to enter dangerous gaseous areas such as ships' holds to unload cargoes of fruit, or to fight fires in confined spaces, an air-tube apparatus was later developed. First references to this form of protection are found in the early nineteenth century. This air-tube equipment severely restricted the distance into which the user could enter a hazardous atmosphere. The development of portable or self-contained compressed air breathing apparatus took place about the middle of the nineteenth century.

The next step was to increase the duration for which the self-contained breathing apparatus would maintain protection. This resulted in the evolution of the closed circuit oxygen breathing apparatus. At the end of the century, breathing apparatus and dust and tube-type respirators were in use.

Further development of a filtered-air type of respirator did not take place until the First World War, when protection was required for soldiers during gas attacks. Although modifications to design and materials have been made, the working principle is still the same as that in use today, both in the Services and industry.

The evolution that has taken place over the years has produced a wide variety of respiratory protective equipment, which can be broadly classified into two groups: those which filter impure atmospheric air, such as dust and gas respirators, and those which provide clean air, such as self-contained breathing apparatus. These classifications, which are based on the degree of protection required, are further subdivided by their method of operation.

Equipment filtering impure atmospheric air

One of the major applications in industry is for the removal from the atmosphere of suspended particles created by the manufacturing process. This is achieved by using a respirator which covers the nose and mouth of the wearer and filters the inhaled air; the exhaled air is returned to the atmosphere through a non-return valve.

Dust and light-fume respirators

The nose and mouth are covered by a rubber or plastic moulding to which is attached the filtering agent. The unit is retained in position by an adjustable harness which clips around the back of the neck or head. When the wearer inhales, the air is drawn through the filtering material via an inspiratory valve, thereby ensuring that all the air required is filtered. The exhaled air passes freely to the atmosphere through another valve, usually sited at the base of the moulding. Non-return valves are used in both instances to create a breathing circuit. It is essential that the exhaled air does not pass back through the filtering media, as the moisture content of the exhaled air would seriously reduce the effectiveness of the filter.

The operational efficiency of this equipment depends entirely on the fit of the moulding to the face, and on the filtering media. The essential close facial fit is achieved, either by the use of soft rubber pads or by a reverted edge seal, both methods reducing the ingress of air between the face and the mask to a minimum. Moreover, these methods ensure the optimum fit on any facial shape. A beard or even a poor shave can adversely affect the efficiency of a mask of this nature.

Filtration of the inspired air is achieved by the use of several materials of varying degrees of efficiency. The choice of materials is dependent on whether it is required to filter dust or light fumes or a combination of these hazards. The best known materials are resin-impregnated merino wool, paper, and wool felts. The medium most commonly employed in this country is resin-impregnated merino wool. The resin impregnation increases the electrostatic charge already present in the wool. This promotes a dual action—mechanical filtration and filtration by precipitation.

In most modern respirators the filtering agent is contained in a plastic or metal capsule, in such a manner that the filter cannot be reversed when fitted into the filter holder. This ensures that the filter cannot be inadvertently reinserted the wrong way round, causing the wearer to inhale the dust particles already present in the filter.

A large proportion of the dust particles to be filtered are likely to be of a coarse nature and, to extend the life of the encapsulated filter, a pad of wool and gauze is inserted in front so that the coarser particles are trapped before the air reaches the filter cartridge. Frequent replacement of the prefilter is desirable and, because of its low cost, is extremely economical. The frequency of change

Fig. 32.1   Puretha and Gaspro respirators being worn during the handling of toxic chemicals in the open air.

of the filter is dependent on the environmental conditions and will only become apparent with experience.

For the removal from the atmosphere of very fine dusts, it is advisable to utilize a respirator fitted with twin cartridges—one on either side of the face moulding. These are usually equipped with a high efficiency filtering agent (filtering particles as small as 0·5 μ) and have the advantage, because of their increased filtering area, of decreasing the velocity of the air as it is drawn through the filters. This prolongs the life of the cartridge and greatly reduces the resistance to breathing.

In addition to dust cartridges, filter cartridges are available which can be utilized with the same respirator but which offer protection against a wide range of fumes, gases, and vapours. In this instance, the filtration medium is a mixture of chemicals specially prepared to maintain their activity during a specified usage life determined by the nature and concentration of the impurities in the atmosphere. Combined cartridges are also obtainable which filter a mixture of dust and fumes, etc. A common application of this type of cartridge unit is for persons engaged in paint or wax spraying, or using agricultural insecticides.

When it is necessary to provide both eye and lung protection, due to the presence of irritant gases or dusts or the possibility of spattering of chemicals, combined protection is afforded by the provision of a full-vision face mask to which is fitted the single or twin filtering agent, such as those already described. Protection against radioactive dust is afforded by a Porton filter with a full face mask. Finally, a PVC cape and hood can be utilized with the full-face mask and filters, giving complete head protection.

It is prescribed under the Factories Act that within particular industries, employees carrying out certain operations are required to wear respirators. The respirator that is provided must be approved by H.M. Factory Inspectorate and comply with the requirements of:

B.S. 2091: Respirators for protection against harmful gases and dusts.

Respirators issued to agricultural workers should comply with:

B.S. 2617: Respirators for agricultural workers using toxic chemicals.

The equipment already described is for protection against dusts and light concentrations of fumes, vapours, or gases of approximately 1000 p.p.m.

Where concentrations of gases greater than 1000 but below 10000 p.p.m. are likely to be encountered, such as in refrigeration and chlorinating plants, more efficient equipment is required to safeguard the employee. The most commonly used is the canister type of respirator.

Canister respirator

In this form of protection the impure air is drawn into a sealed canister and the toxic gases are absorbed or converted by a chemical compound. There are two basic versions of canister respirator: one where the canister is sufficiently light to be fitted to a full-face mask, the other where the chemical contents give longer duration and the canister has to be supported by means other than the face mask. Longer duration and greater efficiency of this latter respirator are achieved at the expense of increased weight.

The supported canister respirator consists of four essential parts: face mask, canister, connecting tubes, and carrying harness. The face mask offers panoramic vision and comprises a rubber moulding into which a visor is fitted.

At the bottom of the visor a valve housing incorporates the non-return exhalation valve and the breathing tube connection. The mask is retained in position by means of a moulded rubber head harness and has a breathing tube of corrugated format so that if bent or compressed there is still a free air passage through the tubes. A fitting on top of the canister is attached to this tube by means of an easily removable clip.

The canister is mounted in a harness which is produced in several forms—for slinging from the neck, wearing the canister on the back, or carrying at the side. The harness material is impervious to oils, greases, or contamination by chemical splashing.

On the mask-mounted version of this respirator the canister, which is smaller and has a reduced life, is attached directly to the face piece. The face piece can either be of the wide-visor variety or of the twin-eyepiece type, reminiscent of the gas masks used during the Second World War.

The canister is constructed from metal or plastic with ridges which serve not only to strengthen the container but also to ensure that channelling of impure air up the sides does not occur. A non-return inlet valve is sited at the bottom of the canister and the chemical absorbent and dust filter, if fitted, are retained between metal screens held in place by compression springs. This prevents the formation of channels or displacement of the material during transit and use.

There are different canisters to combat various gases or combinations of gases and dust and it is important that the correct canister is used. The canisters are all colour coded for easy identification and there is an appropriate canister to counteract practically every known gas in industry or agriculture. Unlike filter cartridges, the life of a canister is specified in the majority of cases as giving protection for 30 min against 1 per cent by volume of the relevant gas. Most canisters are approved by the Factory Inspectorate and comply with the relevant British Standard specification. Other non-approved canisters are available and should be used in accordance with the manufacturer's instructions.

The canisters are sealed to preserve their shelf life and these seals must be removed when fitting to the respirator.

The smaller and lighter mask-mounted canister, while not having the life of the harness version, nor providing protection against so many gases, is a particularly compact unit for emergency use for limited periods, or in the agricultural industry where it has been found especially useful for crop spraying.

While all these respirators provide adequate protection for low concentrations of dusts and gases, they cannot be used in atmospheres deficient in oxygen, or which contain high concentrations of gases. Moreover, under no circumstances should they be worn in confined spaces, such as vats, tanks, and stills where there is no free flow of air. If these situations are encountered it is necessary to use equipment to which clean air is supplied to the user.

Equipment using clean air or oxygen

Equipment of this nature affords complete respiratory protection against all toxic gases and vapours. It can be divided into two categories. First, airline apparatus which uses air supplied to the wearer by atmospheric pressure, manually-operated blowers, or bellows, or, at a higher pressure, from power-operated compressors or high-pressure storage cylinders. Second, self-contained breathing apparatus in which the breathing mixture is carried by the wearer.

Airline breathing apparatus

The short distance breathing apparatus is most commonly used in the U.K. and is the simplest form of airline equipment.

Fig. 32.2 Compressed airline apparatus working from storage cylinders affords complete respiratory protection for men engaged in tank cleaning. The two men wearing self contained breathing apparatus are on emergency stand by duty as a safety precaution, and, as a final safety measure, an oxygen resuscitator is present.

The wearer has a full-face mask with a breathing (equalizing) tube, to which an inhalation valve is fitted at the inlet end. This valve, which is secured to a waist belt, is coupled to a length of heavy duty air hose. To avoid any extraneous materials blocking the air hose a simple perforated metal filter is fitted on the intake. When in use this end of the hose is secured in fresh air by means of the spike or hook provided. It is vitally necessary that this securing is carried out so as to avoid the wearer drawing the hose into a contaminated area should he attempt to proceed further than the length of hose permits. The apparatus is usually supplied with a 30 ft (9 m) length of hose. Lengths in excess of this are likely to increase breathing resistance beyond the acceptable limit.

If greater penetration into a contaminated atmosphere is necessary a hand-operated rotary blower or foot-operated bellows must be fitted to the hose. The maximum safe distance allowed with forced manual means of air supply is 120 ft (37 m); above this distance a compressed air supply is necessary.

Constant-flow airline respirators. For working at distances greater than 120 ft or where long periods of work in toxic atmospheres have to be undertaken, such as in steel works or large chemical plants, the constant-flow compressed airline respirator system, utilizing the works compressed air system as the source of supply, is the most popular.

This equipment consists of a full-face mask, or half mask, into which is inserted an exhalation valve. Light-weight pressure tubing fitted to the mask connects to a combined filter and control valve secured to a waist belt. The filter control valve permits the wearer to regulate the flow of air into his mask according to his needs. A length of pressure tubing connecting the air supply with the filter control valve on the waist belt completes the equipment. The compressed air can be obtained from the factory airline system or direct from a mobile compressor.

It is extremely important that the air to the user is free of oil, fumes, vapours, and particulate matter. A combined mechanical and chemical filter unit is therefore generally inserted between the source of air and the wearer. Finally, a reducing valve ensures that air is supplied at the correct pressure to the user. The operating pressure of the reducer depends on the length of hose and the number of men operating from the same source of air.

Should the atmosphere contain chemicals or dust that attack or can be absorbed by the skin, then a hood covering the head and shoulders can be used, the air being introduced above the visor in a downwards direction. This downward diffusion of air serves to keep the visor from misting and to provide the wearer with a cool and comfortable environment. The air escapes to atmosphere through the skirt of the hood thereby preventing the ingress of toxic matter. It is vitally important with this type of equipment that a sufficient volume of air is maintained otherwise the air barrier is nullified. A good example of this form of protection is the commonly used shot blast helmet.

The constant-flow airline system depends on a continuous and unlimited supply of air. There are occasions, however, when it is necessary to protect persons working in hazardous conditions where there is no airline and it is not possible or convenient to use a mobile compressor. In such conditions, the demand valve type of compressed airline apparatus has many advantages.

Demand valve airline system. The major advantage of the demand valve equipment is conservation of the air. As air is only supplied when the wearer inhales, the duration of protection is considerably extended when compared with the duration achieved with a constant-flow system and the same volume of air.

The sources of supply for this equipment are cylinders of compressed air of large capacity which can be hired from a supplier of compressed gases or purchased from a breathing apparatus manufacturer.

For this source of air the full-face mask is fitted with a demand valve and connected by pressure tubing to a reducing valve screwed into the neck of the storage cylinder.

The reducing valve is equipped with two pressure gauges, one indicating the pressure of the air in the cylinder and the other indicating the pressure of the air being supplied to the wearer. The demand valve fitted to the face mask is operated by the inspiratory effort of the wearer and automatically supplies him with the volume of air he requires. To prevent the weight of the tubing being taken by the demand valve connection, it is attached to the wearer's harness. Once again, should further skin protection be required a hood can be fitted to the mask.

Airline apparatus, as described above, is the most appropriate equipment for providing complete respiratory protection for prolonged periods. It is particularly suitable for personnel involved in hazardous production-line processes or maintenance. However, it is not recommended for use in emergencies where persons have to be rescued from a dangerous environment. The limitations are engendered by the presence of the air line. The line could become entangled around machines or supporting columns and, as a result of its restricted length, may not allow the wearer sufficient penetration. Furthermore, should the hazardous area rapidly expand, the wearer's attachment point to the source of supply could become engulfed. This lack of manœuvrability on the part of the wearer makes it impracticable for this type of equipment to be utilized under emergency conditions.

Self-contained breathing apparatus

In emergencies the only equipment to be considered is self-contained breathing apparatus which enables the wearer to be independent of the surrounding atmosphere while having complete freedom of movement.

In a self-contained breathing apparatus the air or oxygen supply is usually compressed into steel alloy cylinders at pressures of 1 980 or 3 000 lb/in^2 (137 or 207 bar), and carried by the wearer as an integral part of the apparatus. The apparatus works on either the open-circuit principle where the exhaled air passes directly to atmosphere, or the closed-circuit principle where the exhaled air is purified and recirculated.

Open-circuit. In open-circuit apparatus, air is the most common breathing mixture. It is contained under pressure in one or two steel alloy cylinders retained in a frame carried on the back. Air from the cylinder is fed through a reducing valve, where the pressure is reduced, and then into a flexible rubber pressure tube. This is connected to a second stage reducing valve fitted to the visor of a full vision face mask.

The second stage valve is operated by the inspiratory effort of the wearer. On inhalation, the valve opens and supplies a sufficient volume of air irrespective

Fig. 32.3 Air Master self-contained compressed air breathing apparatus affords respiratory protection for thirty minutes.

of the task being performed by the wearer. A non-return exhalation valve fitted into the mask ensures the disposal of exhaled air to atmosphere.

A constant-flow or by-pass valve is incorporated in the first stage reducing valve to enable the wearer to pressurize the mask in the event of accidental damage occurring to the visor, such as from collision with an obstruction. Pressurization of a cracked mask prevents the entry of any toxic or hazardous gases.

The components of the equipment are mounted on a frame attached to a terylene harness. The harness comprises shoulder straps and a waist belt all easily adjusted with quick-acting buckles. The harness can be readily adapted for lifting purposes by the addition of a life-line attachment to the frame, and a chest strap.

The open-circuit compressed air equipment is very simple in operation and requires the minimum of training and maintenance. On the other hand, the cylinders are comparatively heavy and the duration of such sets is limited to approximately 40 min. Cylinders giving a longer duration can be provided but are so heavy that they seriously hinder and tire the user.

Closed-circuit. Closed-circuit breathing apparatus can use compressed gaseous oxygen, liquid air, or liquid oxygen, but the most common in use are those using compressed oxygen. Oxygen is retained under pressure in a steel alloy cylinder worn on the back.

All closed-circuit breathing apparatuses comprise an oxygen cylinder, reducing valve, purifying canister, cooling medium, breathing bag, and breathing tubes, fitted either to a mouthpiece or a full vision face mask. The positioning of these components around the body is determined by the duration of the equipment, the method of mounting and carrying, and individual manufacturer's design.

In closed-circuit oxygen breathing apparatus, the wearer inhales from the breathing bag via a chemical cooler. On exhalation, the impure air passes through a chemical absorbent, where the carbon dioxide is removed, and returns to the breathing bag. There it mixes with the fresh oxygen which is delivered into the bag from the cylinder at a flow of $2-2\frac{1}{2}$ litre/min through the reducing valve. This cycle continues for the duration of the equipment.

There are times when the flow is insufficient for the effort being made by the wearer and, to enable him to supplement his supply, a by-pass valve is fitted which allows the flow rate to be increased. Similarly, there are occasions when the wearer does not require the full amount of oxygen being supplied and a relief valve fitted to the breathing bag enables him to jettison the excess oxygen to atmosphere. Non-return valves in the breathing tubes ensure the correct circuit flow.

Closed-circuit sets are generally lighter than open-circuit sets since they incorporate smaller capacity cylinders. It is possible, therefore, to obtain sets with durations of up to 4 h without exceeding an acceptable weight limit. On the other hand, these sets tend to be hotter and, once again, increase the fatigue factor. Recent developments have greatly improved on this drawback and modern oxygen breathing apparatus is now greatly simplified, cooler in operation, and economic to maintain.

Pressure gauges are fitted to both types of self-contained breathing apparatus to indicate to the wearer the cylinder contents. A further refinement is a low pressure warning whistle which automatically indicates when the pressure has fallen to 42 atm for compressed air and 30 atm for closed-circuit breathing apparatus, thereby allowing 10 min for evacuation to fresh air.

Statutory requirements

It is a requirement of the Factories' Act 1961, Section 30, that when persons have to work in a confined space where dangerous fumes may be present, suitable breathing apparatus of an approved type must be provided. Such apparatus should be subjected to a thorough examination by a competent person at least once a month. Reports of these examinations must be kept available for inspection by the appropriate authority (Breathing Apparatus, etc. (Report on Examination) Order 1961, S.I. 1345).

It is prescribed under the Gas Cylinder (Conveyance) Regulations 1947, that all cylinders associated with breathing apparatus that have to be conveyed by road for recharging, must be subjected to a hydraulic test every 5 years.

Personnel who are liable to wear breathing apparatus in an emergency should have frequent training sessions in which they spend as much time as possible wearing the apparatus.

Training and stowage

At least one person, in every organization where breathing apparatus is used, should be trained to a particularly high standard of proficiency. He would then be competent to carry out the required inspections and also any minor adjustments and repairs that become necessary. Emergency breathing apparatus should be in position around the works in easily identifiable containers which must preclude access to unauthorized persons and yet be easily removable in the event of urgent need. One method, favoured by a number of large organizations, is to house the equipment in a wooden container with a lock, the key being displayed in a glass-fronted compartment secured to the box.

Selection of equipment

Very careful consideration should be given to the choice of equipment. Manufacturers of industrial safety equipment employ highly trained staff who are fully competent and more than willing to offer guidance on the precise equipment required to combat the many and varied hazards which face the worker in industry today.

33. Medical services: first aid and casualty treatment

Dr J. D. Cameron

Group Medical Adviser,
Pilkington Brothers Ltd

The two chief aims of any safety organization must be the prevention of accidents by the elimination of causal factors and, when accidents do occur, the prevention of injury by the use of appropriate safety measures. If injuries are caused, the third line of defence is to minimize their effects by prompt and efficient first aid and treatment.

The aims of first-aid and casualty treatment

These can be listed as follows:

    (a) to ensure as rapid and complete a return of function as possible;

    (b) to retain the injured person at work in his own job or, failing this, in some suitably modified or alternative occupation;

    (c) to reduce to a minimum any unavoidable period of absence from work due to injury;

    (d) should permanent disability result and prevent a return to the previous occupation, to arrange any necessary retraining and resettlement as expeditiously as possible.

These aims, designed to prevent unnecessary loss of working time and to minimize the risk of permanent incapacity, are of the greatest economic importance in an industrial society. They are also in the best physical, psychological, and financial interests of the individual.

First-aid and treatment services

The nature and extent of the first-aid and casualty treatment facilities in any given industrial situation will vary with the particular local needs. A first-aid box may suffice for the smallest factory. Most large industries have a comprehensive occupational health service. Many require facilities between these two extremes.

First-aid provisions at any place of work must at least comply with the statutory requirements of the appropriate legislation. In Britain the existing first-aid, together with other health and safety law, is to be progressively replaced by a system of regulations and approved codes of practice as provided for in Part 1 of the Health and Safety at Work, etc., Act 1974. Until this is done the legal requirements set out in the Factory Act 1961, the Offices, Shops and Railway Premises Act 1963, and the various other Acts relating to particular industries still apply. These regulations are summarized in the Health and Safety at Work Booklets published by H.M.S.O.[1, 2]

First-aid service

First aid is the emergency treatment of a casualty, provided immediately after injury or the onset of sudden illness. Conventionally it includes only those measures which can be taken by those quickly available until more expert help can be organized. In serious situations, such as asphyxia and massive bleeding, correct action at once can be life saving. In major injury it is easy to understand how efficient first aid can help to reduce disability and promote recovery (a) by preventing aggravation of the local condition, and (b) by preventing deterioration in the general condition.

The great majority of industrial injuries are fortunately of a minor degree and here the importance of a high standard of first aid is sometimes not as readily appreciated; it is no less important. The prevention of infection in a minor wound, for example, may avoid serious and disabling complications.

First-aid equipment

This can be divided into that which is of general use and that which is required to meet the needs of particular situations.

First-aid box or cupboard. This is a statutory requirement in all factories (Factories Act 1961, Section 61 (First Aid)). No matter how small the factory a first-aid box has to be provided. Where the number employed exceeds 150 an additional box is required for each 150 or part of that number.

All first-aid boxes must be in the charge of a designated responsible person. Where more than 50 persons are employed the responsible person must hold a qualification in first aid approved by H.M. Chief Inspector of Factories. The name of the responsible person must be displayed in the work place served by any particular first-aid box or cupboard. The minimum contents of first-aid boxes are prescribed for factories employing less than 10, between 11 and 50, and over 50 persons.

Structure. Different types of first-aid boxes are available to meet the needs of most situations. The box or cupboard should be of a strong and durable construction. The lids of some metal boxes tend to bend and make closure difficult and unsatisfactory. Fixed boxes are best made of wood. Portable first-aid equipment is most conveniently carried in a satchel or attaché-case type container.

Location. If provided to meet the needs of a particular hazard the first-aid box should be suitably placed for that purpose. If provided for the general needs of the working population it should be situated so as to meet the convenience of the majority. Cases of minor injury are often reluctant to leave their job and seek first aid if this involves a journey. A situation near the entrance and exit is often most convenient and encourages these cases at least to seek treatment when leaving the factory.

The box should be in a well-lit area near to washing facilities and drinking water. A chair and a table are desirable. If no other working surface is available the inside of a first-aid box lid, hinged at the bottom, which lets down to form a level tray, is a useful substitute. All working surfaces should be of plastic or other easily cleaned hygienic material.

The box should be clearly marked and in a conspicuous position. Those working in the area should know where their first-aid box is situated but others may need to use it in an emergency.

The box must be readily accessible at all times and should never, during working hours, be under lock and key. The problem of pilfering is often advanced as an excuse to lock the first-aid box and place it in an inconspicuous position. Fixed boxes or cupboards can with advantage have a toughened glass panel in the door or lid. In this way the contents can readily be seen by all passers-by and the danger of finding the box deficient of first-aid material in an emergency is avoided.

Contents. The basic contents should be the statutory requirements. Additional first-aid material should be included to meet the needs of special hazards. Nothing other than first-aid material should be permitted in a first-aid box.

Stretchers. In larger organizations it is often advisable to have a stretcher or stretchers available. These are best kept, with at least two blankets, in clearly marked wall cupboards. If the blankets are stored in large polythene bags they are always clean and dry when required in an emergency.

Where problems of evacuation from inaccessible situations are likely to arise one of the special stretchers designed for this purpose should be provided.

Special first-aid equipment

Varying arrangements have to be made to meet the needs of the different problems of individual industries. The provision of first-aid boxes is a management responsibility. Where a medical or nursing service exists it will be responsible for first-aid arrangements. It is sometimes difficult to decide what is a *safety* and what is a *medical* responsibility. This applies, for example, to some rescue and resuscitation apparatus. It is imperative that there should be no possibility of confusion. Responsibility for the provision and supervision of this often vital equipment must be clearly defined. Asphyxia and chemical injuries are quoted as examples of situations in which special equipment is necessary.

Resuscitation equipment

Wherever a serious risk of asphyxia exists, be it from gassing, other poisoning, or electric shock, facilities for rescue and resuscitation must be readily available. Professional advice should be sought on the precise needs of any particular situation.

Once breathing has stopped seconds count, and the sooner artificial respiration is started, the better the chances of recovery. If not established within 3 to 4 min, the prospect of a successful outcome becomes more remote. The mouth-to-mouth technique which requires no special apparatus may be adequate for the immediate situation. It should only be used, however, until the arrival of the necessary expert assistance and the necessary equipment.

Manually operated resuscitators are available and to some of these oxygen cylinders can be attached. The most efficient methods involve the use of mechanical respirators which are adjustable to provide pure oxygen or a mixture of oxygen and air. These usually have the added advantage of having a suction attachment which is invaluable in keeping the airway free from obstruction. It is useless to provide this type of equipment unless personnel are fully trained in its operation and their efficiency maintained by regular practice.

The periodic testing and maintenance of this type of equipment is also most important. It should be inspected regularly every week, and details of the inspection recorded. If this is not done apparatus used infrequently is liable to become unserviceable.

Special wall notices which include a description of mouth-to-mouth artificial respiration are available and should be displayed. They are statutory under the Electricity Regulations (S.R. & O. 1908, No. 1312).

Chemical hazards

Where certain dangerous chemicals are handled special first-aid measures must be available even in the smallest unit. Where strong acids, alkalis and other corrosive agents are used, or where poisonous chemicals which can be absorbed through the skin are handled, provision must be made for the immediate removal of skin contamination. Small areas can be dealt with by washing under a tap, but special showers or baths should be immediately available where any risk of more serious spillage exists.

Where corrosive or other chemicals irritant to the eyes are used 'on the spot' facilities for eye irrigation are essential. Irrigation must be commenced at once and continued for at least 10 min before urgent evacuation to medical care. The normal undines and eye baths are inadequate for this purpose. A special irrigating bottle or, better still, a larger wall container with a flexible pipe and clip or tap should be provided. All those exposed to, or working in the area of, this type of risk should be instructed in the necessary first-aid treatment.

Professional advice should be sought as to the methods and solutions which should be used in any particular situation. The use of water has been assumed in the above descriptions of both skin and eye irrigation because the mechanical removal of contaminants is the urgent essential treatment. Special neutralizing solutions can be provided for most chemicals but these must be immediately available and delay in obtaining them is never justified. They are probably best used after a preliminary washing with water.

Where strong acids or corrosive liquids are handled, the display of a cautionary notice is mandatory under the Chemical Works Regulations (S.R. & O. 1922, No. 731).

First-aid training

Responsible person. It is a statutory requirement that, in a factory where more than 50 people are employed, the person in charge of the first-aid box must be in possession of a current first-aid certificate.

In practice this usually means a certificate issued by one of the three voluntary first-aid societies, St John Ambulance Association and Brigade, the British Red Cross Society, and the St Andrew's Ambulance Association. These organizations have standardized their teaching and examinations. They use a common text book[3] and provide three grades of qualification. The statutory requirement is the holding of the basic First-aid Certificate. This certificate is valid for 3 years and then lapses unless it has been renewed by re-examination. In larger organizations consideration should be given to the training of first-aid attendants for the more advanced qualification known as the Higher First-aid Certificate or for the highest qualification, the Certificate of Advanced Knowledge of First Aid. Full-time first-aid ambulance personnel in particular will benefit from this training.

Other organizations are authorized by the Health and Safety Executive to provide instruction and issue certificates of qualification. Periodically a Certificate of Approval listing these approved bodies is issued under the First Aid

(Standard of Training) Order 1960. (See also Health and Safety Booklet No. 36[1]).

The St John Ambulance Association and Brigade also publish a textbook *Occupational First Aid*.[4] Following a course of instruction based on this book and after examination, a certificate in Occupational First Aid is issued to successful candidates. This course of training is advised for all practising first-aid at work, and is particularly recommended for full-time first-aid attendants and those having responsibility for ambulance rooms.

There are great advantages in arranging first-aid instruction within the organization. In this way emphasis can be placed upon the industrial aspects of the subject and special instruction given on the problems peculiar to a particular industry. This is, however, usually only possible in those industries large enough to support their own medical services.

As an alternative, the voluntary first-aid societies alone or in conjunction with other health and safety organizations provide short intensive courses of first-aid instruction especially designed to meet the needs of industry. These courses usually involve 4 days' full time instruction with the qualifying examination on the last day.

Hospital casualty departments are often willing to provide experience for industrial first-aid workers.

Self help

No workman should be left in ignorance of any possible safety or health risk at his work. It is now a statutory duty on the employer to provide such information, instruction, training and supervision as may be necessary to ensure, so far as is reasonably practicable, the health and safety at work of his employees, as set out in the Health and Safety At Work, etc., Act 1974, Part 1, General duties (2)(c). In this context the workman should be trained in the initial first-aid measures which may be needed in the event of accidental injury resulting from any hazard peculiar to his work. All electricians should, for example, be trained in emergency resuscitation techniques, and the chemical worker should know what to do at once if his skin or eyes become contaminated with a dangerous chemical. These important aspects of training have been described as self help.[5]

Duties of a first-aid attendant

The main responsibility of the first-aid worker is to provide help in the event of injury or sudden illness occurring at work. He must be responsible for maintaining the first-aid boxes and all other first-aid equipment. He must keep accurate records. These should include the name, address, and works number of the injured person, particulars of the accident, the date and time of injury, and details of the treatment provided.

Many industrial injuries are of a trivial nature. If, in addition to his recognized first-aid duties, the first-aid attendant is permitted to treat minor injuries he can do much to lighten the load of overworked doctors and nurses. He can also save an enormous amount of working time which would otherwise be wasted in obtaining treatment outside the factory. Any first-aid attendant given this responsibility must be fully aware of his own limitations and trained to recognize what is in fact a minor injury within his capacity, and what must be referred for more expert opinion. He must maintain a close liaison with the family doctors and local hospital services.

The first-aid or ambulance room

In certain industries an ambulance room of a prescribed standard is a statutory requirement. Smaller factories may not be able to devote a room to first aid and treatment, and first-aid attendants will operate from their first-aid boxes.

Wherever possible, however, a first-aid room should be provided. The first-aid attendant can work much more efficiently with the added facility this provides. Such a room is a minimum requirement for a visiting nurse or doctor.

Situation. It must be convenient to the majority of the workers. The advantages of a central position must be compared with those of a situation near the entrance. It must be freely accessible so that stretchers can easily be carried in from the factory or out to an ambulance. It should be in a quiet and clean location.

Construction. Walls, floor, and ceiling should be of a material which does not collect dust and is easily cleaned. Tiles form an ideal wall covering. Floors should be of a non-slip impervious material which is easily washed. Floor-to-wall junctions should be curved to permit easy cleaning.

A door at least 1·1 m wide should be provided to admit a stretcher. The width of outside passages should be considered. In difficult situations a sliding door may be best.

Privacy must be ensured, by obscuring windows if necessary. A high standard of illumination is required. The temperature should be maintained at 18–21°C.

Adequate washing facilities and a supply of drinking water are essential. Toilet facilities should be communicating or at least adjacent.

Plenty of cupboard space must be provided. Built-in cupboards surfaced with one of the plastic materials are ideal. A domestic sink unit, built in so that the whole can be easily kept clean, is to be recommended. Cupboard tops at sink height provide a satisfactory working surface and save space in a small room.

Equipment. The major articles of furniture are: (a) a special treatment chair, and (b) a couch and suitable screen. The number of additional chairs will depend upon the size of the unit.

Most modern casualty departments are equipped to use sterilized prepacked dressings and instruments. In this way all the requirements for a particular treatment procedure can be made readily available. Disposable materials are widely used. The type of sterilizing arrangements will depend upon the dressing technique adopted.

A portable adjustable lamp, an electric kettle, and a hygienic disposal bin for dirty dressings are also needed.

A day-book in which to record information similar to that specified under the duties of a first-aid attendant must be provided, together with facilities for the recording and filing of other records.

The full and accurate recording of casualty treatment has obvious medical and medico-legal importance. Relevant information concerning accident causation, frequency, and location, is of great value in the accident prevention field. A scheme for the regular communication of this information to the appropriate management representative should always be arranged.

The person responsible for providing treatment and recording this information should not overlook an opportunity this presents for safety instruction.

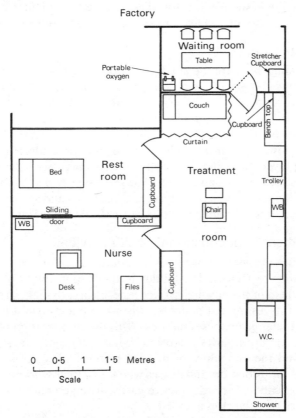

Fig. 33.1 Plan of Medical Department—650 employees.

First-aid, medical, and nursing personnel in industry can often teach as they treat.

A plan of a treatment room, rest room, and nurses' duty room are shown (Fig. 33.1). This serves a working population of 650. The nurses' duty room is also used as a consulting room by the visiting doctor.

Occupational health services

Whilst all occupational health services have many common features, each develops to meet the needs of its own particular industry. In the early days of industrial medicine many were started to fill a need for first-aid and emergency treatment. Most of these have developed into the comprehensive medical services now found in the larger industries. The emphasis today is rightly on preventive medicine. The main objectives being prevention of disease by the elimination or control of health hazards and the maintaining of health, by the provision of the best possible working conditions. All occupational health services also have a responsibility for the organization of first aid and the emergency care of injury or sudden illness occurring at work.

Group services

Factories in Britain which are too small to support their own medical services can with advantage join one of the group services. These are non-profit making organizations providing a comprehensive occupational health service for factories in their area. Most were initially sponsored by the Nuffield Foundation. The first was started at Slough in 1947. Since then other services have been organized at Harlow, the Central Middlesex Hospital, Rochdale, and Dundee. A group of employers have independently started their own scheme to serve the needs of West Bromwich and district. The service provided by these organizations includes the supervision of first-aid arrangements and at least emergency casualty treatment. It is to be hoped that these are the first of many such schemes.

Casualty treatment in occupational health services

The extent to which provision is made for the continuation of treatment at work varies from factory to factory. It is ethically the responsibility of the family doctor, with the help of the hospital service if necessary, to make arrangements for the treatment of his injured patients. It is not economic, in the absence of special reasons, to duplicate these facilities. With the co-operation of the family doctors, however, minor injuries can often be more efficiently treated at work. Time is saved and the difficulties of divided responsibility are avoided. The injured are cared for by doctors and nurses who know their working conditions. Temporary modification of work is more easily arranged and in this way unnecessary absence from work is avoided.

More extensive treatment services are often desirable to meet the special needs of an industry. The big safety problem of flat glass manufacture is the danger of wounds due to accidents when handling the product. This industry in Britain has had a casualty treatment service for more than 50 years. It serves a group of factories situated within 1 mile of each other and together employing approximately 13 500 persons.

Throughout the various works first-aid boxes are placed at strategic points. A worker trained in first aid and employed in the area is responsible for the local box.

Each factory has its own ambulance room staffed by a qualified nurse. Cases are referred to her by the first-aid men and she accepts responsibility for the treatment of minor injuries. When necessary, cases can be referred to the main medical department which is situated near the most central and largest factory in the group.

In this main department all the facilities necessary to treat industrial injury are provided and only selected cases are referred to hospital.

The casualty treatment section includes a reception and waiting area, two treatment rooms, an X-ray department, operating theatre, recovery and consulting rooms.

Individual treatment cubicles have been substituted for the more usual open treatment room. A consulting room and three cubicles are arranged side to side. The partitions are made of plastic laminates. The patients enter the cubicle from one end. The nurse operates from the other end which is an open communicating corridor. On the far side of this opposite each cubicle are a built in wash basin and storage cupboards.

Each cubicle contains a special treatment chair. This is made of glass fibre reinforced plastic for durability and ease of cleaning. It has a special non-slip surface and is designed to hold the average workman wearing his working clothes. It is raised from the floor on a pedestal to provide the optimum working height for the dressing of hand and foot injuries. A special foot support is available.

Prepacked dressings and instruments and other requirements are contained in racks on either side of the chair. Glass shelves provide working surfaces on each side of the chair. Disposable bags are provided to collect dirty dressings and other containers are used to collect used instruments and other recoverable material.

Three other cubicles are provided with a couch instead of a chair and are used for surgically clean cases.

Special arrangements are made in the consulting room for the care of eye injuries.

Among the advantages of these arrangements are clean working conditions and privacy. Unnecessary movements by the nurse are obviated and her good working position reduces the risk of back strain and fatigue.

If possible, after initial treatment, cases are returned to their own work or found suitable alternative work in the factory. If this is impossible but they are fit to attend work, they are employed in a rehabilitation workshop. Here they are given suitably modified work, which can be adjusted to provide any necessary exercises, until such time as they are able to resume work in the factory. Physiotherapy is provided at the main department or in the rehabilitation workshop.

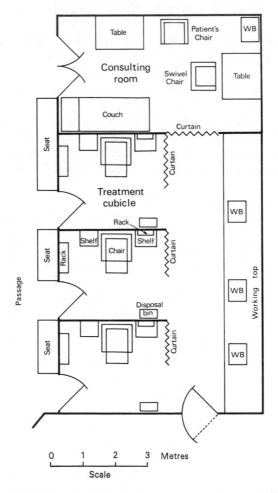

Fig. 33.2 Plan of special treatment room arrangements in large unit.

Other industries have developed casualty treatment services, including in some cases special rehabilitation workshops, to meet their own requirements.

The majority of non-occupational injuries are treated by family doctors or within the hospital service. Any special facilities which industry has provided can often, with advantage, be made available to these agencies, for the use of employees who are their patients.

Rehabilitation

Cases of industrial injury require the help of some, if not all, the following services; first aid, surgical and perhaps medical treatment, convalescence, physiotherapy, occupational therapy, retraining, and resettlement. These are not separate entities but form parts of one continuous process which can be described as rehabilitation. The modern concept of rehabilitation is not to regard it as something which follows the completion of treatment. It should include the entire management of a case of injury. Rehabilitation in industry starts at the first-aid box and is only completed when the injured person resumes his previous job or is permanently resettled in a suitable alternative occupation.

The Department of Employment provides rehabilitation facilities and Employment Rehabilitation Centres and Skill Centres are to be found in most industrial areas. A workman does not need to be on the unemployed register to avail himself of these benefits. The Employment Medical Advisers of the Health and Safety Executive have responsibilities for rehabilitation and are available for advice to government, employer or disabled persons.

Most of the difficulties experienced in the supervision of cases of industrial injury and the avoidable delays in their return to productive work are due to a lack of liaison and, in particular, to inadequate communication between industry, the hospital and doctor services and the officers of appropriate Government departments. To rehabilitate the injured workman efficiently it is important that these different bodies establish the closest possible co-operation. This will ensure that the facilities which each provide are used in the best possible way.

References

1. *First-aid in factories*. Health and Safety booklet No. 36, 4th impression with amendments. H.M.S.O., London, 1974.
2. *First Aid in Offices, Shops and Railway Premises*. Health and Safety at Work booklet No. 48. H.M.S.O., London, 1973.
3. *First Aid*. The St John Ambulance Association and Brigade, The St Andrew's Ambulance Association, and the British Red Cross Society, 3rd edn., 4th impression 1975.
4. *Occupational First Aid*. St John Ambulance Association and Brigade, Macmillan Journals, London, 1973.
5. Ward Gardner, A. and Roylance, P. J. *New Essential First-Aid*, p. 17, Pan Books, 1967.

34. Medical services: liaison with the hospital emergency department

Dr M. H. Hall

Consultant-in-Charge,
Emergency and Accident Dept,
Preston Royal Infirmary

Patients requiring urgent hospital attention are first seen in the Emergency and Accident department.

During the past few years the place of these departments in the treatment of the sick and injured has been subjected to considerable discussion and it has finally been accepted that they occupy a vital place in the treatment of persons who are injured in accidents whilst at work, or on the roads, or who are taken seriously ill under conditions where practitioner assistance is impractical.

New departments are beginning to be opened throughout the country and the quality of the equipment provided is now at a higher level than ever before. In spite of the financial crises that arise with depressing regularity, the past decade has produced improvements in the Emergency Services of a greater degree than has ever occurred before.

The importance of these Units has now been finally accepted by the medical profession, who have agreed to the appointment of consultants to take charge of the organization and the day-to-day running of these departments. As a result, the quality of treatment provided is now equal to that in other hospital units, and the introduction of modern methods of treatment has made it possible to preserve the life of patients suffering from severe injuries or acute sickness in cases for which this would have been impossible a few years ago.

Unfortunately the departments are still overloaded by 'casual' attenders who tend to look upon the Unit as a medical supermarket providing treatment for trivial complaints for which, in many instances, medical attention is not indicated or for conditions for which they should have consulted their own doctor. As a result the number of attenders at these Units increases year by year and the management of the more major cases is severely hindered by the large numbers of unnecessary attenders who do not realize the problems that their visits cause.

The functions of the Unit are to treat:
- (a) patients with recent injuries,
- (b) patients suffering from serious conditions of sudden onset,
- (c) patients referred with a letter to the department by a general practitioner.

Patients with recent injuries

It is impossible to define a recent injury to a degree which will avoid causing hardship to any patient and, at the same time, prevent unnecessary attendances. Fortunately the general principles involved are straightforward. The condition should arise following an injury or accident. It should require hospital attention and facilities for treatment which a general practitioner may not possess. There should also be some incapacity associated with the disorder. Lacerations, fractures, foreign bodies in the eye, and acute strains following muscular effort are examples of conditions which should be seen in the department.

Patients suffering from conditions which have not arisen as the result of a recent injury should not be referred to the department. Examples of this type of condition include backache of several weeks duration, even though it originally followed a lifting strain; insecurity of the knee without recent locking; skin rashes, and septic conditions of all types. Patients suffering from disorders of this nature should be treated by their own doctor. Some may require hospital treatment but the Accident Department is not the correct place for them to receive attention. If necessary their own general practitioner will refer them to the appropriate Outpatient Clinic.

Many patients are referred by first aiders at work for a 'check-up'. Some patients are referred because of the fear of possible litigation arising in the future. On other occasions patients are forced to come to the hospital against their will by a works foreman or a first aider after sustaining a very trivial injury. It must be emphasized that medical grounds alone are the only adequate reason for a person to attend the Accident and Emergency Department. It does not exist to provide legal cover for deficiencies in the first-aid services.

Patients suffering from serious conditions of sudden onset

In general, this group of disorders is easier to define than the previous one. In most cases there is little doubt as to the severity of the conditions, and the necessity for hospital treatment is immediately apparent.

Acute medical and surgical disorders occur fairly frequently especially in the larger firms, and coronary thrombosis, perforation of a stomach ulcer, or acute appendicitis are all conditions which require hospital attention. Acute glaucoma, and hypoglycaemic attacks occur from time to time and, on occasions, attempts at suicide are made whilst the patient is at work.

Patients who do not require hospital attention are those with long-standing complaints which temporarily develop an acute phase, similar to a previous one which has been treated by the patient's general practitioner. To send such a patient to hospital does not help in any way because his previous medical history is unknown to the departmental staff. Generally the patient's own doctor is in a much better position to deal with the illness because of his detailed knowledge of the patient's background.

It is also questionable whether a known epileptic need be sent to the hospital. In properly staffed and equipped works-surgeries there is no reason why the patient should not be detained until he has recovered and then sent home to rest. Only if the attacks become frequent and do not settle need the patient be referred to hospital.

Patients referred with a letter to the department by a general practitioner

These are usually patients who require X-ray examination or a second opinion after an injury, and patients who require minor procedures under a general anaesthetic.

The use of the department to provide a quick entry to the formal consultative clinics is to be deprecated. If there is a waiting list for these clinics it is unfair to the patients already on the list for someone to jump the queue.

The quality of the treatment given to patients referred to Accident Departments is primarily controlled by the available staff, both medical and nursing, by the available facilities and by the time that the staff have to deal with them.

It is to be hoped that the developments taking place at the present time will lead to improvements in the first two, but unless the number of patients referred is kept to reasonable levels the staff will be overloaded and unable to deal adequately with the problems. Inevitably this will lead to a lower standard of treatment and a prolongation of the disability time.

A good relationship between industrial medical services and Accident Departments can be built up only by personal contact between the two services. This is of considerable importance in improving the overall efficiency of treatment and it is surprising how it enables apparently unsurmountable problems to be overcome. A shortage of time is the main obstacle to achieving a close liaison, but an annual meeting between the hospital staff and the personnel from the works services would be an acceptable substitute. Apart from a general discussion and an interchange of views, it would afford an opportunity for contributions to be given by members of the various hospital specialist departments about industrial problems which arise in their speciality.

Communications

An efficient system of communication between the two services is essential. Any methods adopted must be quick, simple, and free from unnecessary detail. Paper work in the works surgery and in the Accident Department must be kept to a minimum.

Forms used to refer patients to Emergency Departments should be simple and straightforward. Unnecessary writing can be avoided and much time and effort saved by the use of boxes which merely require ticking in the appropriate section. If the factory medical service wish to continue with the patient's treatment it is helpful if this can be incorporated in a form. An expeditious reply from the hospital is much more likely if it merely involves ticking a simple printed request, e.g., please continue with this patient's treatment.

The removal of sutures, and the treatment of eyes after removal of foreign bodies are conditions which can be dealt with on a routine basis if the works surgery knows the hospital requirements. If this approach is adopted it should be clearly understood that unless the patient's progress is satisfactory he must be referred back to the hospital immediately. It is only in this way that the hospital will be able to control the efficiency of its treatment.

When the works' surgeries are staffed entirely by first aiders without any trained supervision it is questionable whether they should be allowed to carry out any further treatment. Their facilities and knowledge will be restricted and it is better, under these conditions, for the management of the patient to be carried out by the hospital.

If the injury is relatively minor in nature and the patient can give an adequate history there is little point in sending a form to the hospital unless there are special points of relevance concerning the history of the injury, the treatment given or the possible diagnosis. This would apply for example when there might be an intraocular foreign body. A note sent with the patient would help to ensure that this is not overlooked by a relatively junior doctor. Similarly, the nature of any chemicals to which the patient has been exposed should be pointed out.

When the history is of great importance the patient may be unable to give the details himself because of unconsciousness or illness. In these circumstances the history is especially relevant to the diagnosis and full details should be sent with the patient. These should include his name and address, his works number, and his past medical history as far as is known. A history of diabetes requiring insulin treatment for instance or of a previous coronary thrombosis can be of great value.

After head injury, the following observations are required if the diagnosis of a treatable intracranial haemorrhage is not to be delayed:

    (a) the time of the accident,
    (b) the circumstances of the accident,

(c) the observations of any witnesses, e.g., workmates, first aiders, and professional staff with especial reference to:

  (i) the level of consciousness since the injury and any changes that have occurred,
  (ii) the duration of the amnesia—retrograde or post-traumatic,
  (iii) the state of the pupils—constricted, normal or dilated and, if possible, their reaction to light,
  (iv) the pulse rate.

The level of consciousness can be assessed by observing the patient's manner; whether he is alert, drowsy, rational, orientated, confused, or reasonable, and his response to various stimuli, e.g., a normal voice, tickling, slight or severe pain.

In these patients the importance of maintaining an adequate airway is so great that it must receive priority over other injuries, and the training of first aiders must be such that this becomes a routine matter in all cases of unconsciousness. Associated injuries, e.g., wounds and fractures, should receive as much attention as is practical in the circumstances existing at the time but treatment of these conditions must never be allowed to prejudice the maintenance of a good airway.

Valuable though written messages are, in certain conditions where speed is important they are time consuming, and the telephone remains the best method of avoiding delay. The information can be given quickly to the staff who will be treating the patient and at the same time the hospital is alerted and is ready to deal with him as soon as he arrives.

When the patient is unconscious or gravely ill it is preferable for the most highly trained person available to travel with the patient. This ensures that the airway is under constant supervision and that alterations in the patient's general condition are noted. No discredit arises if the patient's condition when he reaches hospital is not so serious as it was thought to be. This is inevitable in about 90 per cent of apparently serious incidents; it is only by constant practice and training that full efficiency can be gained, but it is still much better to have many false alarms than to allow the reception of one urgent case to be delayed because the Accident Department were unaware of its impending arrival.

The training of first aiders

First aid for the injured is no longer a matter for the disinterested. A skilled first aider can do much for the patient and, especially in cardiorespiratory emergencies, can be the means of saving a life. Unfortunately in many instances the treatment given is inadequate and of no help to the patient.

The possession of a first-aid certificate from one of the voluntary organizations is a necessary requirement for a person to be recognized as possessing the skills of a first aider. Unfortunately, the possession of such a certificate does not automatically mean that the person is competent to deal with the many difficult problems that can arise. Each factory has its own problems and whilst first-aid training gives an overall view of basic treatment it cannot and does not deal with the many variables that are encountered in industry.

Less serious injuries such as lacerations and certain minor fractures are, on the whole, dealt with fairly well but the more major injuries, especially those giving rise to interference with breathing, are not handled as well as they should be. To some extent this arises from a lack of familiarity on the part of the first aider with the potential seriousness of the situation. To recognize that an unconscious patient, who is bleeding into the pharynx, is at risk from respiratory obstruction is not appreciated when the patient is breathing quietly and is of a good colour. Appropriate positioning, under these circumstances, is usually badly carried out but it is difficult to see what can be done to remedy this until first aiders are given an adequate clinical training to help them to develop a sense of anticipation. Numbers alone rule out any prospect of this being possible on a large scale but there is no reason why a selected number should not be encouraged to spend some time in hospital Emergency Departments to be instructed in the practical management of certain injuries and disease.

The training programme for first-aid is largely theoretical, though it can be supplemented by the use of photographs, cinematograph films, models and simulated injuries. Unfortunately none of these give the familiarity that attendances at a hospital would produce.

In the past, hospitals have welcomed and indeed benefited from the voluntary assistance given by members of the National Hospital Service Reserve as part of their training. Although this service was disbanded some years ago, there is no reason why persons responsible for first aid, particularly in the larger factories, should not be offered similar opportunities for attending hospitals in order to improve their ability in the handling and the treatment of patients.

It is unlikely that the larger departments, with consultants in charge, would offer any objection to limited numbers of first aiders attending the departments. Indeed this could only improve the standard of first-aid in industry and thus allow the hospitals to use their improved facilities to greater effect.

There are many items of equipment that are in routine use in hospitals which would be of considerable value to first aiders and especially to those working in industry. Training in the use of this equipment is necessary as it cannot be adequately demonstrated on models. It is from this aspect that the staff of hospital Accident Departments have an increasingly important part to play in first-aid training. The basic syllabus is within the competence of most doctors, but when it comes to instruction in the use of such items as suction apparatus and airways it is best for this to be dealt with by members of the hospital services who are using these regularly in their normal work. It is easy to distinguish between those lecturers who are familiar with the apparatus and those who are treating the subject academically. It is essential that first aid is taught from a practical point of view by doctors who are interested in the subject. Only in this way can a proper perspective be given.

The hospital Accident Department has a vested interest in ensuring that the equipment and methods used outside the hospital are adequate and up to date. They should be prepared to take an active part in the efficiency of the local first-aid services. This is particularly true of the ambulances where the drivers and attendants are professional first aiders of the highest order.

The use of airways and suction apparatus in ambulance vehicles is increasing largely as a result of action by hospital medical authorities. These items and a knowledge of their use should also be available to first aiders in industry. The cost is small but the value is great.

Inflatable pneumatic splints are open to certain objections but with adequate training and knowledge of the risks they can be of great value.

The use of apparatus to administer oxygen either continuously from a cylinder or through a portable respirator requires a great deal of care if it is to be of value. The standard of instruction in the use of this apparatus is high but the standard of use is low. It is used under conditions where no benefit can be expected; it is used inadequately when it could be of great value.

In recent years two important developments are becoming apparent when industry is looked at from the hospital point of view. There is an increasing use of highly complex and toxic chemical material, and a rise in the incidence of coronary thrombosis, amongst all grades of industrial workers, can be seen.

The survival of patients suffering from the results of exposure to toxic substances depends on skilled first-aid at the scene of the incident, followed by complex hospital treatment to deal with the medical problems raised by the chemicals. All first-aiders should be aware of the hazards raised by any chemical substances used within their factory. They must be provided with detailed instructions setting out the appropriate first-aid measures and be thoroughly proficient in the application of these measures to the patient. This is particularly relevant to substances which may interfere with breathing.

There are many types of equipment now available which will provide efficient and simple methods of maintaining respiration. Familiarity with such apparatus is essential if the patient is to survive. Reliance cannot be placed upon the ambulance service for this help; many patients will die before this service arrives and it is upon the industrial first-aiders that the survival of these patients depends. Factory management cannot be divorced from their responsibilities towards first-aid, because they must ensure that those given this responsibility are fully trained and provided with the necessary equipment and knowledge of the specific local hazards that they may meet.

When patients who have been subjected to toxic substances are brought to hospital, it is essential that the person accompanying the patient should bring with him the necessary information about the nature of the product, the toxic reactions that might be expected and methods that are recommended for the management of the patient. However well-equipped an Emergency Department may be with works of reference, there can be appreciable delays in identifying the precise results of exposure to chemical hazards and if this information is available in the factory it should automatically be sent to hospital with the patient.

Coronary thrombosis is on the increase and deaths are much commoner than they were some years ago. After the age of 40 survival after apparent death from coronary thrombosis diminishes rapidly with increasing age but in the younger age groups, provided the circulation to the brain can be kept working by the use of assisted respiration and external cardiac massage, the patient has a reasonable chance of living.

These methods of life support are well recognized and taught on all first-aid courses, but it is depressing to see how often, in practice, the application of these methods fail.

The majority of first-aiders are able to perform external cardiac massage without difficulty, but very few appear fully competent in maintaining the ventilation of a patient's lungs. There is still a lack of appreciation of the necessity to maintain oxygenation of the blood in these patients, and many patients are seen in whom the first-aiders go through the motions accurately but fail to produce any movement of the chest indicating inflation of the lungs. External cardiac compression without ventilation of the lungs is useless, and the importance of being able to maintain this still needs a greater emphasis on the practical side than it already receives.

The 'quality control' of first-aid measures can only be carried out in Accident and Emergency Departments. This is a strong reason for these units playing a greater part in the training of first aiders than they have done hitherto because it is essential that the standard is raised to a much higher level than has so far been accepted. It is also necessary that those persons who are responsible for first aid in industry, where there is a constant high accident risk, must have a higher level of competence than the occasional first aider. The gaining of an advanced first-aid certificate, valuable though it is, is not the final answer to the problem. The degree of knowledge required for this certificate is high but, unfortunately, the possession of such a certificate does not necessarily imply competence when using the knowledge in the real life situation. One answer to this problem might be to restrict the granting of the advanced first-aid certificate to those persons who have undertaken a short period of attendance in hospital Emergency Departments. This would go far to ensure that the possessor of an advanced certificate was not only knowledgeable in theory but also reasonably skilled in applying it to the injured person.

Hospital emergency team

From time to time in factories an injured workman is trapped either in machinery or under fallen debris. If his release is liable to be prolonged or difficult and his life is considered to be in danger, most hospitals are prepared to supply the services of an emergency team. Such a team has been functioning in the Preston area since 1963. Its main value has been in dealing with the victims of road accidents, but it has also answered calls to works incidents. These have all involved employees trapped following an accident and, on some occasions, the team has been of considerable help.

The team consists of a senior doctor and a senior nurse. It is usually called out by the ambulance authorities though it would be prepared to go out at the request of a responsible person in the factory.

The aim of the team is to collaborate with the other rescue services in freeing the patient from the machinery as quickly as possible and, at the same time, treating shock and other conditions which might be present.

The team is able to leave the hospital within 2–3 min of receiving a call and has a full set of equipment which will allow any necessary procedure to be carried out. The equipment is carried in small fibre boxes for ease of handling. It is in two basic sections. The resuscitation equipment includes plasma and dextran with the appropriate giving sets and transfusion stands. The Stephenson minuteman has been modified to enable oxygen/fluothane or oxygen/trilene anaesthesia to be given using face masks or intubation equipment. The tubing allows anaesthesia to be given up to 15 ft (4·5 m) away from the minuteman. An extension tube is carried in the lid of the minuteman to enable the sucker to be used at a similar distance away from the machine.

The other equipment consists of dressings, drugs and syringes, surgical instruments and inflatable splints, and torches. Protective clothing consists of duffle coats, elastic topped golfing trousers for the nursing staff, fluorescent jackets and protective headwear.

There is no point in calling out the team if the patient can be freed within 2 or 3 min. The situation is usually assessed by the hospital doctor before the team goes out; in the case of road accidents the radio-link which connects the department to both the County and Borough Ambulance Services is used for this purpose.

There are several points about the management of trapped victims that are of importance. The free use of morphine or pethidine as a pain reliever is very dangerous. If the patient is in any way shocked the drug will have little beneficial effect unless it is given intravenously. If, because of an apparent lack of response when given by other routes, several doses are given within a short time, as soon as the patient's condition improves following adequate resuscitation, all the doses will be absorbed simultaneously. This results in marked depression of the respiration, masking of abdominal symptoms and signs, and makes the diagnosis of certain intracranial lesions impossible until the effect of the drug has worn off. On two occasions patients have been admitted to the Accident Department with severe injuries and associated morphine poisoning, due to the administration of unnecessarily large doses of the drug. Our experience suggests that the majority of patients immediately after injury do not require more than minimal sedation if, indeed, any is needed at all.

The release of a person from machinery is not a medical matter. The doctor's function, apart from giving comfort and necessary treatment to the patient, is to advise and to assist, as required, the other services who are involved in the rescue. Heroic surgery, e.g., amputation, should not be undertaken to release a patient unless there is a high risk of further danger to both the patient and his rescuers as, for example, by debris falling on top of them.

Patients trapped in machinery can be given a blood transfusion before extrication. When this has been done the patient's condition on arrival at hospital is very much better than it would otherwise have been. Because the patient is under treatment the return journey to hospital can be made at a very slow speed. The seriously injured do not stand up well to a high-speed journey which entails jolting and swaying in the ambulance, and one of the values of the team is that this harmful and unnecessarily traumatic procedure can be avoided.

The Emergency Team is being called out with increasing frequency to persons who have collapsed, frequently following an attack of coronary thrombosis. If the patient is receiving external cardiac massage and adequate ventilation, calling out the team may cause delay in transporting the patient to hospital. When possible this delay should be avoided. However, there are patients who have suffered a major coronary attack, who are in a state of severe shock and who are suffering great pain. In these patients the team is able to provide relief from pain and supportive treatment with drugs and oxygen to allow transport to be carried out with less risk of harm to the victim. They are monitored during the journey, which can be made slowly, and the risk of ventricular fibrillation is diminished. It is in this latter group of patients that the Emergency Team can be of the greatest assistance, and in several areas of the country coronary care teams, whose sole purpose is on-site treatment and transport to hospital, have been formed.

The incidence of coronary thrombosis, particularly in the younger age groups, is increasing and if a coronary care team is operative in any area it is advisable to discuss with them the indications for the use of the service and the 'call out' arrangements.

From time to time patients collapse from causes which are not immediately clear, for example a worker may take a drug overdose at a factory. It may be thought that the patient's transport to hospital would be benefited by medical attention before and during the transport. In these circumstances it is always advisable to discuss the problem with the hospital doctor responsible for the Emergency Team.

The final decision as to whether or not the team can be of assistance must be made by the hospital, but it will be found that requests for assistance in unusual circumstances are always dealt with sympathetically.

Conclusions

The hospital Accident and Emergency services and the Industrial Medical services are complementary to each other in the treatment of the injured workman. Lack of personal contact between members of the different services produces a gulf which can be overcome by regular meetings between the two authorities. The Accident Department has a large part to play in assessing the standard of treatment given by first aiders, and endeavouring to see that the standard of this treatment is constantly raised. The hospital can assist the works services with advice about equipment and methods and can also be of great help in assisting with certain aspects of the training.

Reduction in the death rate following major accidents can come from a greater knowledge of first aid on the part of those people who initially deal with the patient. The recent advances in all branches of medicine and surgery are now beginning to be applied in the Accident and Emergency Department but, unless the patients can be brought alive to hospital, these facilities will be of only limited value.

References

1. *Casualty departments and their setting.* Published for the Nuffield Provincial Hospitals Trust by Oxford University Press, 1960.
2. *Report of the subcommittee on accident and emergency services.* Ministry of Health Standing Medical Advisory Committee of the Central Health Services Council. H.M.S.O., London, 1962.

35. Setting standards for safety

R. Berry

Group Manager,
British Standards Institution

A distinction needs first to be drawn between personal protective equipment which can rightly be termed 'safe' in the absolute, and that which can achieve only the comparative description 'safer'. The nature of a possible accident and the degree of severity of the consequences to the individual can, in most cases, be fully assessed. In these cases, protective equipment properly designed, constructed, and maintained, will prove adequate and safe. However, there are a great many operations or activities in industrial practice for which a claim of providing absolute safety or complete protection cannot be justified. A combination of adverse circumstances, or extreme intensity of accident conditions, can reduce or nullify the effectiveness of the protection given. Fortunately these extremes are the exception but they underline the vital need to provide the safest possible equipment which will, even in the extreme case, considerably lessen the degree of injury suffered.

This is the challenge to research scientists, designers, and manufacturers in the safety equipment field, in the same way that their counterparts who are concerned with the means of production, endeavour to reduce the possibility, the level of incidence, and the effect of operation fault, or human error, which could produce accident conditions. Self contained processes and other forms of automation have in some industries brought danger to a minimum, but the main field of industry still needs to invoke the four fundamental barriers against accident and injury: training and instruction; intrinsically safe design of machine or process; protection devices; personal protective equipment. Protective equipment is the last barrier against injury. Such equipment needs to be as efficient in fulfilling its purpose as design and manufacture can make it.

The user or wearer must be assured of the quality of his equipment—his helmet, safety harness, or respirator—but he himself does not normally buy the equipment. A supplies officer, advised by an experienced safety engineer, may find great difficulty in selecting, on a firm basis of knowledge, from the designs of different manufacturers with a wide range of prices. In many cases it is impossible for the purchaser to test the equipment in accident conditions. The test apparatus is often specialized and costly, and the test likely to be destructive or render the equipment unfit for use. He often, justifiably, relies on a manufacturer's reputation, but not all buyers are experienced or discriminative. The protective equipment industry, like others, is highly competitive and keen salesmanship must play its part. Price, whether high or low, is not a scientific criterion of quality and performance. The purchaser needs another basis by which to assess the product offered, in a word—a standard. If he can stipulate when ordering that the protective equipment must be made to meet the requirements of a recognized standard he can form his own assessment beyond that point.

The use of such a basis will have further significance both to the purchaser and the manufacturer of protective equipment. In cases of accident and injury submitted to the courts, the requirements of a nationally accepted standard may be referred to in legal proceedings as a guide to the difficult problem of determining whether adequate protection, as far as possible, had been provided.

In the U.K., standards of quality and performance are set by the British Standards Institution for many items of personal safety clothing and equipment. These specifications are not mandatory either to the maker or the user, but they are fully recognized as authoritative documents both in industry and the departments of government.

The philosophy behind the preparation of a British Standard is to base the work on the concept of absolute protection, even though this may in the ultimate be unattainable, and to set the requirements of performance at the highest level that existing knowledge of materials, design and construction will allow. By this philosophy the target remains unchanged and manufacturers making equipment to the Standard are free to improve the product by research, development, and production techniques. A purely dimensional Standard would tend to restrict such advances, though minimum dimensions are often specified to ensure quality for features where performance testing would be difficult, for example, when tests are not reproducible, or on balance are uneconomic.

One difficulty faced by a national standards organization lies in constantly reviewing the positions reached by one or more manufacturers in improvement of a product, and the timing of the introduction of higher levels of requirement in the Standard. Such improvements may be the result of considerable research and expenditure by one maker, but with a reasonable period unavoidably allowed, during which time the advance is proved in practice, individual manufacturers in the safety equipment industry recognize the desirability of resetting the performance in the national Standard with as little delay as possible. The change to the Standard is often introduced to become operative on a forward date, by which time supplies to the new standard requirement will be available to purchasers.

In setting standards for protective equipment, there are certain fundamental requirements which must be taken into account. The weight of the equipment, whether worn or held, must not be such as to cause fatigue, or the equipment will just not be worn. The Standard must not preclude attractive design. Materials and design must not only provide the required performance initially, but should have a reasonable wear factor. Even for safety equipment, due thought must be given to the price of the product. Realism dictates that an item of equipment made to give marginally higher performance only by using excessively costly materials will not protect anyone if it is not bought and if it is not worn.

The primary concern of those setting the standards is a high standard of protection, as some of the limiting factors may be transient in effect. New materials can reduce weight without loss of strength. Designs can improve in appearance and become accepted on a popular trend. Surface treatments and construction techniques can be developed to improve the wear factor. Price can be affected by the high quantities sold; the result is reciprocal.

The level of performance is seldom set only on the basis of the strength of the product. Strength may easily be achieved in the case of helmets or harness; the true requirement is absorption of the force applied. The object is protection of the human body which has its own limits of resistance to impact or deceleration. A helmet lacking sufficient absorption quality may save a head but not prevent a broken neck. A safety belt may injure the wearer if the level of performance has been set 'out of balance'. Vision may be reduced by use of a goggle lens which has great strength but lacks optical quality. The Standards are set by specialists in each type of equipment.

British Standards are prepared by Technical Committees which represent all the interests concerned. Members of individual firms, both the makers and the users of the equipment, are appointed to represent relevant trade associations; scientists attend from research laboratories; interested government departments are represented; specialist advice is contributed by the medical profession. The work of committee members is financed by their organizations, the British Standards Institution providing the administration, secretariat facilities, and knowledge of the technique of drafting and producing standards. Copies of draft Standards are circulated to organizations all over the world for criticism before the Committee completes its work.

The British Standards Institution (B.S.I.) was founded in 1901 and has operated under a Royal Charter since 1929 as the sole national standards organization of the U.K. The organization itself is non-profit making and is financed from four main sources: contributions and membership fees from industry and large-scale purchasers; a grant from central government; revenue from sales— more than 3 million copies of standards are sold each year; and receipts from services of testing and approval. It is essential that the policies and practices of the B.S.I. are not open to influence from financial pressures from any source. Standards are published on the basis of their value to the country as a whole and on their technical merit. The size of the organization increased five-fold between 1950 and 1968—the total number of active and standing committees exceeds 3 000—whilst the demand for standards, in all fields of industry, has outrun even this expansion.

The same period saw a similar expansion in the work of the two major international standards organizations: the International Organization for Standardization (I.S.O.) and the International Electrotechnical Commission (I.E.C.), both based in Geneva. These organizations prepare internationally agreed Standards for the national standards bodies (numbering some 55 nations) to reproduce in whole or in part in their own Standards, or to sell as I.S.O. or I.E.C. Standards. The value of this work lies in the interchange of technical and scientific knowledge and in reducing barriers to international trade. The B.S.I. is the member body for the U.K. and plays a very great part in the activities of both I.S.O. and I.E.C.

International standards work is organized through Technical Committee Secretariats held by member bodies, each dealing with a particular activity, which may be an industry—textiles, steel, aircraft—or an individual product or concept—gears, safety colours, welding. The B.S.I. holds the Secretariat of the I.S.O. Technical Committee on 'Personal Safety: Protective Clothing and Equipment' and is the Secretariat of many Sub-committees and Working Groups, such as 'Safety Footwear', 'Industrial Safety Helmets', 'Safety Belts', and 'Life Jackets'. The I.S.O. Committee structure of TC94, for the different subjects, is similar to that of the B.S.I.'s own technical committees on personal safety equipment; but the I.S.O. committees are of course made up of the various national member bodies most directly interested. U.K. delegations from B.S.I. committees attend all the meetings concerned in this work. Improvement in the protection potential of safety equipment is the main purpose of this pooling of international knowledge and research.

Recently, in the national sphere, the need has been recognized for much greater co-ordination of research and development work on protective clothing and equipment. Not only between the various research laboratories—which may be within a trade association, individual company, government department, or university—but also between the manufacturer and the 'user'. The personal protective equipment industry, with so many quite distinct products, is by nature fragmented in its activities, and national sources of research knowledge are not utilized on a collective basis.

The financing of research projects sometimes presents problems if the project is too large or expensive for one manufacturer or a group of manufacturers to undertake. B.S.I. has an arrangement with its sponsoring Government department whereby, in fully justifiable cases, the cost is shared between Government and industry.

The results of research and development are introduced into British Standards in the simplest form digestible by the industry—in standard methods of reproducible tests and in the specification of higher performance, made possible by the work of the laboratories.

Most British, and some foreign, manufacturers of personal protective equipment make their products to conform to the British Standard specification. Criticism is sometimes made by manufacturers that the level of performance requirement in the British Standard has been overtaken by developments in their companies, or that new techniques of production, previously unforeseen and resulting in improved performance, are not permitted by the Standard. The Standard may have had to lay down methods of construction to ensure lasting quality as distinct from test results on the product as new. Amendments improving the Standard can, in fact, be issued as soon as the Committee has the information and can accept the claims as viable. It is interesting that in the case of safety boots and shoes, both rubber and leather, some 90 per cent of sales are of footwear made to comply with B.S. 1870. The specification is successfully kept up to date even though the Standard deals necessarily with details of construction.

The purchaser of any item of safety equipment, anxious to ensure that the product is well designed and made, can order to the relevant British Standard, selecting sizes and types and perhaps adding special requirements. He will be assured, and so will the wearer, by the mark of the British Standard number which appears on the product, as required by the Standard. This is a claim by the maker that the product complies with the Standard's requirements. To make this claim the tests in the Standard must have been carried out successfully.

Two problems arise, as regards testing and as regards the guarantee of the quality of every product. Testing in modern practice sometimes demands sophisticated and costly equipment, e.g., the fluorescent test for permeability of materials against toxic liquids, the dynamic testing of vehicle seat belts, the efficiency of respirators. Many of the manufacturers of safety equipment are small firms and cannot afford the capital expenditure for the necessary tests.

Guarantee of every item produced becomes virtually impossible for equipment which can only be properly tested in a manner which renders the product unserviceable or, in some cases, destroys it. It is vital that the materials, design, and construction of a prototype of safety equipment should be tested in accident conditions, though not necessarily at the extreme of possibility, against which no known equipment may protect. Tests which harm or destroy the product present a problem as to how many items, as a sample from a production run, are necessary to provide reasonable assurance as regards the quality of the remainder.

The first difficulty, of each manufacturer providing specialized test equipment at his own works, suggests the need for centralized test facilities. Tests of materials can usually be done by industrial test laboratories, for example textiles, leather, or plastics; but the tests of the finished product in accident conditions are special to each kind of equipment and may be better concentrated at a national test centre which can provide the equipment, test, and certify the product on an authoritative basis.

The extent of sample testing, following prototype approval, needs to be on a statistical basis—even though this must recognize a level of failure possibility; but the sample testing must be co-ordinated with an inspection and quality control scheme, applied during production, which will reduce that level to a minimum whilst still keeping the product economically viable.

The B.S.I. provides centralized test equipment and carries out the tests required for a number of different kinds of personal safety equipment at the B.S.I. test centre. These are tests required by the British Standards for certification of the product under the 'Kite Mark' scheme, a scheme of approval registered with the Board of Trade. For Kite Mark schemes applied in certain industries, such as safety footwear, testing is done by the trade's research association on their test rigs, again on a centralized basis.

The B.S.I. scheme is voluntary on the part of manufacturers, except in the case of vehicle seat belts, and involves: prototype testing; quality control; inspection, both by the manufacturer and independently by the B.S.I.; statistical random sampling for examination and test; and an agreement regarding changes in materials and design. The maker of approved equipment can use the mark itself on the product to denote this independent approval. The purchaser—and the wearer—has evidence that the claim of compliance with the British Standard is supported by the B.S.I.

The range of British Standards for personal safety protective clothing and equipment is second to none in the world. Specifications are available for safety boots and shoes; for helmets; goggles and face screens; respirators and dust masks; safety belts; gloves; flameproof and intense-heat clothing; industrial life jackets; foul-weather clothing; welder's safety equipment. Others are in preparation. The Standards, which are recognized throughout the world, are listed for reference in the *B.S.I. Yearbook* and in a special leaflet separately.

There will always be a need in the provision of personal protective equipment for the product to be made safer and more readily accepted by the person at risk. In the U.K., manufacturers generally in this field are conscious of their special responsibility: the level of co-operation towards this end is high. The work over the years to establish standards for tests and performance, prepared and agreed by all interests; the constant effort to improve those standards; international co-operation; the centralized facilities available for testing and independent approval; and now the harnessing of research and development to this effort, are all indicative of the co-ordinated endeavour being made towards the increased effectiveness of this last barrier against injury and all that injury entails.

36. The role of government in occupational safety

W. G. Symons

Formerly H.M. Superintending Inspector of Factories,
North Western (Liverpool) Division and Midlands (Birmingham) Division

In the very early days of industrial legislation, it became evident that legislation by itself was not enough. The Factory Commission of 1833 made a survey of conditions in the developing textile areas, and found that earlier factory legislation had been 'almost entirely inoperative with respect to the legitimate objects contemplated by it'.[1] It recommended the setting up of a professional inspectorate, appointed and paid by the government and responsible for enforcing the law. As a result, the Factory Act of 1833 authorized King William IV 'by warrant under his sign manual to appoint during His Majesty's pleasure four persons to be Inspectors of Factories'[2]—a daring innovation at that time.

The first four 'H.M. Inspectors' so appointed were a vigorous group of men, who had to work out a new role without precedents or guide-lines. They made mistakes, but they also made a strong impact on public life; their reports became vital social documents of the industrial revolution. The Hammonds summed up the importance of the inspection provision in the otherwise very limited Factory Act of 1833 in these words—'The Government had wanted to do as little as possible in the way of a Factory Act; they had done a great deal more than they intended'.[3]

The development of government labour inspection

From these pioneering beginnings the British Factory Inspectorate initiated a form of government action which is now to be found in almost every country in the world, and which received international support in 1947 by I.L.O. Convention.[4] In the newly developing countries, as well as in those with a longer

industrial history, we now expect to find qualified government 'labour inspectors' (to use the I.L.O. phrase) maintaining organized inspection of workplaces and working conditions as part of the effort to improve health and safety
at work. A striking feature of these various national inspectorates is how they
have come to share a common professionalism, in spite of wide differences in
their origins, and in the social, political and legal systems of the countries in
which they operate.

In Britain, this system developed until recently in a characteristically prag-
matic and piecemeal fashion. The factory inspectors, originally concerned
solely with textile factories, were given wider duties, especially after 1867,
until they became responsible for inspecting factories of all kinds, building
work, civil engineering and dock work. At the same time, other bodies of law,
with their own enforcing inspectorates, came into being to cover further areas
of work and risk, such as mining, railways, explosives, petroleum and, much
later, agriculture, offices, retailing and nuclear installations.

After the Second World War, it became evident that this piecemeal approach
was inadequate for the speed and scope of technological developments, and to
meet growing social awareness. In 1970, the Robens Committee was appointed
with wide terms of reference to review provisions made for the health and safety
of persons in all kinds of employment (except in transport operations) and for
protecting the public against hazards from industrial and commercial activities.
The Report of this Committee[5] became the basis of the Health and Safety at
Work, etc., Act 1974, which came substantially into force on 1 April 1975.

In addition to its new legal provisions (which are discussed in Chapter 39),
this Act makes sweeping changes in government organization, including
arrangements for inspection. At the time of writing, these organizational
changes are not yet complete, but the present period of change provides opportunity for taking stock—for looking at past experience and at hopes and prospects for the future. For reasons of space, we shall confine ourselves almost
exclusively to the experience of factory inspectors, but other inspectors (such as
Inspectors of Mines and Quarries) should not be forgotten. These each have
their own bodies of experience and will also be involved in the changes which
are in hand.

The professional work of a factory inspector

The first duty of factory inspectors was to enforce the law; but, from 1833
onwards, there has been a continuing debate as to how the word 'enforce'
is to be understood. Should inspectors simply look for breaches of the law, and
prosecute when they find them? Or should they give their main efforts to advice
and persuasion, having recourse to the courts only as a last resort? On the whole,
the British Factory Inspectorate has followed the latter line; it has measured
its success not by the number of convictions secured but by how far the sub-

stantial aims of the law are achieved. An important discussion of this subject is to be found in Chapter 9 of the Robens Report, which explores the scope and limitations of strict legal enforcement as a means of improving working conditions.[6]

Inspectors have come to accept it as part of their duty not only to explain what the law says, but also to advise as to how its aims can be best achieved. As a result, the Inspectorate now provides a technical advisory service about health, safety and welfare at work. Its headquarters office has become a clearing house for information. It publishes, through H.M. Stationery Office, a wide range of advisory material for the use of employers and employees. It operates an Industrial Health and Safety Centre at Horseferry Road, London. Inspectors are heavily in demand as technical lecturers for societies and institutions concerned with occupational health and safety. More important, though perhaps less obvious, inspectors during their ordinary work receive countless queries which they have to try to answer out of their own experience or from the wider resources of the Inspectorate.

Underlying all this is a continuous responsibility to be alert for anything (whether mentioned in the law or not) which may affect the health or safety of employed persons—to be, in effect, a skilled watchdog on their behalf. The word 'skilled' is used advisedly. One of the most distinctive skills of an ordinary inspector in the field is the ability to recognize a risk or matter needing attention, often in an unfamiliar form and an unexpected place. This is an essentially detective or diagnostic skill, which is quite different from that of applying detailed regulations.

It would be rash to claim that factory inspectors, either individually or as an organized service, have always performed these functions as well as they should. Inspectors are but human, and the Inspectorate has not always measured up to the demands made upon it. However, a study of the way in which labour inspections has developed in many countries gives strong ground for believing that its success depends upon a combination of these three complementary elements—of effective legal *enforcement,* of sound technical *advice,* and of skilled *watchfulness.*

These three elements have to be kept in balance. As the Robens Report wisely reminds us, the primary responsibility for preventing occupational accidents and disease 'lies with those who create the risks and those who work with them'.[7] The wrong sort of free advisory service could have the result of encouraging industry to unload on to government the task of finding answers to all the health and safety problems which industry itself has created—which would be not only wasteful, but damaging. Even to-day, there are still too many managements who are content 'to leave it to the inspector' to tell them what to do about safety, without taking any initiative themselves. Advice needs to be linked with sanctions which can be used, when necessary, to bring home to managements and employees their direct responsibility for tackling their own health and safety problems. Once that responsibility is recognized, an adequate inspectorate can do much 'to help industry to help itself', and so can contribute to the broader effort for health and safety at work.

The development of the work of factory inspectors over the past 142 years has therefore resulted in a wide expansion of their areas of professional concern. They are still responsible for regulating hours of work, but in addition have had to develop specialized expertise in three broad areas—in accident prevention, in the prevention of occupational disease, and in matters of welfare and of the general working environment.

Accident prevention

Factory inspectors had at first no direct responsibility for safety. They were soon aware, however, of the dangers of unprotected machinery in textile mills, and of accidents which occurred; and they reported these things to the Home Secretary. The first amending Act, in 1844, contained specific safety provisions. Inspectors became increasingly involved in discussions on safety standards— discussions which led to some heated controversies with employers and their organizations.[8] This early example illustrates the two-way relationship between inspection and legislation which has characterized the whole process. The inspectors were not content to be simply neutral enforcing officers; they saw it as part of their professional duty to inform and to stimulate legislation to deal with the practical conditions they discovered in their work. From this beginning, occupational safety has become one of the main concerns of inspectors.

The basis of sound accident prevention must be experience of previous accidents. For this purpose, any accident to an employee in a place subject to the Factories Act has to be reported if it causes more than three days lost time. These reports are used by the Inspectorate for statistical studies and for selective investigation. It should be emphasized that inspectors are not directly concerned with compensation or civil liability; their investigations are directed solely to enforcing safety law, and to gaining experience for preventing accidents.

Accidents are of many kinds. Some can be prevented by sound regulations properly observed; others can hardly be prevented by the most elaborate legislation. Different kinds of accidents have differing 'severity potentials'— that is, certain types, though not necessarily causing severe injury in every case, are more likely than average to inflict severe injury. In factories, three broad groups of accidents have relatively high severity potentials:

(a) accidents from power-driven machinery, including mechanical lifting and handling;

(b) the 'technical' hazards of fire, explosion, gassing and electricity;

(c) accidents due to persons and articles falling from above ground level.

Accidents of these three types cause about one quarter of the lost time injuries in factories (and therefore of total reportable accidents), but they produce about a half of the severe injuries (resulting in lengthy or permanent disablement) and 90 per cent of fatalities. The continuing concern of inspectors with machinery safety has a sound basis in experience, which is sometimes obscured by using total accident figures without taking account of severity.

The approach of inspectors has therefore been selective. They have not felt themselves to be concerned equally with preventing every type of accident which may conceivably happen; they have concentrated deliberately on a limited sector—on those things which may cause the graver and more insidious risks, and which may be more effectively controlled by legislation. To make this distinction is not to ignore accidents of other sorts—e.g., from hand tools, manual handling and falls on the level. These have, however, to be tackled in other ways, and may not involve inspectors so directly.

Methods of approach need reassessment as circumstances change. Traditionally, inspectors have worked mainly by periodic surprise visits to premises, a method which concentrates attention on things which can be checked visually during a short visit. They have tended to be chiefly interested in the state of the buildings and hardware. Is this approach still appropriate to-day? Much of our modern and sophisticated plant presents few obvious risks during routine production when all is going well, but may cause serious hazards on occasions— when loading, cleaning or from some breakdown in the process. Only too often, these hazards receive attention only after there has been a major accident or disaster. Safety standards need to be worked out between technical specialists, mid-management and operatives, taking into account the whole cycle of production, and possible emergencies and deviations. These and similar developments point to the need to supplement traditional surprise visits, by methods of inspection which can probe more deeply into techniques, processes and systems of work, and which can involve higher technical management, especially at planning and design stage. How this is to be organized with limiting inspecting staff, operating in a wide and diversified field, is one of the problems of the future.[9]

Fire hazards deserve special note, as they are of two distinct types:

(a) from semi-explosive process fires, which may involve persons in flames in a few seconds;
(b) from persons being trapped by smoke and fire spread after a building has caught fire.

Factory inspectors have been closely concerned with risks of the first type, which are almost certainly an increasing risk today, with the widespread use of petrochemical and other organic compounds, many of which are highly flammable. Risks of the second type are similar to those in buildings of all sorts, and are the primary concern of the County Fire Authorities, who are normally responsible for certifying safe means of escape and other precautions. Changes introduced by the Health and Safety at Work Act are likely to simplify this division of responsibility, although there will still be need for collaboration between factory inspectors and fire officers.

Occupational diseases

We have known for a very long time that certain diseases have been linked with certain occupations—names like 'potters' rot' and 'painters' colic' bear witness to this. These recognizable occupational diseases were mainly due to work-people absorbing cumulative poisons by breathing dust or fumes. During the early years of the present century, the Factory Inspectorate were much concerned with identifying and controlling the use of a fairly limited range of 'industrial poisons' known to cause such diseases. These poisons included lead compounds, white phosphorus, silicious dusts, and chromic acid. They were mainly encountered at that time in certain 'dangerous trades', which were made subject to special regulations—such as those covering potteries, chrome plating, asbestos processes, grinding of metals and vitreous enamelling. During this period, there was a spectacular reduction in the incidence of some of these diseases—for example, from 1900 to 1938, the number of reported cases of industrial lead poisoning fell from over 1 000 per year to under a hundred; during the same period, phosphorus poisoning and industrial arsenical poisoning were virtually eliminated. These successes, when contrasted with the slow changes in accident figures, sometimes created a sense of complacency.

More recently, and especially since about 1950, the subject has opened up on a far wider front. New synthetic materials are increasingly coming into use, not merely in the previous dangerous trades, but in work of all sorts—as solvents, adhesives, cleansing materials, coating agents, sterilizers, etc. Some of these are quite safe, but others may create new health hazards in unexpected places. We have come to realize that some materials and conditions of work may increase liability to diseases which are not exclusively occupational in character—such as bronchitis, dermatitis, allergic conditions and some cancers. The public generally are becoming increasingly aware of problems of health and disease. Consequently, many workpeople are puzzled and apprehensive as to whether their work may be affecting their health.

These developments have coincided with large changes in the health services, both on the curative and on the prevention sides. During this process, there has been widespread debate on how the health services and medical expertise should be related to people at work and to industrial activities, and as to the role of government in this relationship. This debate still continues. Four elements in the relationship have been of particular concern to factory inspectors in dealing with occupational disease:

(1) The first relates to the role of *medical practitioners* in this work. Prior to 1972, the Factory Inspectorate had a small Medical Branch, comprising

inspectors with medical qualifications, and also made use of Appointed Factory Doctors who were mostly part-time and had limited duties under the Factories Act. Since 1972, these have been merged into the Employment Medical Advisory Service, distinct from the Factory Inspectorate, but maintaining close liaison with it. On the non-governmental side, many firms employ Works Doctors or Company Medical Officers of various types. The Robens Committee estimated that about 600 doctors were then employed full-time and possibly 2000 part-time by private and nationalized industry, though it was plain that the duties of these doctors differed widely in different undertakings.[10]

(2) Sound preventive work depends on knowledge of cause and effect, and this in turn depends on *medical research*. Fundamental research on the effects of working conditions and materials on the human body is essentially a part of organized medical research. The Factory Inspectorate is concerned to encourage such work in areas of special urgency and to keep liaison with results of research, both nationally and internationally.

(3) *Occupational hygiene* is a modern specialization concerned with assessing and controlling hazards from atmospheric contamination, skin absorption, radiation, noise, etc. Standards depend on results of medical knowledge but hygienists need not be medically qualified, and normally have qualifications in physics and chemistry, including techniques of atmospheric analysis. Tables of threshold limit values (T.L.V.s) are available, giving quantified indication of the degree and kind of hazards associated with a very wide range of materials used at work.[11] The informed and intelligent use of these data is an essential part of the skill of an occupational hygienist. The Factory Inspectorate set up an occupational hygiene unit in 1966, which has since been much developed. Hygienists are employed by some larger firms, and units are attached to some universities and other bodies.

(4) A special problem of the Factory Inspectorate is the *primary inspection* of areas where occupational health risks are unlikely, but may conceivably occur. In the past, primary inspection of occupational health hazards has been based on inspectors' knowledge of the various trades, assessing particular risks by visual inspection, with the possibility of calling in specialist help in urgent cases. To-day, health hazards are far more widely dispersed; they are not necessarily linked with particular trades, nor even confined to industry. It is hard to ascertain the absence of risk simply 'by sniff and by guess'; yet it would be virtually impossible to try to apply the whole range of sophisticated occupational hygiene techniques universally in every workplace. Techniques of detection and of rough assessment are therefore needed, and some are being developed, to help to identify areas of potential risk, which may need more thorough investigation subsequently.

These aspects of the work of factory inspectors are developing rapidly, and are likely to be the subject of increasing public concern.

Welfare and general environmental conditions

The Factories Act requires suitable welfare arrangements (washplaces, cloakrooms, etc) and a healthy and comfortable working environment (as to temperature, ventilation, lighting, etc). Most of these requirements are straightforward and non-controversial, but they sometimes receive less attention than they deserve, perhaps because defects do not make themselves evident in accident figures or in cases of recognizable occupational disease. Yet, for most people most of the time, discomfort at work matters more than what may seem a remote risk of accident or disease.

Factory inspectors are only too aware of much needless discomfort at work, not only in factories but also in offices and commercial premises. Many workrooms are uncontrollably too cold or too hot for parts of the year. Ventilation in parts of a room may mean discomfort in other parts. Some of our newer buildings are among the worst offenders in these respects. Unfortunately, inspectors normally discover uncomfortable conditions only when buildings are complete and when remedies will be makeshift and costly. They develop some skill in trying to make intolerable conditions tolerable; but they have little opportunity to concern themselves with these things at design stage. This subject may come into greater prominence in the future.

Hours of work

The regulation of hours of work was historically the first responsibility of factory inspectors, and is still one of their duties. The Factories Act restricts the hours of employment of young persons (under 18) and of women, but not of men over 18. The form of the law results in sharp differences between restrictions applying to factory work and to other comparable employments. This subject was not within the scope of the Robens Committee, and is likely to receive further consideration along with the wider question of the protection of children and young persons in industrial and non-industrial employment generally.

Prospects for the future

We have so far been looking, in very broad outline, at some of the experience of the British Factory Inspectorate up to the present time. What of the future, in view of the large changes which are now taking place? Doubtless much will continue as before; but there are some points where significant developments are in prospect. All that can be attempted here is to pick out a few of these, recognizing that it is far too early to make any reliable forecast as to how they will work out.

The new structures

The new administrative machinery set up by the Health and Safety at Work Act has been well publicized and will be familiar to most readers. The Health and Safety Commission is responsible for general policy, including the initiation of research and the preparation of regulations and codes of practice. The Health and Safety Executive is the operational arm, which will deal with individual employers and workplaces. The staff of the Executive includes certain inspectorates which were formerly attached to various government departments—i.e. Inspectors of Factories, Mines and Quarries, Pipelines, Nuclear Installations, Alkali and Clean Air and Explosives. It includes the Employment Medical Advisory Service and scientific support services such as the Safety in Mines Research Establishment.

The Factory Inspectorate therefore becomes one of the component inspectorates within the Health and Safety Executive. At the present time, these inspectorates maintain their distinctive functions and organizations, although it is plainly intended that they will work together more closely. The local District structure of the Factory Inspectorate is being reorganized into larger areas, as recommended in the Robens Report.[12]

Unless otherwise provided, enforcement is by inspectors attached to the Executive, but there are powers to allocate enforcement duties by regulations to other bodies, including local authorities. It is apparently intended to allocate inspection of some types of 'non-industrial employment' to District Councils, but no regulations have been made at the time of writing.

Safety regulations and technical change

The Health and Safety at Work Act does not contain a new code of detailed safety provisions; instead it sets up an administrative organization for continuously generating and revising detailed codes as needed. This feature is based on the belief that, with the rate of technical change, detailed safety provisions are out of place in an Act of Parliament. At present, 'existing statutory provisions'[13] are continued in force, the Commission having responsibility to initiate revision by regulations as changes are needed.

Regulation-making under the Factories Act was often a lengthy process, especially where a number of interests were involved. The Commission has a wide field, covering almost the whole employed population. Many matters with which it is concerned will impinge on the work of various government departments and trade interests. Increasingly, national safety standards need to be harmonized by means of international consultation. On the face of it, therefore, the drafting and revising of regulations and codes of practice seem likely to prove a very heavy and continuing load of work for the Commission and its staff.

Involvement of employees

The Factories Act exists to protect employed persons, but it has given virtually no place to employees or their organizations in its local working. The statutory recognition of 'employees safety representatives' in the Health and Safety at Work Act is therefore a very important innovation.[14] This follows broadly similar arrangements in the Mines and Quarries Act, 1954, for the rather special conditions of coal-mining. Similar arrangements exist in Swedish law.

The primary aim of these provisions is to encourage direct consultation between employers and employees as to local safety arrangements. There are provisions for inspectors to provide relevant information.[15] This development is of great potential and may have a strong impact on safety methods—and, incidentally, on the work of inspectors. The vital question is how trade unions and employers generally will use the opportunities it creates.

Safety of persons other than employees

Most previous safety legislation has been divided sharply between legislation to protect people at work and that to protect the public at large. Normally, the two areas have been covered by separate law, enforced by different authorities. The Robens Committee recommended a reorientation of approach; it proposed that legislation and enforcement be unified to protect all persons, whether employed or not, from hazards arising in workplaces or from working processes.[16] The responsibilities of the Commission and Executive have been drawn accordingly.

One aspect of this subject which received close attention was the need to protect the public, as well as employees, from the very large-scale hazards which sometimes accompany modern industrial operations.[17] Recent events have underlined this danger, and it became one of the first concerns of the Commission after its appointment.

Another aspect which has so far had less attention is the effect of the new legislation in workplaces where a minority of employed persons are involved along with a greater number who are not employees. Typical examples are schools, colleges and hospitals. Any attempt to consider employee safety in these places, apart from the safety of school children, students or patients would be quite unrealistic (except for some specialized matters, like clinical risks to patients). The only practical approach seems to be to deal comprehensively with the safety of the workplace and its activities as a whole. This will have a substantial effect on the responsibilities of the Executive. Somewhat similar considerations may arise in some retail stores and in places of entertainment.

The field of inspection

One of the main features of the new legislation is its comprehensiveness covering all employment (excepting only transport operations and domestic service).

This brings 8 million more persons within the scope of inspection, as well as covering aspects of public safety. Perhaps more significant, the scope is based on employment and not on premises. Inspection under most earlier legislation could be directed to cover all premises listed as subject to the appropriate Act; the present legislation has no such boundary. Practically all buildings, apart from dwelling-houses, are work-places; and in addition much work is done away from fixed workplaces. Even allowing for division of duties between the Executive and other authorities, the need will remain to devise a realistic and comprehensive system of inspection to cover this wide diversity of work. Inspection will have to be directed not only towards examining workplaces, but to a much greater extent towards checking systems of work.

Technical information

As we have seen, the various inspectorates became involved in assembling and providing technical advice by pressure of events rather than by conscious decision. The Health and Safety Commission has been set up from the start with explicit powers and duties to initiate research and investigation, and to circulate and disseminate technical information on matters of occupational health and safety.[18] Trade and academic bodies will doubtless carry out much of the actual research, but the Commission will be in a position to keep in touch with results, and to be the official focal point for information on these subjects. Because of its comprehensive responsibility, the Commission will be in a far stronger position to fulfil this role than were the separate headquarter organizations of the various inspectorates in the past. This therefore is likely to be an important area of development.

In some ways, the problems of disseminating and applying technical information are even greater than those of collecting it. In this context, it needs to be realized that the inspection of areas where 'technical' risks are infrequent but occur occasionally (which is the condition in many non-industrial employments to-day) often makes greater demands on the technical knowledge of the inspector than does the inspection of work where technical hazards are more obvious and therefore more familiar to managements. The intention seems to be that local 'Area' offices of the Executive should be developed as local resource centres. A special problem will be to maintain practical liaison between the Executive and Local Authority officers who may be allocated enforcement duties for certain types of employment. For this purpose, the location of Area offices will be important.

References

1. *Parl. Papers* (1833), XX, p. 36.
2. 3 and 4 William IV, c. 103 (Section 17).
3. Hammond, J. L. and Hammond, B. *Lord Shaftesbury,* p. 40, Pelican, 1969.

4. *Labour Inspection.* International Labour Office Convention No. 81.
5. *Safety and Health at Work,* Cmnd. 5034, H.M.S.O., London, 1972.
6. *Ibid,* Chap. 9, especially paras. 261 and 262.
7. *Ibid.* para. 28.
8. Charles Dickens' reference to the National Association of Factory Occupiers as 'the Association for the mangling of operatives' is of this period!
9. *Annual Report of H.M. Chief Inspector of Factories,* 1972. (Cmnd. 5398) Introduction, paras. 25–31.
10. Robens Report, para. 366.
11. *Threshold Limit Values.* Technical Data Note No. 2. Dept. of Employment. These notes are mainly based on values adopted by the American Conference of Governmental Industrial Hygienists.
12. *Safety and Health at Work,* Cmnd. 5034, H.M.S.O., London, 1972, para. 223.
13. Acts of Parliament listed in Schedule 1 of the Health and Safety at Work, etc., Act 1974, and regulations made under them.
14. Health and Safety at Work, etc., Act 1974, Sections 2(4)–2(6).
15. Health and Safety at Work, etc., Act 1974, Section 28(8).
16. Safety and Health at Work, Cmnd. 5034, H.M.S.O., London, 1972, para. 290.
17. *Ibid.,* paras. 294–8.
18. Health and Safety at Work, etc., Act 1974, Section 11(2).

Bibliography

The Early Factory Legislation. M. W. Thomas, M.A., LL.M. Thames Bank Publishing Co., Ltd.

Factory Inspection in Great Britain. T. K. Dyang, Geo. Allen & Unwin, Ltd.

Safety and Health at Work. Report of the Committee 1970–72. (Chairman: Lord Robens) Cmnd. 5034, H.M.S.O.

Annual Reports of H.M. Chief Inspector of Factories, H.M.S.O.

37. Management responsibilities for safety and health

R.W. Hearn

Director of training,
RoSPA

Senior management responsibilities fall into three broad groups: economic, humane and legislative.

Economic

Accidents invariably result in loss or damage to people, plant or product. It is axiomatic therefore that the cost of loss, temporary or permanent, of employees, production or goodwill must be considered. This consideration must include the cost of introducing and maintaining a safety organization to reduce and eliminate accidents.

Another item under this broad heading is insurance, including employee liability, public liability, product liability, fire, etc. It is not realistic to attempt to balance the cost of accident prevention against any insurance premium which a good safety programme may achieve. Although a 'no claims bonus' may show a considerable and acceptable saving.

The whole question of compensation is at present being considered by the Royal Commission on Civil Liability and Compensation for Personal Injury, which in the light of experience gained in Canada and New Zealand could have a major effect on the economics of an accident-free work situation.

Humane

The humane responsibility of management is an extension of that centuries old query 'Am I my brother's keeper?'. Some managers may be able to turn an entirely deaf ear to such appeals, but with today's general trends towards social liberalization, to worker cooperation and so on, it is a rash employer who does not have his moral obligations well to the forefront of his mind.

An employer may not feel an economic or a statutory obligation towards the wife and family of his employees, but he will usually accept a moral responsibility to the widow and dependants.

Statutory

In general terms legislation lays down minimum standards, but the prudent employer always ensures that these minimum standards are adequate to meet the economic and humane needs of a particular situation.

Legislation in this field is devised primarily to protect the employed person and the general public from the hazards of the work activity rather than to protect the business. The recent Health and Safety at Work, etc., Act 1974, may raise this minimum standard to a more generally acceptable level by the introduction of Codes of Practice, which will set out, in non-legal terms, the best current methods of compliance with legislation.

Management responsibilities under the new Act embody items from previous criminal legislation and others from common law. This does not mean that the employer's responsibilities are necessarily greater, just that he has them all simply arranged in the new Act: however Section 1 does state that it is the intention to maintain or improve previous standards of health and safety at work, so that it is unlikely that the responsibility will be less or lighter.

The employer's general duties can briefly be summarized as requirements to provide as far as is reasonably practicable:

(a) a safe and healthy system of work and to maintain plant accordingly;
(b) absence of risks in conjunction with the handling, storage and transport of articles;
(c) provision of information, instruction, training and supervision;
(d) provision of a safe place of work, including access and egress, and, very important, a healthy and safe working environment.

Before going on to list other responsibilities, it is worth while commenting on the above duties. Firstly the full meaning of these provisions (Health and Safety at Work, etc., Act 1974, Section 2(2)(a)–(e)) will be seen more clearly when regulations and codes of practice are issued, but one factor is clearly seen now, and that is that these responsibilities are not absolute but are clarified with the phrase 'so far as is reasonably practicable'. This will of course raise many problems in quantifying the risk and relating the cost of its elimination— in assessing, in other words, how much value the employer places on safety! The courts will no doubt decide about this as time goes on, but it is worth noting that Section 40 of the Act lays down that it is the duty of the accused to prove it was not practicable to do more than he had actually done.

In the past there have been cases where workers have suffered diseases many years after handling certain substances. Now the employer has the responsibility of telling employees about the characteristics of the substances being used: this may be seen by some as a step towards open management, but it will require a great deal of enquiry and research to make it effective.

The term 'working environment' is a somewhat fashionable term, but is often glossed over as referring only to such things as the cleaning of windows and the painting of walls. Although these activities will certainly be deemed to be included, more important aspects will doubtless be dealt with, e.g. the provision of atmosphere free from dust and fumes, and the control of noise. The word 'welfare' is used here and though it is not defined in the Act will certainly embrace a wide variety of requirements for hygiene and sanitation facilities, from lavatories to emergency showers.

Further duties of the employer include the provision of a written statement of general policy with respect to health and safety; this policy must be kept up to date and must show how the objectives are to be achieved, i.e. it must explain the safety organization in the company and the allocation of safety responsibilities to specified individuals. In the past, some companies have had no written policy and others have had epigrammatically short ones. In future, these statements will have to be full, precise and clear, and, most important, they must be brought to the notice of all employees.

Many companies have had safety committees for a long period, but in future these committees will be a statutory obligation in 'prescribed cases'. At the time of writing, the regulations to bring these ideas into practice have not yet been issued, so detailed guidance cannot be given, but the general principles on which safety committees have worked tentatively in the past are discussed later.

A further duty is to ensure that the health and safety of the general public are protected. There are several cases that come to mind in which the health and safety of the public have been infringed, or at any rate jeopardized, such as by the falling of scaffolding into the street, the emission of obnoxious dust over the neighbourhood and the explosion of oil tanks in a thickly populated area. The difficulty of foreseeing danger to the public will make this a somewhat complex and continuing responsibility for management, but obviously one in which eventually his own self-interest is served.

Another new responsibility for management will be the inclusion of a report on health and safety throughout the company, to be embodied in the company's annual report, as required by the Companies Act. Details of such requirements of this duty are not yet published, and indeed it may be some time before it is possible to establish a pattern of reporting which is suitable for large companies and small companies, and in innocuous processes and hazardous processes.

Full acceptance of these responsibilities requires management to draw up a strategic plan to cover the whole safety field: if it appears that the cost is high, then there must be a strong implication that in the past, savings have been made at the expense of safety.

Policy

It is incumbent on all employers employing more than five persons to prepare and issue a written policy statement outlining the company's safety and health philosophy. It will be necessary to keep the contents of this document under review, to revise and re-issue it as often as is necessary in the light of changes in that philosophy, and to bring it to the notice of employees.

Organization

It will be necessary to have a clearly defined picture of the line and functional responsibilities of the people responsible for health and safety. Consideration will have to be given to having a member of the board of directors nominated as having responsibility for the health and safety organization. A director with personal responsibility for safety and health is recommended. The nominated person is not alone in his responsibilities and it will be necessary to tabulate clearly the responsibilities of all persons, management and employees, in the arrangements to carry out the company's policy.

Responsibilities

Line management

Line managers have the task of carrying out the safety policy approved by the board. They should regard accident prevention as a prime contribution to their productive effort, and should promote and encourage an awareness and willing acceptance of the importance of health and safety among their staff.

Senior managers have a similar responsibility for safety as for productive output and should ensure that this responsibility is accepted and shared throughout all supervisory levels. They must provide personal leadership in carrying out the policy, in participating in safety committees, in introducing training programmes, considering all accident investigations and setting a personal example in the observation of safety regulations in their day-to-day work.

Theirs also is the responsibility for keeping the board informed of the progress in implementing the policy, of supplying data or other material on safety for the guidance of the board, and making recommendations when necessary for improving safety standards. They should also be prepared to take disciplinary action for negligence or breach of the company's safety regulations; equally important, they must recognize particular effort and interest by staff who seek to improve safety.

Foremen and others who are in immediate contact with shop floor employees have almost the last opportunity of preventing accidents. They must themselves practice correct procedures at all times and ensure that those under their supervision know what to do and do it. They have a special responsibility for the supervision of new employees and young people. Responsibility rests on foremen to ensure that all plant, machinery, tools, equipment, etc., are in a safe condition and they must report all defects or breaches of safety discipline, beyond their own power to rectify, to their immediate superior.

All line management must ensure that plans are worked out to deal with emergencies; that they are clearly promulgated, understood and periodically tested; that all statutory requirements are met; and that every effort is made at all times to ensure correct employee attitude to safety.

The safety officer

In small or medium-sized units a safety officer may be appointed for duties which cover industrial accident and fire prevention, road safety, security, welfare, personnel, etc., but in large units a safety officer, specializing in safety and hygiene, is essential.

He should advise on the formulation of a company's safety policy and guide all employees on the implementation of this policy, and he must be given the status necessary to enable him to carry out his duties vis à vis all levels of line management.

Duties of safety officer

The duties of the safety officer will depend to a great extent on the size and nature of the works. In general they will consist of advice on safety measures to all members of management, and to specialist employees such as architects, designers, purchasing agents, etc.

He should try to ensure that basic safety principles are incorporated in the design stage of buildings, plant, processes, storage and distribution, not only to provide a safe working environment, but also to facilitate production. He should advise on the incorporation of safety measures in all operational procedures, machine usage, use of hoists and other lifting equipment; on the provision and usage of personal protective equipment; on the preparation of safety and emergency instructions; on reporting and investigating accidents and the preparation and analysis of records; on training, safety propaganda, safety incentive schemes, etc. He must at all times work in close harmony and collaboration with line management and the trades unions to ensure that no aspect of safety is neglected.

Qualifications

The qualifications necessary vary, to some extent, in different circumstances, but he will need to be a man of energy, intelligence and strong personality, with a good educational and technical background. A qualification in engineering or other technically appropriate subject and corporate membership of appropriate professional institutions are desirable, but equally important is personal enthusiasm and the ability to convey that enthusiasm to others at all levels.

Probably the most outstanding quality that he must possess is that of being able to communicate with all grades within the works. He must be personally acceptable to everyone because, if his job is to be well done, people of all grades within the works must at times place in his hands information which appears dangerous to their interests, and they must be able to trust him with this knowledge. In short, he must be a person of standing and integrity. He must be able to bring cogent argument to bear towards the attainment of any object he has in view and he must be able to vary his argument to suit his audience: at the same time, he must be able to appreciate other people's problems and their arguments.

The safety officer must be competent to interpret correctly the requirements of the Health and Safety at Work Act and other safety and health legislation, and to point out when the requirements have been breached, and he must appreciate the implications of common law as applied to his works. Another aspect of his work is that he must be able to visualize the effects of changes of methods, materials and machinery in order to ensure that they comply with statutory requirements. He must ensure that the most effective and economical measures are taken in order to keep changes within the law and, even when this is achieved, he should only consider it to be the foundation of his work and should go on to consider what steps above the minimum requirements of the law should be taken to render it difficult, or even impossible, for an accident to take place in the working environment.

Whilst it is in the long run probably advantageous that a safety officer should be a competent technician in the work of his firm, undue familiarity with a technique can make it difficult to distinguish hazards. The main hazards may in any case be dissociated from the actual special techniques involved, e.g., falls of persons are common to all processes. The safety officer's technical attainments should, however, be such that he can talk to technicians who are competent in the various operations carried out in the works, and at least understand them sufficiently well to grasp exactly how hazards arise in particular processes. He should be able to judge the effectiveness of suggested preventive measures and follow closely the significance of events which have led to an accident. He should, in other words, know enough to be able to use the more expert knowledge of others to his own ends.

He must understand plans and drawings sufficiently well to be able to visualize the objects which they represent and to judge whether they will give rise to hazards of any kind, so that he may call for the right type of alteration to be made. The drawing-board stage is the stage at which alterations cost least and can be most effectively carried out. Vision and applied imagination of the safety officer is most usefully employed at this stage. He must, in addition, keep abreast of the technical developments affecting safety work as a whole, especially the work of his own plant, and he should keep in touch with sources of information on similar subjects and developments in safety devices and equipment of all kinds, so that he can properly influence design and the purchase of stores and equipment.

The safety officer must also be able to collaborate with a training officer to ensure that the various training schemes in operation, whether for the new worker of lowest grade or for the final year of technical apprenticeship, or for senior management, contain adequate safety material. He should be able to work closely with the medical officer and those responsible for the nursing or first-aid service. The work of the safety officer and of those responsible for occupational health are closely interlocked, and cooperation on this subject must therefore be close and continuous. The constant exchange of information on accidents and accident-producing conditions between medical staff and safety officer is essential to a highly efficient accident prevention programme. The medical staff's specialist knowledge may combine with the safety officer's knowledge to lead to a different conclusion concerning an injury than either would arrive at independently.

He should be able to express himself clearly and concisely in language which will be properly understood and will, at least, not give offence to his audience, whatever grade it may be. His powers of persuasion must be such as to enable him to overcome resistance based on either self interest or financial stringency— or both.

In larger works he must have the ability to carry out executive duties in his own department. These may include the supervision of assistant safety officers, clerical staff, statisticians, safety foremen, inspectors, etc. He must be able to present to his board of management, in a clear concise form, essential informa-information on safety matters, and this requires that he should be competent to assemble, collate and analyse reports and statistics received from line management.

He must be progressive and flexible in his thinking and in carrying out his duties, so that his recommendations are seen to be an aid to productive output and, therefore, more readily acceptable.

Joint consultation

Joint consultation between management and shop-floor employees can make a contribution to a better safety standard. With the Health and Safety at Work Act this will no longer be a voluntary exercise, though at the time of writing the regulation defining the requirements have not been seen. There will be four distinct steps towards achieving joint consultation.

Firstly, the appointment of safety representatives. These may be appointed by the trade unions, or possibly elected from among the workers by the workers.

Secondly, having got the safety representative it will be a requirement for management to consult that representative on matters affecting health and safety in the work place.

Thirdly, the safety representatives may make a request for the formation of a safety committee, which will become the channel of consultation.

Fourthly, the representative will have the right to receive factual information from factory inspectors after they have carried out inspections.

The precise role of the representative has not yet been prescribed, but it will be necessary to allow the representative sufficient time to carry out his duties, and these will almost certainly include facilities to make inspections of the work place.

The safety committee

Composition

The committee should be composed of representatives of management supervision and the nominated or elected safety representatives.

Size

It is impossible to lay down any hard and fast rule regarding the number of safety representatives required to represent the shop floor, although this may be prescribed. The ratio of safety representatives to management representatives also cannot be rigidly laid down, but it should never be less than equal representation with management. If the works is very large or is formed of a group of smaller works, it may be found desirable to divide it into sections and to form a committee in each; adequate representation may otherwise be unobtainable without making the main committee too large.

Chairman

The choice of chairman is important. An influential member of the management as an active chairman is essential, as his presence at the meetings will prove that the interest of the management in accident prevention is real and practical. Such a choice also ensures that the committee's recommendations are already known to an influential member of the management when they come up for sanction.

The exact managerial status of the chairman may sometimes be of less importance than his ability to conduct a meeting. If there are always interesting discussions by a committee composed largely of elected or nominated representatives, the reason is usually to be found in the personality and methods of the chairman.

Secretary

The secretary may be elected or ex officio; the latter is more practical. The personnel officer is possibly the ideal person in that he has the facilities for preparing minutes and agendas. He also has the lines of communications available for their distribution. The personnel department is generally available to all employees and therefore personal communication between the members of the safety committee and the secretary is easily accomplished. It is often found that the safety officer is given the task of secretary, but this should be avoided whenever possible, so that he can provide his full attention to his more important function of advising on safety.

The secretary's duties will include convening the meeting and the taking of minutes. It may be an advantage to have a minute secretary making a verbatim record of proceedings. It will also be his duty, with the safety officer and others, to see that the committee's resolutions are implemented or otherwise properly disposed of.

Management representatives

Management representatives may be either ex officio or chosen personally by the management. There is something to be said for each system. On the one hand, by choosing positions rather than individuals it is more easy to ensure adequate representation of all departments. On the other hand, by this method some very useful and active individuals might be passed over. In many works it will be possible to get over this difficulty by having some management representatives acting ex officio and others picked for their personal qualifications. Among those who might well be ex officio members are the works engineer, the personnel manager and the medical officer.

Supervisors

The supervisor is a key man in regard to safety as well as production, and his active help is essential. Supervisors must be told about the safety committee and its functions before any general announcement is made that one is to be set up. Moreover, it is most important that supervision be kept continuously in touch with the committee's work. The method used for bringing him into contact with, and securing his interest in the committee's work depends almost entirely on the nature of the existing works organization.

Workers representatives

These may be nominated by the trade unions from among the employees or elected by their colleagues. It is desirable to have as many worker representatives to serve as possible, and this can best be done by limiting service on the committee to one year.

Meetings

Meetings should be held regularly, and monthly meetings are the most usual intervals. It is good practice where possible to have the meeting at the same time and day each month, e.g. 3.30 p.m. on the third Thursday of each month, so that members can ensure they are available for meetings. Emergency meetings may be called when serious and urgent matters arise. Safety committees should not be used as a platform for complaints or matters other than health and safety.

Arrangements for meetings

It is worth while giving some care to the way in which meetings are arranged. A slovenly or haphazard way of getting the committee together gives the impression that the work is considered not very important. In contrast, a little formality makes members feel that the committee has some prestige and that it is a privilege to belong to it.

Notices of meetings, preferably accompanied by an agenda, should be sent personally to each member. This has a better psychological effect than if a general notice is posted on the works notice boards or a verbal message passed round. The custom of sending apologies for *unavoidable* absence should be encouraged and an attendance book should also be kept.

Place of meeting

It is true to say that a meeting takes on the colour of its surroundings, and it is neither fair nor sensible to convene a committee in a dismal room and expect anything but a dismal meeting. On the other hand, if the committee meets in surroundings which are entirely foreign to them, those who are more used to the shop floor than to the boardroom may not react with that naturalness and ease which are so desirable.

Agenda

The method in which the committee gets its business is of vital importance and the agenda can play an important part of this. It is necessary to ensure that items raised at the committee have already been through the normal worker–supervisor–manager line of communication and that the normal system is not being by-passed.

This can be overcome by requiring items for the agenda to be in the hands of the secretary seven days before the date of the meeting. It then gives the secretary time to see what action has been taken. Items which arise within the seven-day period can be dealt with by the chairman asking if there are any emergency items. Properly used, this system can prevent items such as cracked windows, missing electric lamps, etc., coming before the committee unless the normal systems have failed to achieve the desired results. A typical agenda would be:

    Apologies for absence
    Minutes of previous meeting
    Matters arising—progress reports
    New items
    Safety officer's report
    Date of next meeting.

Conduct

The meetings of the committee should be conducted according to the generally accepted rules of procedure but, whilst order must be respected and maintained, enough latitude should be allowed to prevent formality from overwhelming the meeting and having an adverse effect on free and frank discussion.

The chairman should not be too hasty in ruling a matter out of order, without giving a reasonable explanation. Otherwise, the member trying to raise it may feel that he is being deliberately muzzled. An atmosphere of freedom of speech must be created from the beginning; once the foreman or workers' representative feels that it is useless or unwise to voice an opinion the committee has failed.

Minutes

The minutes are taken, prepared and circulated by the secretary. They are of great importance and should be circulated as widely as possible. Each committee member should receive a personal copy, and copies should be sent to the heads of all departments and posted on notice boards. The minutes should be published as soon as possible after the meeting. They must record accurately all decisions taken and must give a clear idea of why each matter has been decided in that particular way. They need not, however, go into every detail of the discussions; a concise and comparatively short account can be more informative than an involved document running into many pages.

Activities

The committee will regularly consider reports of accidents and safety inspections, hear suggestions from its members, discuss training arrangements, poster and other publicity material, and safety requirements for new developments or expansion in the works. It should also have the opportunity of inspecting new protective equipment, and members should be encouraged to give such items personal trial.

Every opportunity should be taken to improve the safety knowledge of committee members by inviting guest speakers to talk on specific subjects. In-company experts who could be used would be the electrical engineers, works medical officer and chemists, etc. Invitations could also be sent to the local factory inspector, fire prevention officer and safety groups.

As safety is an attitude of mind, the committee should include, among its activities, off-duty safety. This has the effect of making workers conscious of the part safety can play in their homes to the advantage of their whole family. It can prove a most rewarding aspect of a committee's work in the contribution it makes to correct worker attitude.

Safety inspection techniques

There are many techniques available for carrying out safety inspections. The best method of inspection is the day-to-day observation of safety needs by supervisors and operatives as they go about their normal work so that safety is an integral part of their activity.

With regard to formal inspections or audits there are five main points for progress.

(a) identification of possible loss producing situations;
(b) the measurement of the potential losses associated with these risks;
(c) the selection of methods to minimize the losses;
(d) the implementation of the selected method within the company;
(e) the monitoring of the result.

A safety inspection

The detailed nature and extent of a safety inspection depends on the type of work carried out and the hazards which exist. These will be known to the safety officer, who will devise a routine accordingly. It will be a routine scheduled inspection. In a small establishment the whole factory may be inspected; in larger factories the inspection may be confined to one department or unit at a time. The inspection team normally consists of one or two members of management accompanied by the safety officer.

In addition to looking at plant, gangways, floors, racking, exits, housekeeping, obstructions, etc., observation should be made to ensure that correct safe working practices and procedures are carried on and the necessary use made of personal protective equipment.

Inspections should be carried out in a formal manner and never be cursory or hurried.

A safety tour

This is an unscheduled examination of a work area. The scope of the tour is less than that of the safety inspection and the inspection team may vary considerably. It may consist of the nominated safety representatives, the safety committee, or members of management.

Techniques

Safety sampling

This is designed to measure by random sampling the accident potential by counting safety defects or omissions.

The area to be sampled is divided into sections and an observer appointed to each. A prescribed route through the area is planned and the observers follow their itinerary in the time allowed—say 15 minutes. During the sample they note safety defects on a safety sampling sheet which has a limited number of points which are to be observed. These points include items such as non-use of personal protection, obstructed fire exits, environmental factors, lighting, heating, ventilation, faulty hand tools, damaged guards, etc.

The staff making the inspections should be trained observers with a knowledge of the procedures being carried on.

The results of the inspections should be collated by the safety officer and presented in graph form. The system monitors the effectiveness of the safety programme.

Damage control

The basis of damage control or damage costing is that non-injury accidents are as important—or more so—than injury ones. The elimination of non-injury accidents removes the others. For example, a brick falls and just misses a man below, no injury results and no accident is recorded. The next brick to fall injures a man! This is recorded. The elimination of the cause of the first brick falling would have prevented the injury when the second one fell.

Damage control aims at providing a safe working environment and calls for the keenest observation and co-operation by all staff who see or experience a condition which may lead to an accident.

Contact scheme

The object of the contact scheme is to give added facilities to supervisors in imparting information or instructions on health and safety to the work people in their control.

Many supervisors are very conscious of the fact that they are not teachers, and even 'tool-box talks' are not always within their scope, quite apart from the fact that these tend to disrupt production whilst everybody sits around and has a talk. If, however, a supervisor walks around and talks to all his men in turn he can usually put over various ideas. The contact scheme enables him to do this and to keep a record of what he has done.

A simple scheme that can be arranged is a piece of squared paper on which the names of the individuals should be entered in the vertical column and the subjects horizontally at the top. Having made out a programme of items to discuss, put the subjects, e.g., mushroom-headed chisels, hammer heads, clearing work benches, wearing eye protection, etc., along the top. Then all the supervisor has to do is to give a little informal talk to each person at his place of work and put a tick in the square denoting the person and the subject.

Incentives

Opinions vary as to the desirability or value of incentives to encourage employees to consider their own safety. Schemes such as the RoSPA Safe Driving Awards for commercial vehicle drivers do much to develop a personal and collective pride in the job. Generally, schemes are more successful where the final result or award is gained by collective effort. However, in many incentive competitions there is a danger that accidents will not be reported; firm application of definitions is necessary.

Participation in regional and national competitions concerning medical first-aid and fire drills often builds up a team spirit, and company awards, preferably of the take-home kind, to supplement those won in the event are well worth while.

Exhibitions of safety and personal protective equipment, the use of posters, literature, films, exhibitions, all play a part. The considerable incentive of correct worker attitude to 'off-the-job' safety is a powerful medium, and families should be encouraged to participate as much as possible in all matters connected with the safety of the breadwinner.

Permits to work

These are essential where temporary work is being carried out on or within a plant. Work of a hazardous nature, or which may create a hazard, should not be allowed without a permit. This should lay down the safety precautions to be taken. It should be valid for a limited period depending on the nature of the work and the hazards. All the methods to be used and precautions to be taken should be agreed beforehand and clearly stated on the permit. The senior supervisory staff authorizing and carrying out the work must sign the permit before it becomes permissive. The number of such permits should be reduced to a minimum conducive to efficient manning of the plant.

Accident investigations

Every effort should be made after every accident, whether injury producing or otherwise, to determine the cause. An accident indicates a failure of preventive measures, and, if the cause can be found, the failure can be rectified.

Purpose of investigation

The prime purpose must be to find the cause so that a recurrence may be avoided, and not to find a scapegoat. It is important that this be clearly understood by everyone. Cause, when determined, may, of necessity, apportion blame, but this is not a function of the person or persons identifying the cause.

Scope of investigation

Ideally all accidents should be considered, but the amount of damage or injuries, and the degree of potential hazard will in practice apply limitations.

The remit

This will also vary, but should always include the chain of events leading up to the accident, the cause and recommendations to prevent a recurrence. Other points may be breach of statutory or company regulations, or other indirect contributory hazards or weaknesses revealed by the investigation.

Preliminary investigation

Again, severity indicates the immediate action to be taken. The safety officer should be informed immediately of an accident so that he may decide the extent of the preliminary enquiry. Photographs, as soon as possible, of the circumstances can be of considerable value.

Formal investigation

The safety officer can advise management on the need or otherwise of a formal investigation, or it may be determined entirely on the nature and circumstances, as in the case of a fatality.

A senior member of management should act as chairman and should have the help of specialist expert advisers. Usually, the smaller the number of persons formally investigating an accident, the better and quicker the result will be.

Witnesses

It should be made clear to witnesses that the investigation is solely to determine cause. Questioning should be conducted in a formal but friendly manner consistent with the nature of the enquiry. Witnesses should not be called to give evidence on oath and should be allowed a witness's friend if so desired.

Report and follow-up

The result of a formal investigation should be promulgated to all concerned and fully discussed by the safety committee. Care must be taken to ensure that recommendations are implemented.

38. Safety administration

M. J. White

Principal Assistant,
Pensions and Social Services Dept.
General and Municipal Workers' Union

The victims of industrial accidents are, as often as not, members of a Trade Union. It is generally accepted that Trade Unions do a lot for their injured members—cash benefits, convalescent treatment, legal advice and assistance—but all this comes after the accident. What are Trade Unions doing to prevent industrial accidents?

Trade Unions exist to represent, protect, advise, and educate their members and to serve their interests generally. The responsibility for providing and maintaining safe working places and safe systems of work and for enforcing the system of work is a managerial responsibility. Any attempt to share that responsibility would create considerable legal and practical difficulties, and uncertainty as to where the responsibility lay would probably lead to more rather than fewer accidents.

The Unions are well aware of the mounting toll of industrial accidents, and of their cost in economic terms and in terms of human suffering and disability. Every year the concern of the Unions is reflected in the proceedings of their Annual Delegate Conferences and of the Annual Meeting of the Trades Union Congress.

Trade Unions co-operate with Government Agencies and with employers in the investigation of the causes of industrial accidents and diseases. The history of British safety, health and welfare legislation is, in part, the history of the struggle of the Trade Unions to win legally defined minimum safety standards for their members.

It is indicative of the Trade Union attitude that the Unions agreed to mark the centenary of the first Trades Union Congress (2 June, 1868) by providing funds to establish a permanent Institute of Occupational Health to discover and attack the causes of industrial diseases.

Trade Unions are in exactly the same position as any other body which provides a service in a free society. They have to order their priorities according to what the 'customer' wants. The accident rate in a particular plant or industry may be bad, but if industrial relations generally are also bad, it is not a bit of use a Trade Union officer talking to his members about co-operation to prevent accidents when his members have wages and working conditions on their minds. The effort to reduce the toll of accidents must be part and parcel of a combined effort to improve working conditions generally. Safety, health, and welfare are inseparable.

Accidents are a reflection of human error in one guise or another. No one wills or designs them. They cannot be tidily explained away. The contributing factors are innumerable and not always immediately obvious. How can one explain the unhappy combination of the boiling saucepan, the temporarily preoccupied mother, and the climbing child? This simple and sadly common domestic situation contains all the ingredients of industrial accidents, the danger point, lack of proper supervision, and the unheeding or inadequately instructed victim.

Statistics published by the Chief Inspector of Factories show that we are most vulnerable in precisely those areas of activity within which mechanical guards, sophisticated rules for the performance of certain tasks, or the provision of protective clothing serve no purpose. In other words, in those areas in which we have to rely most on human intelligence, forethought and caution. That more than half of the reportable accidents fall into this category reflects the failure of management to manage, and says very little for the quality of the training provided for men or management.

A serious preventive programme will inevitably call into question our systems of selection, induction, and supervision. In the altogether wasteful and un-productive business of war a great deal of intelligent effort is devoted to the task of selecting and training the right men for the job. The armed forces demand a high standard of fitness and intelligence, particularly in the case of those who are likely to be placed in positions of command. Many of us know from personal experience that this remains true even in the stress of war.

Modern industry inescapably presents hazards to health and safety, but how much real care is devoted to the task of selecting the right man for the job at shop-floor level in every instance?

Many unskilled, or semiskilled processes require a high degree of mobility of labour. Men are shunted from one job to another according to the exigencies of the hour. It is true that many of the processes are fairly simple, such that a child could be taught in 10 min. The trouble is that workers whose reflexes have become attuned to one rhythm are required to quickly adapt themselves to an entirely different rhythm and to perform the movements at speed. This is unobjectionable so long as there is some margin for initial error and the initiation is properly supervised, but so often the supervisor is preoccupied with production problems and has far too many employees to manage, and even when the supervision and induction are good, it is sometimes resented by the worker.

Supervision is very important but when the question of promotion arises, how much consideration is given to the candidate's physical and temperamental fitness for the job or to the effect he is likely to have on those under his supervision? To what extent are his personal views tested on the relationship between productivity and safety? How much careful and sympathetic study is devoted to the delicate problem of the connection between physical fitness—or the absence of it—and accidents?

A particularly shocking feature of our accident record is the number of young persons under the age of 18 whom we maim. Some people expect boys or girls to become responsible adults the moment they enter a factory gate. So do the youngsters themselves. It cannot be overemphasized that proper training and patient, intelligent supervision is essential in the case of young people if we are not to perpetuate bad working practices. Their initial induction into the world of industry should be an extension of the road drill they were taught in primary school, and we should drop like a ton of bricks on any adult worker who is caught either introducing youngsters to dangerous short cuts or who scoffs at safety rules and procedures in the hearing of youngsters.

One could go on endlessly listing problems of this nature but enough has been said to indicate the field to which attention should be paid if we are to get out of the present dilemma and tackle the accident problem afresh.

It is precisely in this field that the Unions can be of most use. They cannot initiate and enforce safe and efficient working practices, but if employers are to overhaul their procedures for selecting, training, supervising and promoting their employees, they will do so more smoothly and effectively if, at the outset, they consult the Unions and seek their co-operation and active participation. Co-operation implies a degree of partnership, and partners have to be consulted and their opinions and suggestions considered with patience and respect. It would be no use simply to confront the Unions with a programme of radical reform—no matter how well intentioned—and to expect them to endorse it without having had a say in the formative stages. Neither can one expect that such a programme would be agreed upon in its entirety without discussion, disagreement, and amendment. There is bound to be a conflict of ideas and interests, but a workable scheme will emerge if both sides conduct themselves according to the rules of diplomacy and not the rules of war, and are repeatedly reminded that the common aim is to improve standards of safety, health, and welfare.

Once the co-operation of the Unions has been sought and secured, it need not be confined to simply underwriting the good intentions of the employer. Given any encouragement at all the Unions can persuade their members to accept changes which are in their interest and can educate them to be safety conscious. It is part of the built-in defensive mechanism of human nature that one tends to think that disaster is something that happens to the other fellow. The purpose of education in safe working practices is to demonstrate that disaster need not befall any of us if we learn that there are two ways of doing most things—the right way and the dangerous way. Trade Unions have excellent educational facilities and can draw upon a wealth of talent and experience, but here again the emphasis should be on co-operation. Whenever possible training in safe working practices should be a joint exercise relating to specific industrial conditions, because each industry has its own health and safety problems and, to some extent, its own terminology. Joint educational activities will help convince the workers concerned that their employer and their Union really mean business—that the training courses and safety conferences and committees are not just pious exercises which, in the end, lead to nothing.

So far, this discussion has been confined to the role of the Unions as associations of employees representing their members' interests but not directly and personally affected. Accident prevention is a shop-floor problem and no amount of impersonal exhortation is going to solve it. Any scheme designed to make the workplace a safer place directly concerns shop floor personnel and, in some cases, may even affect their earnings or their prospects of promotion. Their willing participation is essential. The most vital part of the Union's function is to secure that participation and to demonstrate that the advantages are real and tangible.

Joint consultation at every stage and at every level is the key to progress. Employees should be diligently consulted and kept informed in the formative stages of the preventive scheme. Thereafter, their representatives should meet representatives of management regularly and be free to discuss every aspect of accident prevention and to air their criticisms and suggestions in the sure knowledge that what they say will receive careful consideration and that the hazards and malpractices to which they draw attention will be remedied.

Once the Safety Committee has been formed and is functioning, the day-to-day tasks of consultation should be the responsibility of the Safety Committee men. The Trade Union officer should remain in the background to be summoned as arbiter only when every other avenue of approach has been explored unsuccessfully. This unalloyed worker/management relationship is really fundamental to a good safety programme because it breeds responsibility, trust, and mutual appreciation of both sides of the problem. It enables employees to influence directly and to understand the significance of managerial decisions in matters which affect their health and safety. It generates a positive will to accept the notion that safety is everybody's business. Joint consultation of this order is good Trade Union practice. Trade Union officers exist to lead and service their members, not to wet-nurse them. There is a great deal more satisfaction to be had from leading active participants than from shepherding the apathetic. We cannot, however, rely solely on voluntary co-operation. It will be effective only if it is allied to and backed by the enforcement of a sound structure of statutory regulations to spur the sluggard and penalize the recalcitrant.

The Trade Union viewpoint can be summarized briefly. Any initiative designed to reduce the toll of accidents in industry deserves the closest possible co-operation of all Trade Unionists. Here at least is one field in which the respective interests of employers and employees undoubtedly coincide. The motive of a safety conscious employer may be largely economic, but should motives concern us so long as the result is the provision and maintenance of a safe and healthy workplace? In the pursuit of industrial safety there is no room for barriers, no place for the 'them and us' mentality. Joint consultation and effective co-operation aimed at curing industry of the accident malaise represents enlightened self interest at its best.

It is doubtful whether anyone would seriously dispute what has so far been said about the fruits of joint consultation and yet, oddly enough, both industry and the Trade Unions continue to accept a complicated and divisive system of compensating those who suffer industrial injuries. Accident prevention is bedevilled by common law claims for damages. It is doubtful whether such claims make any appreciable contribution to the prevention of accidents. There are those who argue learnedly and honestly that they do, but the annually increasing accident statistics and the never ending lists of actions awaiting the judgement of the courts are effective arguments to the contrary. The system is divisive and, therefore, detrimental to the atmosphere of goodwill and mutual trust so essential to preventive measures because it is a contest, and retains something of the air of the medieval joust. It is not the business of the judiciary to prevent accidents but to determine the question of liability in any given case according to a body of legal principles based upon case and statute law. Judgement is given according to the evidence as it is presented in court—usually a long time after the event. Many cases which, on paper, appear to be cast iron certainties, have been lost because the plaintiff faltered under cross-examination or because his witnesses failed to live up to their proofs of evidence. Anyone who has had the delicate task of advising injured workers can sympathize with their bewilderment and their sense of having been cheated.

National Insurance benefits may remove some of the hardship and sting, but it is still questionable whether any useful purpose is served by retaining a legal right when the remedy is dependent upon so many fallible factors.

This is not written in a spirit of disillusionment. It is appreciated that many learned and courageous judges have wrestled with this difficult branch of the law to achieve just and equitable decisions. Nevertheless, in practice, claims for damages remain something of a gamble and it is no disrespect to our legal system to hold that such a method of compensating the victims of industrial accidents does not measure up to the needs of a complex industrial society.

The greatest single contribution that could be made to our efforts to prevent accidents would be to lift the system of compensating the casualties of industry out of the realm of the law of tort and to pay compensation without reference to liability. The dividing line between employers' liability and simple industrial hazard is often a matter of narrow legal technicality and there is no logical reason why injured employees should have to rely upon contentious common law claims to seek restitution of their physical and financial loss.

It is hardly relevant to this discussion to consider alternative schemes. Suffice it to say that others have canvassed similar views, and reform is not only possible—but probable.

In the field of accident prevention the primary function of the Trade Unions is to co-operate and to educate. The verbatim proceedings of the Trades Union Congress confirm that this view is widely and officially held within the Trade Union movement. There is a reservoir of goodwill, but one must guard against creating the false impression that the task is an easy one; anything worthwhile seldom is. There are many pitfalls for the unwary along the way. There are historically rooted suspicions and prejudices to be overcome. Trade Unions can go just as far and as fast as their members will follow. Sometimes their members are cautious and sceptical and have good reason to be. In spite of the acknowledged difficulties there are grounds for optimism. If one disregards much publicized strikes and disputes, the evidence would seem to suggest that the hard logic of post-war economic difficulties is bringing both sides of industry closer together. There is a general movement towards employee participation in the management of industry, a development viewed with some alarm in quarters where the old legal concept of 'master and servant' still lingers. The vigorous pursuit of effective joint consultation in matters of health and safety could be doubly useful as a means of achieving the objects of the Health and Safety at Work, etc., Act 1974, and as a neutral testing ground for the wider notion of worker participation.

39. Legal aspects of industrial safety

W. D. Gold,
Barrister,

Senior Law Lecturer, Blackburn College of Technology and Design

---

'Accidents don't just happen—they are caused.' Generally there is no liability without fault. The only true accident is one that could not be reasonably foreseen beforehand and there is no fault. It is an accepted fact that most industrial accidents can be prevented, and at least the majority of those that do occur can be prevented quite easily by the use of simple safety devices and proper work systems. Apathy on the part of both management and employees, carelessness and the lack of proper training are probably the main contributory factors leading to accidents and health risks. Apart from the usual conditions and terms associated with modern employment, such as the protection of employment and state insurance schemes, the law is concerned with the enforcement of safeguarding employees from risks connected with their employment which could result in physical injuries and industrial health deterioration.

So far as the Common Law is concerned, the remedy for damage sustained is monetary compensation based upon an action for negligence. Thus such an action is necessarily after the event and such damages that are awarded to the plaintiff are almost never paid by the person liable at law for the injury, the cost usually being borne by insurance. It is, of course, compulsory for all employers to take out insurance covering their employees whilst at work under the provisions of the Employers' Liability (Compulsory Insurance) Act 1969, which became effective in January 1972.

The great difficulty in bringing a Common Law action for negligence is the onus of proof. The plaintiff must prove his case upon the balance of probability. To do this, he must prove that the defendant was negligent and that such negligence was the cause of the incident.

Negligence has been defined by the law as 'the omission to do something which a reasonable man, guided upon those considerations which ordinarily regulate the conduct of human affairs, would do, or doing something which a prudent or reasonable man would not do'.—*Blyth v. Birmingham Waterworks Co.* 'A person must take care to avoid acts or omissions which he can (or should) reasonably foresee would be likely to injure his neighbour. Who then in law is your neighbour? The answer seems to be—persons who are so closely and directly affected that a person ought reasonably to have them in contemplation as being so affected when he is directing his mind to the acts or omissions which are called in question.'—*Donoghue v. Stevenson.*

Thus at Common Law there are two elements in negligence, e.g., the 'reasonable' person and that of 'foreseeability'. The degree of care which is required at Common Law is that which is reasonable in the circumstances of a particular case; the standard of care is the foresight and caution of an average prudent person. Where a person, such as a skilled craftsman, professes to exercise some special skill or calling, and is by contract engaged upon work requiring his special skill, the law requires the standard of care of the reasonable or average person in that special calling or trade. These are questions of fact for a court to decide. There is always an element of risk in all walks of life. Because of the emphasis upon reasonableness it is often very difficult for a plaintiff to prove that the defendant was negligent. The plaintiff must prove upon the balance of probability that the defendant failed to take proper and sufficient precautions. It may be pleaded by the defence that, although the accident subsequently showed that there was a risk connected with the employment, the kind of risk was such that it could not reasonably have been foreseen before the incident. However, liability does not depend on the precise nature of the injury or incident being foreseeable; it is sufficient if the incident was of a kind that was foreseeable though the form it took was unusual.

Vicarious liability

An employer is vicariously (superiorly) liable at Common Law for the negligence of his employees who, in the course of their employment, injure third parties or cause damage to third party property. The term 'course of employment' is determined by the question 'Was the employee doing the work he was employed to do at the time of the incident, but did he do it in a negligent manner?' In *Warren v. Henlys* the defendant employer was not vicariously liable when his employee assaulted a customer since the employee was not employed to assault customers; the employee was personally liable. But in *Century Fire Insurance Co. v. Northern Ireland R.T.B.* the employers were liable when their employee, employed as a petrol tanker driver, while delivering petrol at a garage lit a cigarette and the lighted match fell to the ground thus causing the petrol to ignite—here the employee was doing the work he was employed to do, but he did it in a negligent manner, albeit he did light the cigarette for his own benefit. (This latter case surely illustrates the principle of foreseeability and that of a reasonable man.)

An employer can, however, claim at Common Law an indemnity from the employee whose negligence results in his employer having to pay damages to a third party *(Lister v. Romford Ice and Cold Storage Co. Ltd.)*. But an employer cannot claim an indemnity when, due to the negligence of the employer, the employee was engaged upon work for which he was untrained or was physically incapable of performing in a safe and satisfactory manner; in such a case the employer has himself been negligent—*Jones v. Manchester Corporation.* Invariably, an employer who is ordered to pay such damages will call upon his insurance company to indemnify him and to pay the damages to the third party. Where an insurance company pays out in accordance with the policy it is subrogated i.e., placed in the same legal position as the assured as regards any indemnities from the person or persons who caused the damage and can recover against those persons in the name of the assured. Thus, here again, it will be observed that the Common Law can only offer an ex post facto remedy and not a preventive one.

If the plaintiff is awarded judgement direct against the person who caused the damage, he usually has difficulty in recovering, for the defendant may well be a 'man of straw'.

There is no statutory requirement for employees in the course of employment to cover themselves against third party risks, as is required by the Employers' Liability (Compulsory Insurance) Act 1969. One effect of the Health and Safety at Work Act 1974 is that in similar circumstances it is possible for the person injured to seek compensation from the Criminal Injuries Compensation Board if the circumstances of the incident constitute an offence under the Act.

Contributory negligence

Under this defence (The Law Reform (Contributory Negligence) Act 1945) the defendant does not deny that he has been negligent but seeks to reduce the damages payable by him by proving that the plaintiff by his own negligence contributed to his own injury. In assessing damages the Court will reduce the amount awarded to the plaintiff having regard to his respective degree of negligence as the Court thinks fit. Thus, both the plaintiff and defendant may be equally negligent, in which case the amount awarded to the plaintiff, calculated in the first instance as if he had not been negligent, would be reduced by 50 per cent. Problems arise in civil actions where workmen use equipment provided by an occupier of any premises, i.e., fixed installations, and not provided by the workmen's employer. The principle involved is whether the equipment was defective and whether the occupier knew or should have known of it. In one particular case an electrician went to premises to wire up equipment; he switched on and was injured by a moveable electrical saw which the occupiers had left unsecured. Both the occupier and electrician were held partially to blame.

Statutory duties

The Common Law of England, being the ancient law as modified, extended and adapted, by the judges over the years, is clearly insufficient to meet the requirements of all the new safety problems that now arise and will continue to arise in an ever-changing complex industrial society. Hence, Parliament has from time to time found it necessary to interfere with the laissez-faire and to legislate, inter alia, upon matters relating to the safety, health and welfare of employees. A series of Statutes culminating in the Factories Act 1961, and the Regulations that could be made thereunder by the Secretary of State—The Offices, Shops and Railway Premises Act 1963—are but two examples of such legislation. The Acts sought to impose a minimum standard of care (in some cases there is an absolute duty) upon employers and those persons so designated in the legislation, to provide and maintain safe working conditions. The aim of such legislation is preventive rather than retributive by the infliction of criminal penalties, but such penalties are minimal, the real remedy again being compensatory at Common Law and this, as stated previously, is generally recovered from insurance. Such preventive legislation is applied only to those persons and to those premises that fall within the ambit of the relevant statute. If an incident occurred in circumstances not covered by a statutory duty designed to prevent such incidents, the plaintiff in civil cases would be forced to rely upon his Common Law rights and prove, if he could, that the defendant was negligent. Many complex legal problems thus arose.

In consequence of the Robens' Committee on Health and Safety at Work, published in 1972 (Cmnd. 5034), there has now been enacted The Health and Safety at Work Act 1974, which became fully operative from the 1 April 1975. It replaces the hitherto piecemeal statutory legislation by a single comprehensive Act. The Act is intended, with the assistance of Codes of Practice, to ensure a minimum standard of care by those persons identified in the Act. It is penal in its sanctions (with possible imprisonment in some cases), but it is revolutionary in that not only does it make a breach of its requirements a criminal offence, but also in that it is for the defendant to prove 'that it was not practicable or was not reasonably practicable to do more than was in fact done to satisfy the duty or requirement imposed by the Act, or that there was no better practicable means than was in fact used to satisfy the duty requirements' (Section 40). In other words, the defendant is guilty of a criminal offence unless he can prove otherwise. It is submitted that the burden of proof will rest upon the balance of probability rather than upon the standard of 'beyond all reasonable doubt' as required by the prosecution to prove, generally, in criminal cases.

The Act does not, however, immediately repeal the previously existing legislation. Such legislation will be repealed by regulations made by the Secretary of State in due course when Codes of Practice are introduced on the recommendations of the Health and Safety Commission established under the Act. Whilst a failure on the part of any person to observe any provision of an approved code of practice will not of itself render that person liable to any civil or criminal proceedings, such code of practice is admissible in any criminal proceedings for an offence under the Act.

New regulations under the Act will be made by the Secretary of State on the recommendations of the Commission to meet new requirements and to render the law less complex.

The Act does not of itself confer any right of a civil action for breach of any of its requirements (Section 47), but, of course, breach of any duty or requirement resulting in damage to the person concerned will be actionable at Common Law for monetary compensation. Section 47(6) defines 'damage' as including the death of, or injury to, any person (including any disease and any impairment of a person's physical or mental condition). Basically, the Common Law remedy for monetary compensation remains the same. However, it is of importance to make reference to the Civil Evidence Act 1968 (Section 11), whereby a criminal conviction is admissible as evidence in subsequent civil proceedings that the defendant committed the offence in question. Thus, where it is relevant to do so, the plaintiff may introduce the conviction as conclusive evidence against the defendant. In a criminal case it is no defence for the accused to plead contributory negligence on the part of any person injured by the negligence of the accused (in such cases the person so injured may himself be guilty of an offence under the Act) but, in subsequent civil proceedings, he may plead contributory negligence to reduce the quantum of damages payable. Section 1 of the Act provides that its provisions shall have effect with a view to:

(a) securing the health, safety and welfare of persons at work;
(b) protecting persons other than persons at work against risks to health or safety arising out of or in connection with the activities of persons at work;
(c) controlling the keeping and use of explosive or highly flammable or otherwise dangerous substances, and generally preventing the unlawful acquisition, possession and use of such substances;
(d) controlling the emission into the atmosphere of noxious or offensive substances from premises prescribed by the Act.

It will be seen from the foregoing extract that the effect of the new legislation is all-embracing.

These duties apply, inter alia, not only to the employer–employee relationship (Section 2), but also to employers and self-employed persons as regards persons other than their employees (Section 3), to persons who allow other persons who are not their employees to use non-domestic premises as a place of work or as a place where they may use plant or substances provided for their use there (Section 4); to persons who design, manufacture, import or supply any article or substance for use at work (Section 6); and places a general duty on every employee while at work (Sections 7 and 8). The general effect of the Act is thus, in one context, to give statutory effect by the infliction of criminal sanctions to the hitherto common law duties of care.

Irrespective of an incident having occurred, an Inspector appointed under the Act can always take action if he is of the opinion that there is a breach of the requirements of the Act. Enforcement does not necessarily mean a prosecution.

Section 2(1) provides: 'It shall be the duty of every employer to ensure, so far as is reasonably practicable, the health, safety and welfare at work of all his employees.'

At first sight, it may appear that there is not much difference between this standard and the common law standard of reasonable care, but, bearing in mind the requirements of proof as demanded by Section 40, it is for the accused to prove that it was not reasonably practicable for him to have done more. He must show that he has done his best. What is reasonable is a question of fact. A statutory duty is higher than the common law duty. However, the expression 'so far as may be reasonably practicable' is well understood at law and will be interpreted in the light of previous cases. In *Marshall v. Gotham Co. Ltd* it was said that such a duty does not necessarily involve a duty to shore up every piece of roof in every working place against rare and undetectable geological faults. In another case (*Jenkins v. Allied Ironfounders Ltd*) it was said by the House of Lords: 'On the one side, we must put any expense, delays or other disadvantages involved in adopting the preventive system and, on the other, the nature and extent of the risks involved if that system was not adopted.' In other words, the employer must take into account the likelihood of danger if the device is omitted. A safety device costing a small sum may well minimize the danger, but it would be unreasonable to expect an employer to expend vast sums of money in buying a device to rule out a one in a million chance of someone being injured, although it is no defence to say that because of economic difficulties the employer could not afford any expenditure.

If there is a latent defect which could not be detected by reasonable examination, or if, in the course of working, plant becomes defective and the defect is not brought to the employer's knowledge, and could not by reasonable diligence have been discovered by him, the employer cannot be guilty of an offence (*Toronto Power Co. v. Pashwan*). The distinction is between latent defects and those which are, or ought to be, known to the employer.

'So far as is reasonably practicable' is *not* the same as 'so far as is practicable.' An employer is not bound to prevent all risks and is not required to take fantastic risks. It may well be that in some sudden emergency an employer may require an employee to accept some risk in order to save life, but such a situation could only arise in special circumstances (*Watt v. Herts. County Council*). The undertaking of dangerous employment, such as that of a fireman or steeplejack, is not of itself an acceptance of all the risks connected therewith, although the very nature of the work exposes the employee to the hazards and risks connected with such work. The fact that the employment is inherently dangerous requires the employer to take extra precautions not required in a less dangerous employment. The demolition of buildings can be, and often is, extremely hazardous, but the employer is not relieved from taking care 'so far as is reasonably practicable' because his employees ordinarily accept the risk of everyday injuries incident to the trade. Indeed, he is required to take extra care.

Nor will the expression 'so far as is reasonably practicable' lessen the hitherto 'absolute' duty provided by statute, e.g., Sections 12, 13 and 14 of the Factories Act 1961. It will be no defence to plead that the cost of guarding machinery which is obviously dangerous would be financially prohibitive.

Absolute safety is almost impossible to achieve although it is a very desirable standard. Machinery and equipment may be inspected at regular intervals by competent persons but, without warning, an accident may occur through some defect which was neither obvious nor foreseen. If, however, a machine has a known tendency to eject materials the onus is upon the employer to reduce this possibility of injury.

The onus of establishing that there is a proper system of inspection and facilities for reporting defects rests upon the employer. From the point of view of evidence, the employer is well advised to keep and maintain records of inspection, and, further, to adduce evidence that he has provided proper working conditions and maintained the same. Such evidence of the taking of reasonable care is best shown by the production of written instructions and procedures.

Section 2(2) further provides 'without prejudice to the generality of an employer's duty under Section 2, the matters to which that duty extends include in particular:

> (a) the provision and maintenance of plant and systems of work that are, so far as is reasonably practicable, safe and without risks to health"

A safe system of work covers such elements as the physical layout of the job, the sequence in which the work is to be carried out, proper instruction as to the operation of machinery, the issue of warnings and notices, and special instructions as to safety procedure. Simply because nothing has happened in the past does not mean that this by itself provides a proper system of work.

If there is a known danger, say of silicosis from dust, employers must do all that which is reasonably practicable to get the employees to wear the masks provided (*Crookall v. Vickers Armstrong*). In a similar case, the defendant employers had complied with the then statutory duty and had provided suitable goggles for eye protection in their foundry. The plaintiff, in common with the other workers, gave up wearing the goggles after using them for only a few days simply because he had constantly to wipe them clean and this interfered with the speed of his working which, in turn, limited his earnings since he was paid on piece work. The goggles supplied by the defendant had been recommended by the British Safety Council. When, however, the men gave up wearing the goggles the management simply acquiesced in the employees' practice. The plaintiff recovered damages at Common Law, for the defendants had been guilty of negligence in not doing all that was reasonably practicable to enforce the wearing of the goggles—it might well be that management adopted the attitude that the employees had themselves to blame if they did not wear the protection provided, but such an attitude cannot be tolerated, if it ever was, under the provisions of the new legislation. Under the provisions of the new Act, both the employer and the employee would be guilty of a criminal offence. Similar incidents arise when employees are in the habit of deliberately removing guards from machinery.

It must be emphasized that it is not legally possible to contract out of the provisions of the Act—*Ex turpi cause non actio oritur*. If an employer were to offer extra payment to an employee to accept some obvious risk rather than to remedy the situation, thus seeking to reduce cost, then a court would take a most serious view of his action and it is quite possible that the employee would also be open to a prosecution under Section 7, and this in turn could debar him from recovering damages in a civil action. An employee may know there is a risk of danger, but may have no alternative but to accept the situation—in doing so, the employee does not accept the risk due to a failure on the part of the employer to reduce the risk to a minimum *(Bowater v. Rowley Regis Corpn)*.

No liability, criminal or civil, will attach to an employer who has provided proper facilities and given adequate instructions and published notices. He is not required to stand over employees every minute; he need only do that which is reasonably practicable for him to enforce safety measures *(Woods v. Durable Suites)*.

It is, however, essential that management, supervisory and specialist staff must know the safety provisions relevant to their departments and work.

The general conditions in which work is to be done should be suitable. In short, the employer must provide the proper equipment for the job not only when they are working on the employer's premises but when they are sent to other premises. The vehicles in which they are to travel and the equipment provided must be suitable.

There may be a duty to provide preferential treatment for employees suffering from physical defects in that the consequence of injury to them would be more serious than in the case of a normally fit person. It may be that employees are unable to read safety notices and they therefore require explanation of them, It is not, it is submitted, safe to rely upon the premise that everyone can read.

A system of work includes the making of arrangements so that employees would choose the correct equipment for the particular task on which they are engaged. However, where an employee is sufficiently experienced and has the choice of tools or appliances themselves reasonably safe, and is injured through choosing the wrong ones for the particular task in hand, the employer would not necessarily be liable for the employee's carelessness. It could be otherwise in the case of an inexperienced employee.

Injury to health can be caused by exposure to fumes, lack of proper ventilation and, perhaps more important, noise and vibration. Noise and vibration must always be reduced to acceptable levels.

Section 2(1)(b) requires an employer to make arrangements for ensuring, so far as is reasonably practicable, safety and absence of risks to health in connection with the use, handling, storage and transport of articles and substances.

The effect of this section is that employers must ensure that employees are instructed in the proper use of and in the handling of materials. They must be warned of any dangers in the improper use of substances. Whilst in a past case an employer was not liable for the death of an employee who was addicted to inhaling noxious vapours *(Jones v. Lionite Specialities)*, it is now possible that in similar circumstances he might be liable, unless he could show that employees had been warned of the dangers involved. Here again, evidence in a safety procedure published to the workers is most desirable in the interests of the employer. Such materials must be properly labelled and not left lying about. Similar requirements apply to inflammable materials and substances. They must be stored in a safe place. Employees must be properly instructed in how to lift materials and how to load vehicles properly. The use of vehicles is another point that requires attention. No employee should be permitted to travel on vehicles other than in accordance with recognized safety procedures. Vehicles must be properly loaded. Fork-lift truck drivers may need special instruction in driving such vehicles. Lifting and handling techniques must be explained.

Section 2(2)(c) requires an employer to provide such information, instruction, training and supervision, as is necessary to ensure, so far as is reasonably practicable, the health and safety at work of his employees.

The foregoing sub-section is self-explanatory.

An employer must provide proper instruction as to the natural hazards of the work involved. This information must be revised and republished in the light of subsequent developments.

An employee must be properly instructed in the use of machinery and appliances. An employee must certainly not be put in charge of machinery without first ensuring that he not only knows how to operate but also that he knows how to operate it properly and safely. There is a higher duty imposed upon employers to supervise inexperienced or young apprentices than in the case of the more experienced personnel, but it is not sufficient in law to leave the latter to take care of themselves—a degree of supervision is still required to show that the employer has not shown laxity in the enforcement of safety measures and procedures. It would be advisable to have some documentary evidence (almost obligatory in the case of new employees) that such instruction and training has been given. It is insufficient merely to give an employee a copy of the relevant regulations and to tell him to read them *(Barcock v. Brighton Corporation)*. Adequate instruction and training must be given for the person may not understand such regulations, indeed, he may not be able to read. Section 2(2)(d) requires that 'so far as is reasonably practicable as regards any place of work under the employer's control, the maintenance of it in a condition that is safe and without risks to health and the provisions and maintenance of means of access to and egress from it that are safe and without such risks.'

The sub-section is most comprehensive. It involves two aspects of control. Firstly, those premises and places of work within the control of an employer. Secondly, those places of work under the control of an employer when working on contract on premises under the control of other persons.

Dealing with the first instance, the premises involved should be well lit, well ventilated and, so far as is reasonably practicable, free from fumes. Defective flooring and staircases are all affected. It is suggested that handrails, where not already fitted, should be fitted. A proper system of work should be effected and maintained when working on other premises. In the second instance, the independent contractor must likewise devise and maintain reasonably safe working conditions when engaged upon such work on other premises. Not only are employees affected by this provision, but also other persons not employed.

The qualifying words 'so far as is reasonably practicable' do not suggest that, say, the floor must be kept absolutely free at all times from a slippery substance or temporary obstruction, but that all reasonable measures must be taken to keep it free *(Braham v. J. Lyons and Co. Ltd)*. Cleaners should be instructed how to polish floors properly and only non-slip polish should be used.

The statutory requirement applies not only to employees and the duty is extended to all other persons who are lawfully on the premises. It most certainly would apply to the situation where there is a hole in the floor and, pending its repair or during its repair, a mere notice 'Beware—Hole in the Floor' is exhibited nearby. The hole must be properly fenced off.

It is submitted that the sub-section cannot apply where there is a safe means of access and a person chooses another which it cannot reasonably be anticipated he would choose.

The case of *Callaghan v. Kidd and Son Ltd* (decided under the Factories Act, and Common Law) is also illustrative of a typical situation. In this case the plaintiff had tripped over some iron bars lying on the floor in front of a pedestal mounting two grindstones on which he was working, and injured himself. The bars had been brought by a girl to be ground, but instead of grinding them forthwith she had gone away for some purpose. The employer was liable in that upon the evidence he had failed to give directions impressing on all operatives the imperative duty of not leaving bars (or indeed anything else over which a person could trip) on the floor unguarded.

Egress to and from the places of work must, so far as is reasonably practicable, be safe. Presumably, this means the employer must devise methods of keeping passages, corridors, and the like, and on the outside, those places under his control, in a reasonably safe condition and free from obvious dangers. In the winter months, he should devise some system of removing ice from the areas under his control leading to his premises.

Section 2(2)(c) imposes an obligation to provide and maintain a working environment for his employees that is, so far as is reasonably practicable, safe, without risks to health, and adequate as regards facilities and arrangements for their welfare at work.

The foregoing is a general statement as to working conditions. In the absence of any Code of Practice, it is submitted that the standard will be that which is already required under existing legislation, namely, the Offices, Shops and Railway Premises Act; the Factories Act 1961, and similar Statutes.

Section 2(3) requires an employer to prepare and issue a written policy statement outlining the employer's safety and health philosophy. It should indicate the involvement of management and employees in the implementation of such policy. The Health and Safety Commission have issued draft policy statements which employers can easily adapt for their own purposes.

Section 2(4) requires the establishment of a safety committee if so requested by appointed safety representatives. It is suggested that the establishment of such a safety committee is more or less mandatory if an employer is to show that he has done his best.

The real effect of the whole of Section 2 (including the various sub-sections) is to lay emphasis upon proper training for employees, information as to the inherent hazards of the work involved, consultation on safety measures, and the reporting of safety effectiveness through safety committees.

Self-employed persons

Section 3 of the Act imposes a general duty upon every employer and self-employed person to conduct his undertaking in such a way as to ensure, so far as is reasonably practicable, that persons not in their employment who may be affected thereby are not thereby exposed to risks to their health or safety.

Non-domestic premises

Section 4 of the Act has the effect of imposing on persons duties in relation to those who are not their employees but use non-domestic premises made available to them as a *place of work* or as a place where they may use plant or substances provided there for their use, and applies to premises so made available and other non-domestic premises used in connection with them. Thus, universities, colleges, training centres and similar institutions are brought within the general provisions of the Act.

Emission of noxious or offensive substances

Section 5 enacts that it shall be the duty of the person having control of any premises of a prescribed class to use the best practicable means for preventing the emission into the atmosphere from the premises of noxious or offensive substances and for rendering harmless and inoffensive such substances as may be so emitted. It should be observed that the expression used in this section is 'the best practicable means' and in this sense it is an absolute duty. 'Substances' is not the same as fumes. It is submitted that an incident similar to the Flixborough Disaster in 1974 would now be brought within the provisions of the section.

Manufacturers

Section 6 of the Act provides that it shall be the duty of any person who designs, manufactures, imports, or supplies any article or substance for use at work to ensure, so far as is reasonably practicable, that the substance, or the article by design and construction, is safe and without risks to health when properly used; to carry out or arrange for the carrying out of such testing and examination as may be necessary for the performance of the duty imposed on him; and to take such steps as are necessary to secure that there will be available in connection with the use of the article at work adequate information about the use for which it is designed, and has been tested, or in connection with the use of the substance, information about any conditions necessary to ensure that it will be safe and without risks to health when properly used.

The effect of this section is that the end-products of the production lines could be the subject of a prosecution. But this duty is really nothing more than the existing common law duty as applied in *Donoghue v. Stevenson,* save only that the prosecution will involve criminal sanctions.

The case of *Vacwell v. B. D. H. Chemicals* is, perhaps, an outstanding example as to how Section 6 will be interpreted by law. In that case the defendants were held liable for damages caused by their product boron tri-bromide. No warning had been given as to the extreme danger of the chemical when in contact with water. The defendants had not consulted all reasonably accessible information before putting the chemical on the market, and, consequently, failed to warn of the hazard.

Section 6(8) provides that articles may be supplied on the basis of the customer's written undertaking that he will take specified safety precautions. It is submitted that this provision will only relieve the supplier from criminal (and civil) liability if it is reasonable in all the circumstances—it could be that there is a duty upon the supplier to point out all possible dangers that occur to him but possibly not known to the buyer. If the supplier fails to obtain the *written undertaking* in circumstances where it is impossible for the supplier to make the article (or substance) safe, the supplier, manufacturer, designer, would be guilty of a criminal offence and possibly the principle of *res ipsa loquitur* (the fact speaks for itself) would be proof of negligence for civil purposes. A mere oral undertaking is insufficient.

An article or substance is not being properly used if it is used without regard to any relevant information or advice relating to its use which has been made available by the person by whom it was designed, manufactured, imported or supplied (Section 6(10)).

The word 'imports' is not the same as 'importer'. If, therefore, an employer buys his articles or substances abroad and brings them into this country, the provisions of Section 6 would apply to him in addition to any other liability under Section 2.

Duties of employees

Section 7 provides that it shall be the duty of every employee while at work to take reasonable care for the health and safety of himself and of other persons who may be affected by his acts or omissions at work; and as regards any duty or requirement imposed on his employer or any other person by or under any relevant statutory provision, to co-operate with him so far as is necessary to enable that duty or requirement to be performed or complied with. It will be observed that there is now a general duty imposed upon employees while at work in the course of their employment to take care for the safety and health of themselves, their fellow workers, and the general public, and further, to co-co-operate with their employer in ensuring that safety measures and policies are observed and enforced.

The so-called practical jokers are now within the scope of criminal sanctions.

Let us now consider a hypothetical situation. An employee is well known for his liking for horseplay during his employment; he is warned about it but not dismissed. It could be that the employer would be guilty of an offence in that he has not taken all reasonable practical measures for the safety of his employees. The individual would be guilty of an offence under Section 7, and any pressure used to prevent dismissal could be construed as a failure to co-operate in safety measures. It is suggested that in such a situation the Inspectorate would be of considerable assistance.

Apart from possible criminal prosecution, failure to observe recognized safety procedures could always be grounds for a dismissal under Sections 23 and 24 of the Trade Union and Labour Relations Act 1974, provided, of course, that it can be shown that the employee knew of and had been instructed in the safety procedures. Section 8 of the Act further provides that no person shall intentionally or recklessly interfere with or misuse anything provided in the interests of health, safety or welfare in pursuance of the relevant statutory provisions.

The inspectorate and their powers

By virtue of Section 20 of the Act, an inspector appointed by the enforcing authorities have statutory authority:

    (a) to have access to premises at any reasonable time and to take with him a constable, if necessary;

    (b) to make any necessary investigations, photographs and the like;

    (c) to take samples as may be necessary;

    (d) to obtain all relevant information from persons on the premises.

The inspectors have additional power in that they may issue 'improvement' and 'prohibition' notices.

Section 21 authorizes an inspector to serve on a person controlling certain activities an improvement notice stating that he (the inspector) is of the opinion that:

    (a) certain provisions are not being met;

    (b) requiring those conditions to be remedied;

    (c) the time by which those conditions must be remedied.

Section 22 authorizes an Inspector to serve a prohibition notice if an activity is being or is about to be carried out against any statutory provision and may involve a risk of personal injury. The notice shall state the inspector's opinion, specify the nature of the risk, the statutory provision being contravened, and direct the activities referred to to cease immediately. Thus, an inspector can effectively close down a factory, premises or stop activities.

There is a right of appeal against such notices to an industrial tribunal.

Section 25 empowers an inspector to seize any article or substance found by him in any premises which he has power to enter and has reasonable cause to believe is a cause of imminent danger of serious personal injury, and cause it to be rendered harmless (whether by destruction or otherwise), and, if it is practicable, to submit a sample to the person responsible for the premises.

In any proceedings for an offence under any of the relevant statutory provisions the onus is cast upon the accused to prove that he has done all that he could reasonably do in all the circumstances (Section 40). It is a defence for the accused to show that the act or omission was the fault of some other person (Section 36). Where an offence is committed by a body corporate, proceedings may also be taken against any officer of the body corporate who is responsible for the act or omission (Section 37). The section uses the expression 'attributable to any neglect' on the part of such person.

Index

THIS BOOK HAS BEEN SET IN MONOPHOTO TIMES NEW ROMAN
AND PRINTED AND BOUND IN GREAT BRITAIN BY
WILLIAM CLOWES AND SONS, LIMITED, LONDON AND BECCLES